102 Advances in Polymer Science

Polymer Synthesis
Oxidation Processes

With contributions by
B. Amédouri, B. Boutevin, K. E. Geckeler,
R. Grubbs, H. K. Hall, M. Lazár, B. Novac,
M. Okada, A. B. Padias, J. Penelle, W. Risse,
B. Rivas, J. J. Robin, J. Rychlý, H. Tanaka

With 12 Figures and 42 Tables

Springer-Verlag
Berlin Heidelberg GmbH

ISBN 978-3-662-14971-3   ISBN 978-3-540-46690-1 (eBook)
DOI 10.1007/978-3-540-46690-1

Library of Congress Catalog Card Number 61-642

© Springer-Verlag Berlin Heidelberg 1992
Originally published by Springer-Verlag Berlin Heidelberg New York in 1992
Softcover reprint of the hardcover 1st edition 1992

Typesetting: Th. Müntzer, Bad Langensalza;

02/3020-5 4 3 2 1 0 — Printed on acid-free paper

Editors

Table of Contents

Ring-Opening Polymerization of Bicyclic and Spiro Compounds.
Reactivities and Polymerization Mechanisms
M. Okada 1

The Development of Well-defined Catalysts for Ring-Opening Olefin
Metathesis
R. Grubbs, W. Risse, B. Novac 47

Captodative Olefins in Polymer Chemistry
J. Penelle, A. B. Padias, H. K. Hall, H. Tanaka 73

Synthesis and Properties of Fluorinated Diols
B. Boutevin, J. J. Robin 105

Synthesis and Applications of Fluorinated Telechelic Monodispersed
Compounds
B. Amédouri, B. Boutevin 133

Synthesis and Metal Complexation of Poly(ethyleneimine) and Derivatives
B. L. Rivas, K. E. Geckeler 171

Oxidation of Hydrocarbon Polymers
M. Lázar, R. Rychlý 189

Author Index Volumes 101–102 223

Subject Index 225

Ring-Opening Polymerization of Bicyclic and Spiro Compounds. Reactivities and Polymerization Mechanisms

Masahiko Okada

Department of Forest Products Science and Technology, Faculty of Agriculture, Nagoya University, Furo-cho, Chikusa-ku, Nagoya 464-01, Japan

Ring-opening polymerization of heterobicyclic compounds, such as bicyclic acetals, lactones, lactams, and orthoesters provides a convenient and effective route to designing functional polymers containing heterocycles in their main chains. Polycondensation and polyaddition also give some of these polymers, but ring-opening polymerization has the advantage of producing polymers not only high in molecular weight but also stereochemically regulated. The reactivity of bicyclic monomers varies over a wide range, and in addition, highly strained monomers are often ring-opened by two or more different ways. In some cases, ring-opening polymerization accompanies isomerization. These features complicate the ring-opening polymerization of bicyclic monomers. The present article reviews recent advances in ring-opening polymerization of selected bicyclic and spiro compounds, focusing on the structure-reactivity relationships and the polymerization mechanisms.

1 Introduction 3

2 Ring-Opening Polymerization of Bicyclic Acetals 3
2.1 Stereochemical Aspects of Cationic Polymerization of Bicyclic Acetals 3
2.2 Stereoregulation in the Cationic Polymerization of Anhydro Sugar Derivatives 11

3 Ring-Opening Polymerization of Bicyclic Lactones 17
3.1 Selective Cyclooligomerization of Bicyclic Oxalactones Having a Bicyclo[3.2.1]octane Skeleton 17
3.2 Polymerization of Bicyclic Lactones Having a Bicyclo[2.2.2]octane Skeleton 21

4 Ring-Opening Polymerization of Bicyclic Lactams 24
4.1 Polymerization of Bicyclic Lactams having a Bicyclo[3.2.1]octane Skeleton 24
4.2 Polymerization of Bicyclic Lactams Having a Bicyclo[2.2.2]octane Skeleton 29
4.3 Polymerization of Bicyclic Lactams Having a Bicyclo[2.2.1]heptane Skeleton 31

5 Ring-Opening Polymerization of Bicyclic Orthoesters 32

Advances in Polymer Science Vol. 102
© Springer-Verlag Berlin Heidelberg 1992

6 Ring-Opening Polymerization of Spiro Orthoesters 35

7 Ring-Opening Polymerization of Spiro Orthocarbonates 39

8 Concluding Remarks 43

9 References 44

1 Introduction

More than three decades have passed since Hall et al. [1] reported their pioneering work on the ring-opening polymerizations of a variety of bridged bicyclic compounds containing heteroatoms. During this period, significant progress has occurred in this particular field of synthetic polymer chemistry. A wealth of knowledge has thus been accumulated on the reactivities and mechanisms of ring-opening polymerizations of heterobicyclic monomers such as bicyclic ethers, acetals, lactones, and lactams. With these basic researches, a number of functional polymers potentially useful for industrial and biomedical applications have been made available by utilizing ring-opening polymerization of bicyclic monomers.

Heterobicyclic compounds often display specific behavior in their ring-opening polymerization. This is attributed to their rigid and bulky structures which contain, in most cases, two or more asymmetric carbon atoms. Sometimes, stereoelectronic effects involving heteroatoms also play an important role in regulating polymerization processes. For example, racemic bicyclic acetals such as 6,8-dioxabicyclo[3.2.1]octane and its derivatives often undergo stereospecific polymerization even in the presence of conventional Lewis acid initiators.

The present article reviews recent progress in the field of ring-opening polymerization of heterobicyclic compounds. The description does not cover the entire range of bicyclic monomers, because it would make the contents enumerative and divergent. Therefore, we shall limit the kinds of bicyclic monomers dealt with in this article to bicyclic acetals, lactones, lactams, and orthoesters, and discuss their ring-opening polymerizations with particular emphasis on the structure-reactivity relationships of monomers and the polymerization mechanisms. Ring-opening polymerizations of spiro compounds such as spiro orthoesters and spiro orthocarbonates are also included because of their close relevance to the corresponding bicyclic compounds. Synthetic and functional aspects of specialty polymers obtained from heterobicyclic compounds are not discussed in this review, because they will be described in detail elsewhere. [2] So far, the ring-opening polymerization of bicyclic compounds has been reviewed in several articles. [3–7] Hence, in the present article we will make an effort to minimize the overlapping with previous works.

2 Ring-Opening Polymerization of Bicyclic Acetals

2.1 Stereochemical Aspects of Cationic Polymerization of Bicyclic Acetals

Ring opening polymerization of monocyclic acetals such as 1,3-dioxolane and 1,3-dioxepane has been extensively investigated by several research groups for many years. [6] However, there still remains something to be clarified as to the polymerization mechanisms, particularly the structure of growing species. One of the several reasons that have prevented the elucidation of the polymerization mechanisms is the lack of information on the stereochemical course of the polymerization. Spectroscopic methods are the best way for identifying the structure of the propagating species, as is the case in the cationic polymerization of tetrahydrofuran. However, the propagating species in the cationic polymerization of cyclic acetals are in general so reactive that it is not easy to detect them spectroscopically. Therefore, we may resort to a second choice in which we allow a monomer containing an asymmetric acetal carbon to polymerize and examine whether the configuration of the acetal carbon is inverted or racemized during the polymerization. However, in general, the replacement of one or more hydrogen atoms of monocyclic acetals by alkyl groups in general reduces the polymerizability to a great extent, making the monomers not homopolymerizable under ordinary conditions. Thus, this approach is practically inapplicable for monocyclic acetals.

Bicyclic acetals are generally more strained than monocyclic acetals; in other words, they show higher polymerizabilities. Therefore, the stereochemical course of their ring-opening polymerization can be clarified from the configurational changes of the asymmetric acetal carbons of bicyclic acetals during polymerization. Questions may arise concerning the ring-opening polymerization of bicyclic acetals such as 6,8-dioxabicyclo[3.2.1]octane (**1**) and 2,7-dioxabicyclo[2.2.1]heptane (**2**). They are (1) which of the two acetal bonds is broken when the monomer is ring-opened, (2) whether the configuration of the acetal carbon is inverted or racemized on polymerization, and (3) whether the enantiomeric monomeric units are distributed randomly or in blocks in the polymer chain when the monomer is racemic.

Concerning the site of bond cleavage, the compound **1** and its derivatives always undergo acid-catalyzed bond cleavage exclusively at the C(5)–O(6) bond to yield a six-membered tetrahydropyran ring. Things with a more strained bicyclic acetal **2** are not as simple as this. [8] At low temperature, **2** undergoes cationic polymerization by the selective C(1)–O(2) bond cleavage to give a polyacetal containing five-membered tetrahydrofuran rings. This phenomenon is in conformity with the antiperiplanar rule proposed by Deslongchamps, [9] but the rule does not always hold at higher temperature or for the polymerization of substituted homologues. In general, either C(1)–O(2) (a) or C(1)–O(7) (b) bond is broken, sometimes competitively, to yield both tetrahydrofuran and tetrahydropyran rings in the polymer chain. In the polymerization of 1,4-anhydrosugar derivatives having a 2,7-dioxabicyclo[2.2.1]heptane skeleton, the proportions of furanose and pyranose units can be altered, sometimes dramatically, by a selection of substituents and initiators. This point will be discussed in Sect. 2.2.

It has been well established that the configuration of the acetal carbon is in general inverted in the cationic polymerization of **1** and its derivatives. [3] In other words, the propagation proceeds through the S_N 2 type mechanism involving an attack of the monomer on the acetal carbon of a trialkyloxonium ion. Therefore, it was thought that no polyacetals entirely composed of the structural units retaining the original configuration of the acetal carbon could be prepared by the conventional ring-opening polymerization method.

However, Okada et al. [10–12] found that a bicyclic acetal, 3(e),4(a)-bis(benzyloxy)-6,8-dioxabicyclo[3.2.1]octane (**3**), gave a polymer entirely consisting of *cis*-2,6-linked tetrahydropyran rings (hereafter this unit is referred to as *cis* unit or β-form following the terminology of carbohydrate chemistry) (**5**) when it was allowed to polymerize in a dilute solution at or above −60 °C. Interestingly, the same monomer gave a polymer exclusively composed of *trans*-2,6-linked tetrahydropyran rings (*trans* unit or α-form) (**4**), when the polymerization was carried out with a higher initial monomer concentration at −90 °C and terminated at low conversions (Scheme 1).

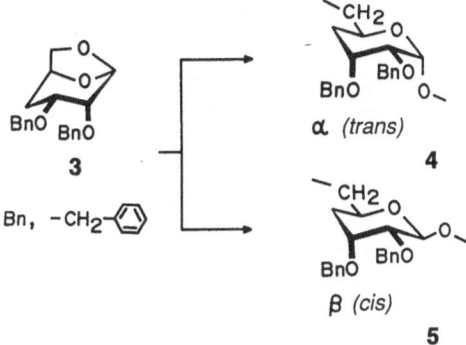

Scheme 1

Some relevant monomers (**6–8**) showed a similar tendency: they gave the polymers predominantly, but not entirely, composed of the respective *cis*-2,6-linked tetrahydropyran units, i.e., (1→6)-β-linked polysaccharide, when they were allowed to polymerize with lower initial monomer concentrations and at higher polymerization temperature. [13, 14] This is contrasted to 1,6-anhydro-2,3,4-tri-*O*-benzyl-β-D-allo-pyranose (**9**), the D-enantiomer of 2(a), 3(e), 4(a)-tris(benzyloxy)-6,8-dioxabicyclo[3.2.1]-octane, which provided a polymer composed of *trans*-2,6-linked tetrahydropyran units, i.e. (1→6)-α-linked polysaccharide (**10**), when cationically polymerized. [15]

It should be noted that the numbering of the atoms in the IUPAC nomenclature differs from that in the nomenclature used in carbohydrate chemistry, which sometimes leads to confusion. The present article follows the IUPAC nomenclature, except for the compound with carbohydrate sources.

The monomers **3** and **6–8** have the following features in common in their ring-opening polymerization behavior: (1) They give polymers rich in the *cis* units at or above −60 °C. (2) The cis unit content in these polymers increases with decreasing initial monomer concentration, as depicted in Fig. 1. (3) Polymerization of these monomers at higher initial monomer concentrations and at lower

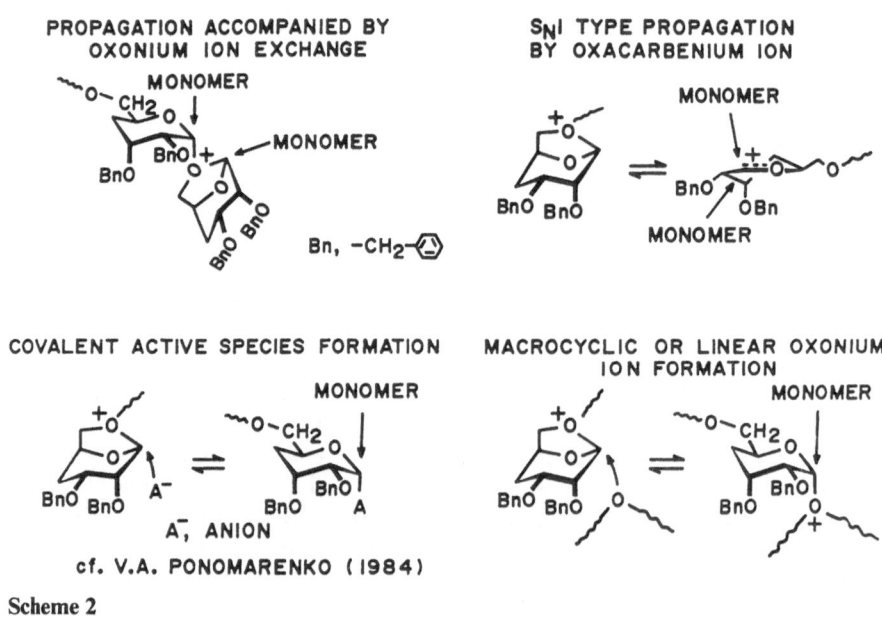

6   **7**   **8**

9   **10**   Bn, $-CH_2-\langle\!\!\bigcirc\!\!\rangle$

temperature gives the polymers predominantly or entirely composed of the trans units. All these monomers possess, as a common substituent, an equatorially oriented benzyloxy group at the position 3. Therefore, this substituent should be the definite key for the steric control of the polymerization.

PROPAGATION ACCOMPANIED BY OXONIUM ION EXCHANGE

S$_N$I TYPE PROPAGATION BY OXACARBENIUM ION

MONOMER

MONOMER

MONOMER

Bn, $-CH_2-\langle\!\!\bigcirc\!\!\rangle$

COVALENT ACTIVE SPECIES FORMATION

MACROCYCLIC OR LINEAR OXONIUM ION FORMATION

MONOMER

MONOMER

A$^-$, ANION

cf. V.A. PONOMARENKO (1984)

Scheme 2

As illustrated in Scheme 2, [12] four possible reactions produce the *cis* unit (β-form). They are (1) the oxonium exchange reaction at a penultimate unit (upper left), (2) the propagation by an oxacarbenium ion (upper right), (3) the propagation by a covalent species (lower left), [16] and (4) the inter- or intramolecular transacetalization via macrocyclic or linear trialkyloxonium ions (lower right). The common features of the polymerization of the monomers **3** and **6–8** described above cannot be accounted for either by the propagation involving oxacarbenium ions or covalent species or by the transacetalization.

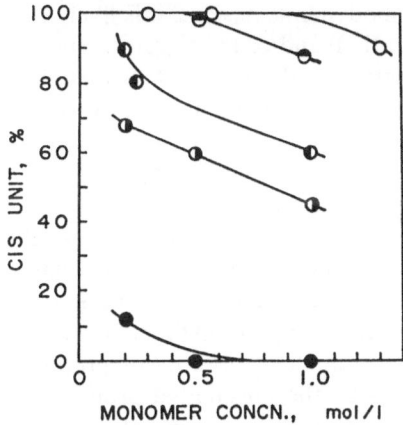

Fig. 1. Dependence of the *cis* unit content of the polymer on the initial monomer concentration.[14] Polymerization conditions: solvent, CH₂Cl₂; initiator, PF₅; temp., −60 °C. ○, 3; ◕, 6; ◑, 7; ◐, 8; ●, 9

A generalized propagation mechanisms has been proposed to explain the polymerization behavior of the monomers **3** and **6–8** (Scheme 3, [14]). When the propagation of a trialkyloxonium ion (**11**) through path A to **12** is re-

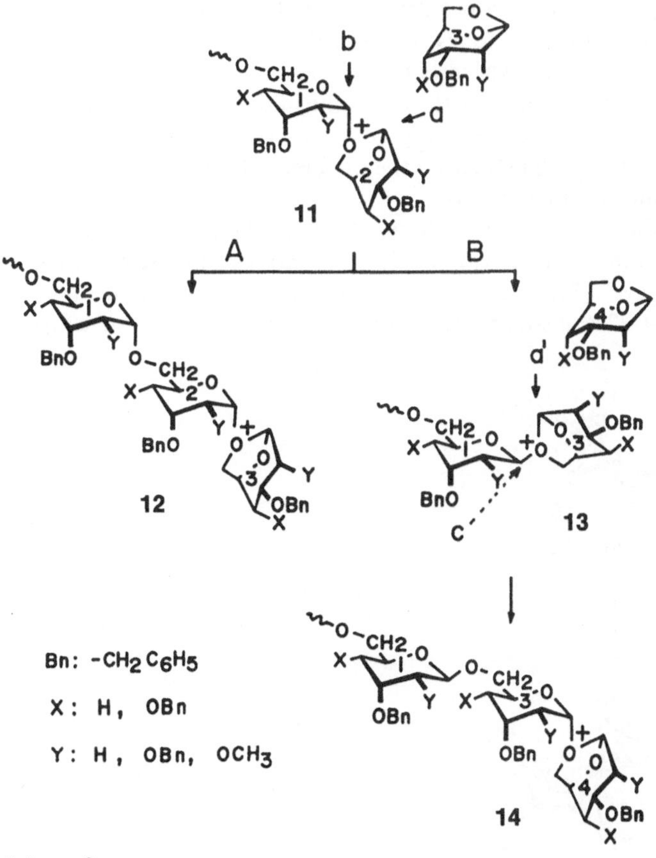

11

A | B

12

13

Bn: −CH₂C₆H₅

X: H, OBn

Y: H, OBn, OCH₃

14

Scheme 3

tarded or prohibited for kinetical and/or thermodynamic reasons, it is likely that
an internal ring-closure reaction (depolymerization) and/or the transformation of
the monomeric unit **1** to a *cis*-type penultimate unit through path B (oxonium
exchange) take place. Such an oxonium exchange reaction frequently occurs
in the cationic ring-opening polymerization of cyclic acetals and ethers. [17] In
the resulting oxonium ion **13**, there occurs a predominant forced monomer
addition to the acetal carbon of the terminal unit **3** (arrow a'). Thus, the
monomeric unit **1** is incorporated into the polymer chain as a cis unit, and the
oxonium ion **11** with a degree of polymerization higher by one (structure
14) is regenerated. Regular repetition of the reaction sequence (**11**→**13**→**14**)
through path B eventually gives a main chain entirely composed of the cis units.

Polymerization with a sufficiently high initial monomer concentration produced
polymers predominantly containing trans units by the preferential propagation
via path A, at least in the initial stage (kinetic control). As the initial monomer
concentration is decreased, the formation of the *trans* units is retarded or prohibited
owing to the increasing contribution of depolymerization, because the *trans*
unit in the polymer chain involves 1,3-diaxial nonbonded interaction between
the exocyclic acetal oxygen and the benzyloxy oxygen at position 3: in other
words, and equilibrium monomer concentration should be relatively high. On
the other hand, the formation of the *cis* unit devoid of 1,3-diaxial nonbonded
interaction takes place even at a lower initial monomer concentration, competing
with the internal ring-closure. Therefore, the lower the initial monomer concentration and the higher the polymerization temperature as well, the higher becomes
the fraction of the *cis* unit in the polymer by the preferential propagation via path
B to via path A (thermodynamic control).

In some cases, neighboring group participation is responsible for the formation
of the β-cnfiguration. In the polymerization of a stereoisomer mixture of 4(a)- and
4(e)-bromo-6,8-dioxabicyclo[3.2.1]octane (42:58) (**15a** and **15e**), a polymer consisting of 78% *trans*- and 22% *cis*-units was obtained even at −90 °C at which a

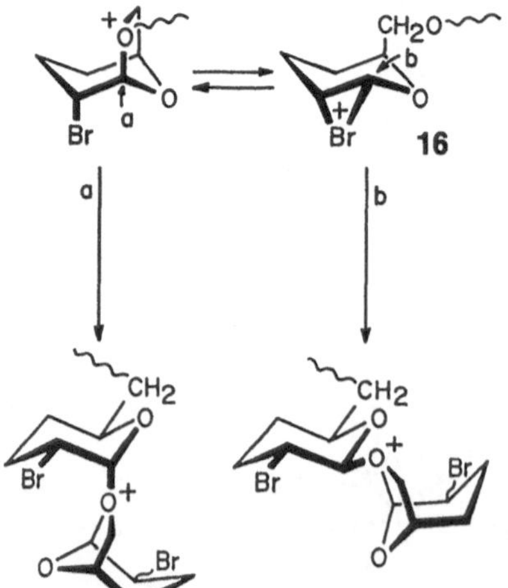

Scheme 4

polymer entirely composed of *trans* units was formed from the pure 4(e)-isomer. [18] This is a clear indication of the effect that the axially oriented bromine atom of the 4(a)-isomer has on the steric course of the propagation by forming a bromonium ion bridge to the acetal carbon, thus preventing the monomer from approaching the acetal carbon from the bromonium ion (16) on the same side (Scheme 4, [18]).

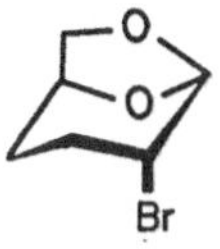

15a 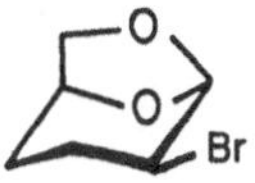   **15e**

By utilizing the neighbouring group participation, $(1\rightarrow6)$-β-D-galactopyranan (18) was successfully prepared in the cationic ring-opening polymerization of 1,6-anhydro-β-D-galactopyranose derivative (17) having a 2-benzoyloxy group. [19, 20]

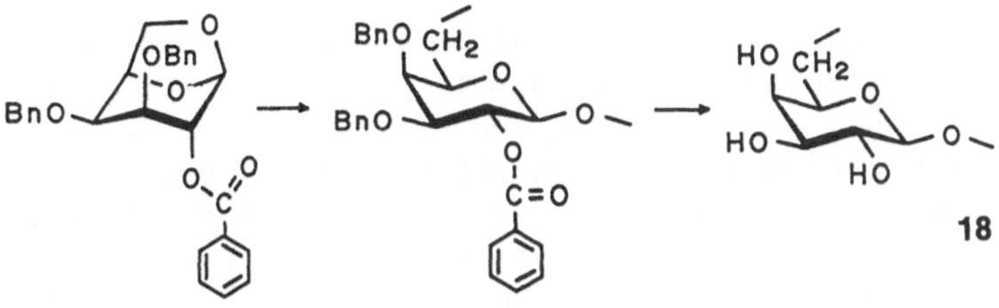

17

As illustrated in Scheme 5 [19], a trialkyloxonium ion is transiently formed at the growing terminal unit, but it is immediately attacked by the carbonyl oxygen of the benzoyloxy group with inversion of the C-1 configuration to form a dioxacarbenium ion. As a consequence, an incoming monomer can attack the reaction center exclusively from the dioxacarbenium ion on the opposite side, that is, the β-side of the pyranose ring, and the trialkyloxonium ion is regenerated. However, the polymerization proceeded very sluggishly even at 0 °C in the presence of 10 mol% phosphorus pentafluoride, and the number-average molecular weight of the polymer was only 2.6×10^3 (GPC, polystyrene standard). This is clearly due to a higher stability of the dioxacarbenium ion. The benzoyloxy group acts as a nucleophile in the first step, but it acts as a leaving group when the monomer attacks the anomeric carbon of the dioxacarbenium ion. Therefore, it is necessary for the preparation of a $(1\rightarrow6)$-β-linked polysaccharide of higher molecular weight to balance these two opposing tendencies of the neighboring group properly.

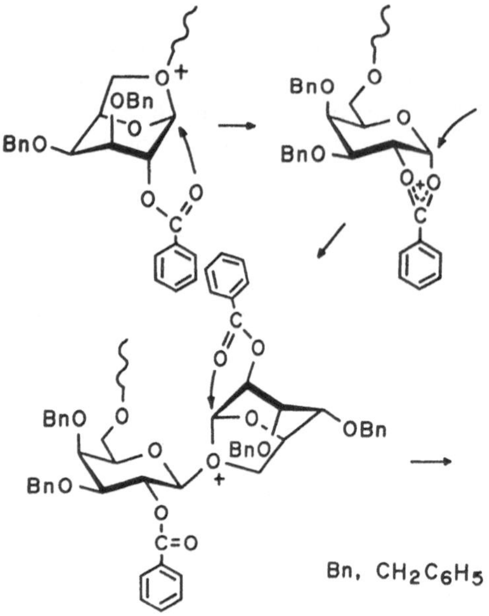

Scheme 5

The third point in the stereochemistry of the ring-opening polymerization of bicyclic acetals is stereospecificity. Since bicyclic acetals derived from noncarbohydrate sources are racemic in most cases, stereospecific polymerization often takes place under properly selected reaction conditions. For example, cationic homopolymerization of racemic 6,8-dioxabicyclo[3.2.1]octane **1** and its 4(e)-bromo derivative **15e** with conventional Lewis acid initiators at low temperature gave the corresponding stereoregular polyacetals rich in isotactic dyads. [21, 22] In particular, the polymerization of the latter monomer at −90 °C in toluene with phosphorus pentafluoride as the initiator yielded a polyacetal having an isotactic dyad content higher than 95%. The bromo-substituted polyacetal (**19**) was reductively debrominated to yield a highly isotactic (1→6)-α-linked polysaccharide backbone structure (**20**).

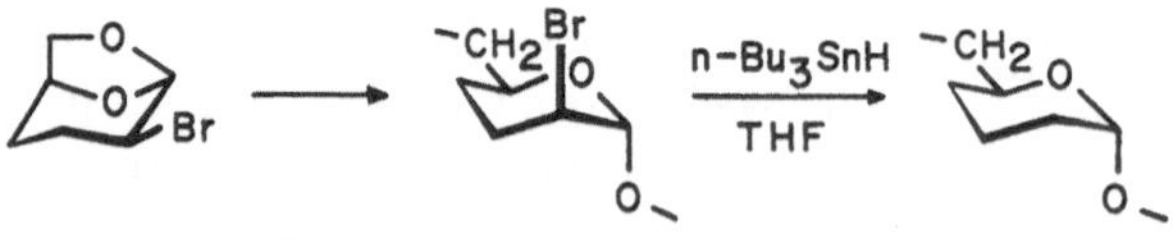

D,L-copolymerization of enantiomerically imbalanced mixtures of 6,8-dioxybicyclo[3.2.1]octane (**1**) has revealed that an isotactic sequence along the polymer chain is preferentially formed by the enantiomer selection at the chiral growing chain end. [21] If this is generally the case, asymmetric copolymerization should occur when a racemic bicyclic acetal is allowed to copolymerize with an optically active monomer without using a chiral initiator. In fact, in the copolymerization of racemic 4(e)-bromo-6,8-dioxabicyclo[3.2.1]octane (**15e**) with an optically active monomer, 1,6-anhydro-2,3,4-tri-*O*-benzyl-β-D-glucopyranose (**21**), the D-enantiomer of **15e** having the same chirality as that of the optically active comonomer **21**, was preferentially incorporated into the copolymer chain. [23]

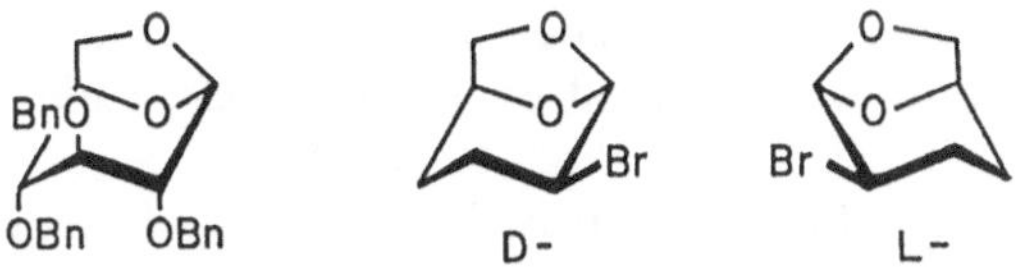

From the copolymer composition dependence of the molar ratio of the D- and L-enantiomeric units of **15e** in the copolymer, the rate of reaction of the growing chain end of **21** with the D-enantiomer of **15e** was estimated to be about four times faster than that with the L-enantiomer. Such asymmetric selection is mainly ascribable to the steric and electronic interactions between the asymmetric environment created by the bulky terminal unit of **21** and the rigid bicyclic monomer having three asymmetric centers and a polar bulky bromine substituent (Scheme 6, [23]).

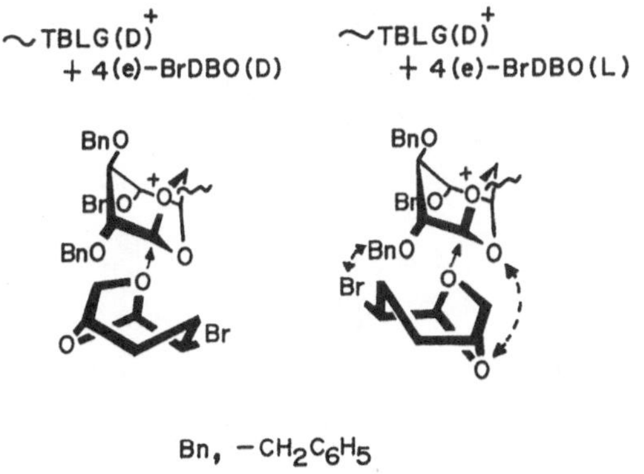

Scheme 6

Similar enantiomer selection at the growing chain end is commonly observed in the cationic polymerizations of bicyclic acetals having bicyclo[3.2.1]octane skeletons. [24]

2.2 Stereoregulation in the Cationic Polymerization of Anhydro Sugar Derivatives

The first stereospecific synthesis of a linear polysaccharide containing only (1→6)-linked α-D-glucopyranosyl residues was achieved as early as 1961 by ring-opening polymerization of 1,6-anhydro-2,3,4-tri-*O*-methyl-β-D-glucopyranose. [25] Since then, chemical synthesis of stereoregular polysaccharides has been a continuing interest of synthetitc polymer chemists who are concerned with biopolymers. [6, 24, 26–28] Synthetic polysaccharides of well-defined structure can be used as model substances to investigate the action pattern of enzymes, the induction and reaction of antibodies, and the effect of structure on biological activities in the interactions of proteins, nucleic acids, and lipids with poly-hydroxylic macromolecules. They should furnish a large variety of potentially useful materials whose properties can be widely varied by appropriate chemical modifications.

In particular, chemical synthesis of cellulose is an attractive target of synthetic polymer chemists worthy of challenge. One of the possible approaches is the ring-opening polymerization of 1,4-anhydro-α-D-glucose derivatives. As shown in Scheme 7, there are four possible structural units in the polymers prepared by the ring-opening polymerization of 1,4-anhydro-2,3,6-tri-O-benzyl-α-D-glucopyranose (**22**). They are the (1→4)-β- and (1→4)-α-D-glucopyranosidic units (**23** and **24**) and the (1→5)-β- and (1→5)-α−D-glucofuranosidic units (**25** and **26**). [28] Synthesis of a polysaccharide of cellulose-type requires controlling the propagation reaction in such a way that bond scission occurs between the anomeric carbon atom and the bridge oxygen atom (1.4-scission), along with the inversion of the configuration of the anomeric carbon by the addition of the monomer.

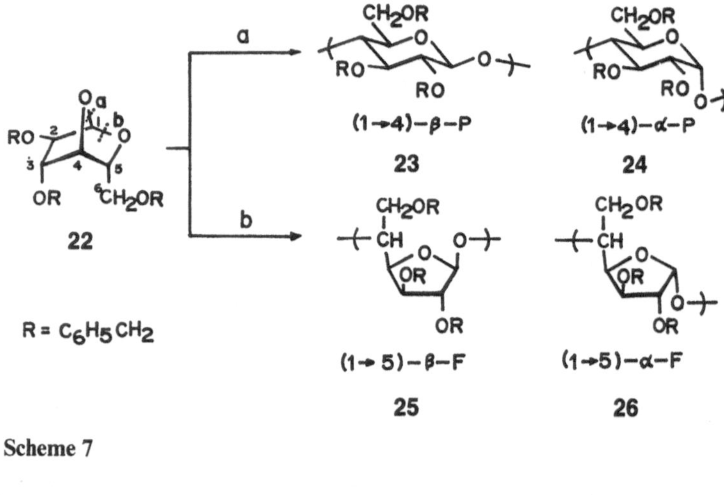

Scheme 7

Micheel et al. [29] claimed that the ring-opening polymerization of **22** by triethyloxonium tetrafluoroborate gave a linear 2,3,6-tri-O-benzyl-(1→4)-β-D-glucopyranan **23**, although it contained some (1→4)-α-linkages. Later, Uryu et al. [30] reexamined in detail the polymerization of the same monomer with a variety of initiators and found that only a phosphorus pentafluoride initiator gave a stereoregular polysaccharide, 2,3,6-tri-O-benzyl-(1→5)-α-D-glucofurans **26** with $[\alpha]_D = +82°$ and a number-average molecular weight of 8.5×10^3 ($DP_n = 20$). Other cationic initiators provided polysaccharides with mixed structures, depending on the polymerization conditions (Table 1).

Table 1. Structure of poly(1,4-anhydro-2,3,6-tri-O-benzyl-α-D-glucopyranose) prepared under different conditions [28]

Initiator	Temp., °C	Polymer structure,[a] %			
		$(1{\rightarrow}4)$-β-P 23	$(1{\rightarrow}4)$-α-P 24	$(1{\rightarrow}5)$-β-F 25	$(1{\rightarrow}5)$-α-F 26
PF_5	−78	0	0	0	100
PF_5	−40	3	4	7	86
$SbCl_5$	−20	0	0	13	87
CF_3SO_3H	−40	17	1	29	53
$(CF_3SO_2)_2O$	0	36	32	11	21
$(C_2H_5)_3OBF_4$	0 → −78	18	15	36	31

[a] Determined from the proportions of C(1) absorptions in ^{13}C NMR spectrum

Scheme 8 [30] shows that a stereoregular structure entirely consisting of $(1{\rightarrow}5)$-α-D-glucofuranosidic units can be obtained by the trialkyloxonium ion mechanism (path c). Probably the selective 1,5-ring opening of **22** occurs from the 1,5-linked

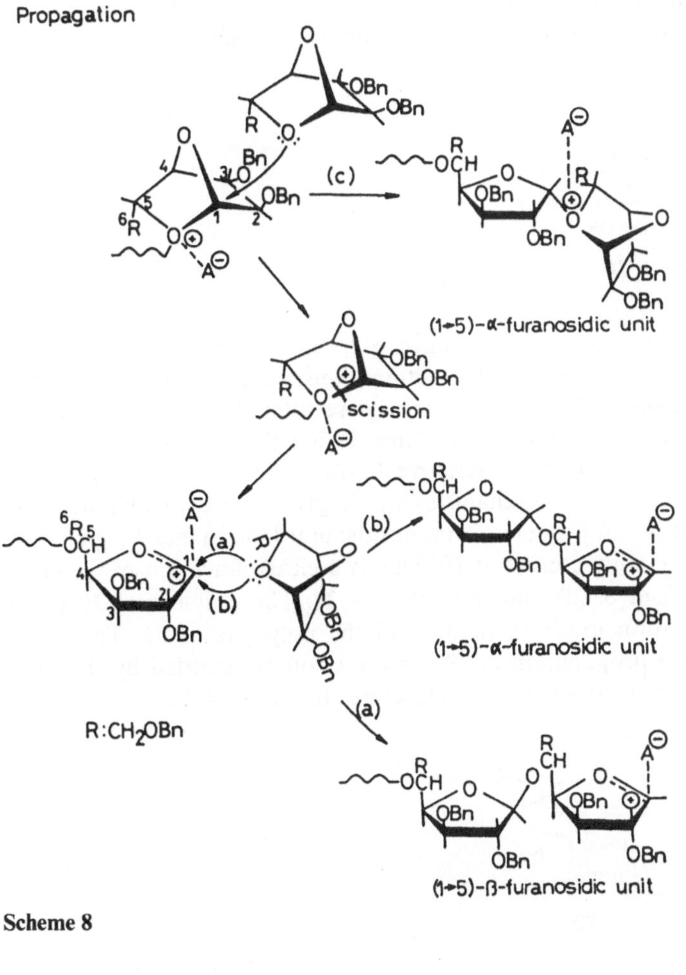

Propagation

Scheme 8

oxygen which has a higher nucleophilicity than that of the 1,4-linked oxygen and/or by the antiperiplanar rule of Deslongchamps et al. [9] On the other hand, a mixed structure consisting of $(1 \rightarrow 5)$-α- and $(1 \rightarrow 5)$-β-D-glucofuranosidic units can be formed by the oxacarbenium ion mechanism shown in Scheme 8 (paths a and b). The polymerization involving the oxacarbenium ion becomes favorable at higher temperature.

It has been demonstrated that the substituents at C(2) and C(3) positions of 1,4-anhydro sugar derivatives influence their stereochemical courses of cationic ring-opening polymerization. [28] Since 22 is different from 1,4-anhydro-2,3-di-O-benzyl-α-D-xylopyranose (27) only with respect to the C-5 substituent, comparison of the polymerization data of these two monomers may give an answer to the question of how much the additional C-5 substituent affects the chain propagation. In fact, boron trifluoride etherate and tin (IV) tetrachloride were found to be effective initiators that produce high molecular weight polymers (28) with the degree of polymerization of up to 477 in high yields. [31] However, they failed to produce high molecular weight polymers of 22 (the DP_n of the polymer obtained under similar conditions was 25). Clearly, the C-5 substituent in 22 hinders more sterically the chain propagation and lowers the polymerizability of the monomer.

27   28

Ring-opening polymerization of 1,4-anhydro-2,3,6-tri-O-benzyl-β-D-galactopyranose (29) gave polymers with mixed structures consisting mainly of $(1 \rightarrow 5)$-β-D-galactofuranosidic units. [30] The structures were altered by relatively minor changes in the polymerization conditions. Since 29 has the configurations at C-1 and C-4 opposite to those of 22, the selective 1,5-ring-opening polymerization of 29 via the trialkyloxonium ion mechanisms would give a stereoregular structure consisting entirely of $(1 \rightarrow 5)$-β-D-galactofuranosidic unit (30). In fact, the polymers of 29 were found to mainly consist of 30. This is a clear indication of the chain propagation mechanism of 29 similar to that of 22. The polymer of 29 had a degree of polymerization lower than those of the polymers of 22. This result implies that the chain propagation of 29 is more strongly retarded by the steric hindrance associated with the monomer structure than that of 22.

29   30

As described above, 1,4-anhydro sugars can be polymerized to two kinds of stereoregular polysaccharides, one having a (1→5)-α- or a (1→5)-β-furanoside structure with five-membered rings and the other having a (1→4)-β-pyranoside structure with six-membered rings. So far, a cellulose-type polysaccharide, i.e. (1→4)-β-glycopyranan, has been obtained only from 1,4-anhydro-α-D-ribopyranose derivatives (31). [32] When polymerized, other 1,4-anhydro sugars such as 1,4-anhydro-α-D-xylopyranose (27), [31] 1,4-anhydro-α-D-lyxopyranose (32), [33] 1,4-anhydro-α-L-arabinopyranose (33), [34] 1,4-anhydro-α-D-glucopyranose (22), [30] and 1,4-anhydro-α-D-galactopyranose (29) [30] yielded (1→5)-furanan-type polysaccharides by ring-opening polymerization but failed to give (1→4)-β-pyranan-type ones. At present, it is not possible to control at will the stereochemical course of the polymerization of these 1,4-anhydro sugar derivatives.

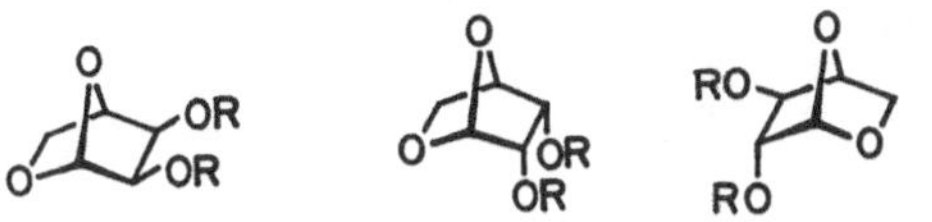

31   **32**   **33**

A variety of branched oligo- and polysaccharides have been found in nature. Those of glycoproteins and glycolipids in cell surfaces play an important role in cell-cell recognition. Comb-shaped branched polysaccharides as model substances for these branched polysaccharides can be synthesized by two different routes via ring-opening polymerization of anhydro sugar derivatives: (1) polymerization of anhydro disaccharide derivatives followed by deprotection, and (2) synthesis of regiospecifically protected linear polysaccharides followed by stereoselective glycosidation and deprotection. The former route is favorable for the synthesis of polysaccharides that are substituted completely and stereospecifically with monosaccharide moieties, though the monomer synthesis is troublesome. Thus, the

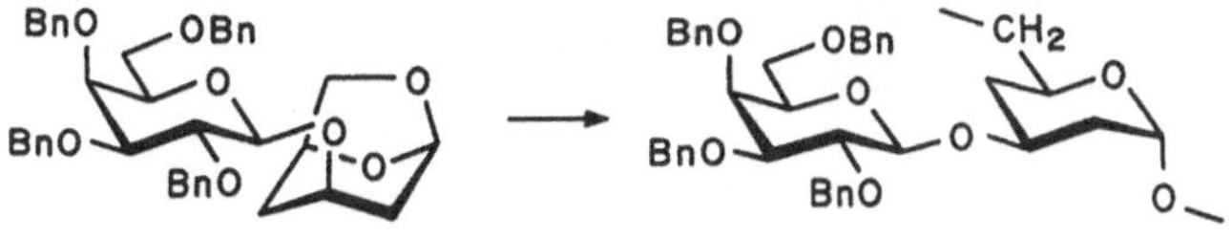

34   **35**

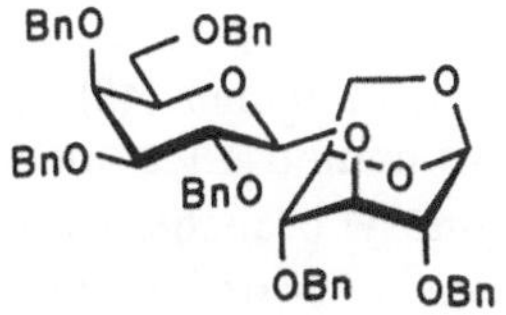

36

polymerization of 1,6-anhydro-3-O-(2,3,4,6-tetra-O-benzyl-β-D-galactopyranosyl)-2,4-dideoxy-β-D-*threo*-hexopyranose (**34**) using 20 mol% phosphorus pentafluoride at −78 °C proceeded rapidly to give a polymer having a glucoside branch in the repeating unit (**35**). [35] In contrast, no polymer was obtained from 1,6-anhydro-2,4-benzyl-3-O-(2,3,4,6-tetra-O-benzyl-β-D-galactopyranosyl)-β-D-glucopyranose (**36**) under similar conditions, but the monomer was recovered unchanged.

The hexabenzyl derivatives of 1,6-anhydro maltose (**37**) and cellobiose (**38**) polymerized very sluggishly at −60 °C. [36] Therefore, it appears that 1,6-anhydro disaccharide monomers with the (1→3)-glycosidic linkage are less reactive than those with the (1→4)-glycosidic linkage.

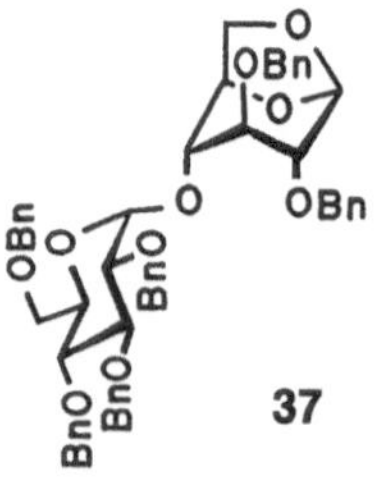

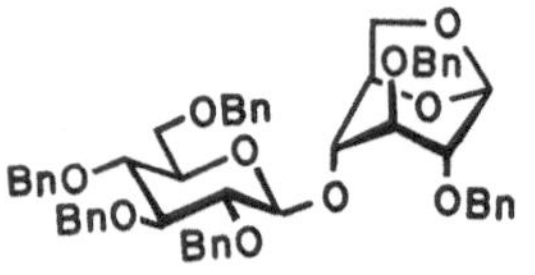

The high reactivity of **34** relative to **36** is probably due to its 2,4-dideoxygenated anhydro structure. Both the acetal oxygen of the monomer and the trialkyloxonium ion of the growing terminal in **34** are less sterically hindered and more basic and nucleophilic than those in the corresponding dibenzyloxylated compound (**36**).

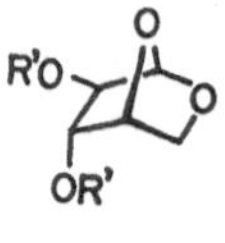

Copolymerization of 1,4-anhydro-2,3-di-O-(*tert*-butyldimethylsilyl)-α-D-xylopyranose (**39**) with 1,4-anhydro-2,3-di-O-benzyl-α-D-xylopyranose (**27**), followed by desilylation with tetrabutylammonium fluoride gave a partially benzylated stereroregular (1→5)-α-D-xylofuranan. [37] A branched polymer (**40**) was obtained when it was glycosylated with 3,4,6-tri-O-acetyl-β-D-mannose 1,2-(methyl orthoacetate). Debenzylation of the polymer having D-mannosyl branches with sodium in liquid ammonia yielded (1→5)-α-D-xylofuranans having 2- or 3-O-α-D-mannopyranosyl branches.

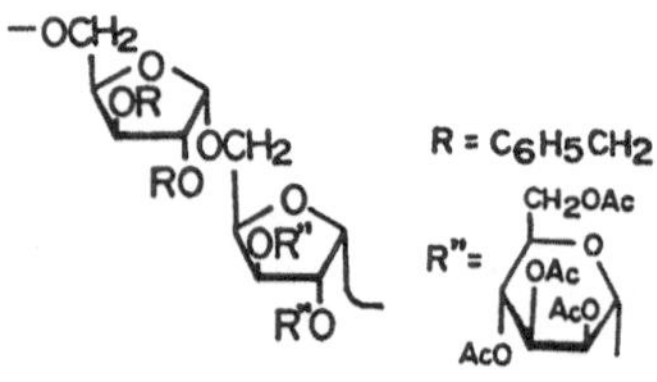

3 Ring-Opening Polymerization of Bicyclic Lactones

3.1 Selective Cyclooligomerization of Bicyclic Oxalactones Having a Bicyclo[3.2.1]octane Skeleton

Cyclic oligomers of various ring sizes are often formed in the ring-opening polymerization of cyclic compounds. [38] For example, in the anionic polymerization of ε-caprolactone by potassium *tert*-butoxide in tetrahydrofuran, fast polymerization to a polyester of high molecular weight was followed by slower depolymerization and yielded a mixture of cyclic oligoesters. [39] The ring size distribution in the mixture was satisfactorily explained by the Jacobson-Stockmayer theory. [40] This section describes cationic ring-opening polymerization of a bicyclic oxalactone, 6,8-dioxabicyclo[3.2.1] octan-7-one (**41**), to cyclic oligoesters of specific ring sizes. In the cationic ring-opening polymerization of **41**, scission occurs exclusively at the C(5)–O(6) bond and is accompanied by the inversion of the configuration of C(5). In this respect, **41** resembles the corresponding bicyclic acetal 6,8-dioxabicyclo[3.2.1]octane (**1**) described in the foregoing section. However, **41** shows specific polymerization behavior in its cationic polymerization.

Scheme 9 illustrates unusual polymerization behavior of **41**. [41] Cationic polymerization of **41** at 0 °C gave a polymer containing both the *trans*- and *cis*-linked 1,3-dioxolan-4-one rings (**42**) in the main chain. [42] A plausible mechanism involving the isomerization of polyester backbone chains has been proposed for the formation of the polymer **42**. In contrast, polyester (**46**) with a number-average molecular weight of 1.1×10^4 was preferentially produced at -60 °C in the initial stage of the polymerization. [43] However, it gradually converted to cyclic oligomers, predominantly 20-membered cyclic tetramer (**44**) and 25-membered cyclic pentamer (**45**) after a prolonged reaction. No polymer was formed from one of the enantiomers of **41**, $(+)$-$(1R, 5R)$-6,8-dioxabicy-

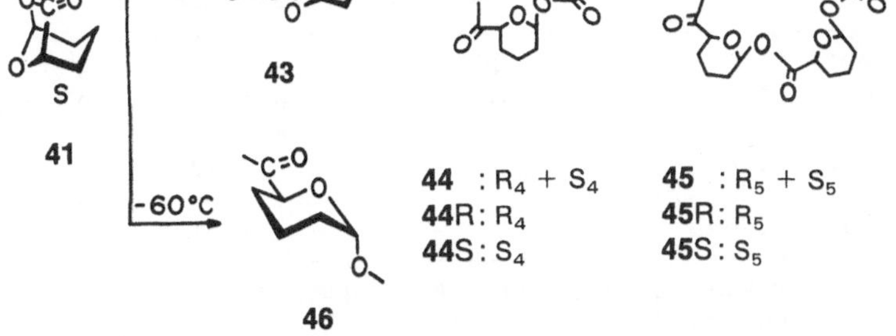

44 : $R_4 + S_4$ **45** : $R_5 + S_5$
44R: R_4 **45R**: R_5
44S: S_4 **45S**: S_5

Scheme 9

clo[3.2.1]octan-7-one (**41R**), under similar conditions. An optically active cyclic tetramer (**44R**) and an optically active cyclic pentamer (**45R**) were instead selectively formed even in the initial stage of the reaction. [43]

In the polymerization of **41** at −40 °C, only cyclic oligomers were formed except in the initial stage where higher oligomers and a polymer were formed. The sizes of the cyclic oligomers depend on the solvents employed for the polymerization. [44] 10-Membered cyclic dimer (**43**), **44**, and **45** were selectively formed in acetonitrile, chloroform, and 1-nitropropane, respectively. The cyclic pentamer **45** once produced preferentially in 1-nitropropane gradually changed to **43** after a prolonged reaction. No cyclic dimer was obtained from the optically active monomer **41R** even under similar conditions where **43** was selectively produced from the racemic monomer **41**, but the optically active cyclic oligomers **44R** and **45R** were preferentially formed at temperatures below −40 °C. The selective formation of these cyclic oligomers of specific ring sizes is closely related to their solubility in the solvents used. For example, the cyclic tetramer **44** was selectively produced in chloroform in which its solubility of **44** was much lower than the solubility of other cyclic oligomers. These results are summarized in Table 2.

The cyclic dimer **43** is a *meso* compound consisting of a pair of different enantiomeric units of **41**. [45] All four substituents attached to the two tetra-hydropyran rings are located in the axial positions. This compound has a center of symmetry and is readily crystallized. The cyclic tetramer **44** is a racemic mixture of optically active enantiomers, that is, **44R** and its enantiomer **44S**. [46] Similarly, the cyclic pentamer **45** is a racemic mixture of **45R** and its enantiomer **45S**. [47] In these cyclic oligomers, every exo-cyclic acetal oxygen occupies the axial position and every carbonyl carbon occupies the equatorial position of the tetrahydropyran ring. The molecules of these oligomers are chiral but not asymmetric (gyrochiral

Table 2. Selective oligomerization of racemic and optically active 6,8-dioxabicyclo[3.2.1]oc-tan-7-one (**41**)[a], [44b]

Monomer,		Solvent,[b]		Conversion,[c] %					
				Dimer	Tetramer		Pentamer		Other
g		ml		**43**	**44**	**44R**	**45**	**45R**	
(*RS*)	1.0	AN	1.0	96	0		0		0
(*RS*)	2.0	CF	2.0	0	91		0		0
(*RS*)	1.0	NP	2.0	3	0		58		0
(*R*)	1.0	AN	2.0	0		41		12	1
(*R*)	1.0	CF	1.0	0		0		46	0
(*R*)	1.0	NP	1.0	0		0		81	0

[a] Initiator, BF_3OEt_2, 1 mol% to monomer; temp., −40 °C; time, 24 h
[b] AN, acetonitrile; CF, chloroform; NP, 1-nitropropane
[c] By gel permeation chromatography

molecules). They form inclusion complexes with polar organic molecules such as acetonitrile and acetone. In addition, they also act as carriers for ion transport through liquid membranes. [48]

In the polymerization of the racemic monomer 41, there is a relatively small probability for the linear growing chain to consist of four or five identical enantiomeric monomeric units unless the monomer of the same chirality as the terminal unit of the growing chain is preferentially incorporated into the polymer chain. Actually, such enantiomer selection at the growing chain end also takes place to some extent in the polymerization of 41. This was proved by the preferential consumption of the enantiomer being in excess in a monomer mixture of 52% optical purity in the polymerization at −40 °C. [44] Therefore, it is likely that sequences of four or five consecutive monomeric units of the same chirality are formed at the growing chain end, from which the cyclic oligomers 44 and 45 are produced by back-biting. This mechanism explains why 44 and 45 are formed competitively with the polyester even in the early stage of the polymerization of 41.

The tendency for the growing chain end to form a cycle of four or five consecutive enantiomeric units is in sharp contrast to the polymerization behavior of the corresponding bicyclic acetal 1. Presumably, replacing the exocyclic methylene carbons of the polyacetal chain by carbonyl groups significantly changes the conformation of the growing chain, owing not only to a wider bond angle of the sp^2 carbons but also to the dipole-dipole interaction of the carbonyl groups with the tetrahydropyran rings.

By utilizing the specific oligomerization behavior of the bicyclic oxalactone described above, it was possible to prepare an optically pure cyclic tetramer and cyclic pentamer from a less optically pure monomer mixture, although the yield was low. [49] The results are presented in Table 3. Needless to say, 43 was concomitantly produced in larger amounts.

Table 3. Preparation of optically active cyclic tetramer (44S) and pentamer (45S) by the oligomerization of bicyclic lactone (41) of lower optical purity[a], [49]

Solv.[b]	S[c] M	Time, day	Conversion,[d] %		
			Dimer	Tetramer (44R : 44S)	Pentamer (45R : 45S)
AN	2	1	54	8 (1 : 99)	11 (0 : 100)
AN	2	3	60	4 (0 : 100)	11 (0 : 100)
NP[e]	1	1	50	0	29 (6 : 94)
NP	1	3	57	1	27 (3 : 97)
NP	1	5	57	1	27 (1 : 97)

[a] Monomer, 41R : 41S = 32 : 68; Initiator, BF_3OEt_2, 1 mol%; temp., −40 °C
[b] AN, acetonitrile; NP, 1-nitropropane
[c] Ratio of the volume (ml) of solvent to the weight (g) of monomer
[d] By gel permeation chromatography
[e] Monomer, 41R : 41S = 36 : 64

The two stereoisomers of 4-bromo-6,8-dioxabicyclo[3.2.1]-octan-7-one (**47a** and **47b**) showed different behavior in their cationic oligomerizations at −40 or 0 °C: [50] The axial isomer **47a** showed a tendency to cyclodimerize, particularly at higher temperature, whereas the equatorial isomer **47e** was much less reactive and gave only a small amount of a cyclic dimer along with other oligomers. Reductive debromination using tri-*n*-butylstannane converted the cyclic dimer **48** of **47a** to the cyclic dimer **43** of **41**, indicating that the former also consisted of a pair of the enantiomeric monomeric units.

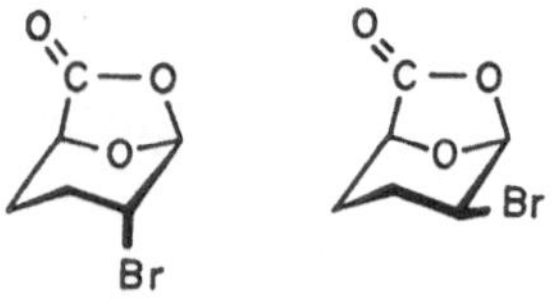

47a   **47e**

The formation of the dimer **48** from **47a** differs from the formation of the dimer **43** of **41** in that **48** is produced even in the early stage of the oligomerization, whereas **43** is formed after the middle stage of the oligomerization. This difference suggests different mechanisms operative in the cyclodimerizations of **47a** and **41**. As illustrated in Scheme 10 [50], neighboring group participation of the axially oriented bromine atom of **47a** facilitates the formation of the cyclic dimer **48**.

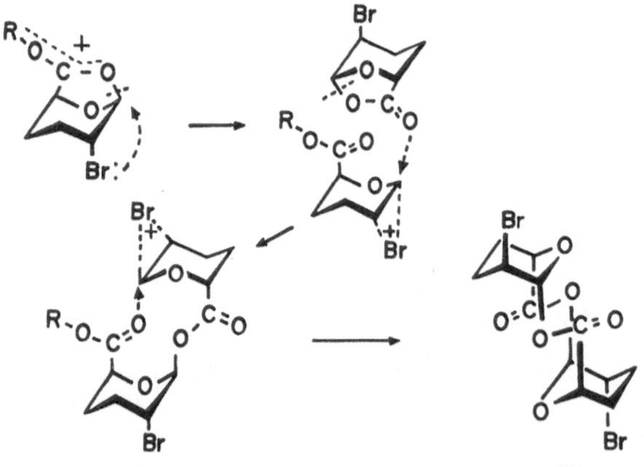

48

Scheme 10

In the oligomerization of **47e**, the neighboring group participation of the equatorially located bromine atom is impossible, unless the oxonium ion is unimolecularly ring-opened to form an oxacarbenium ion. The electron withdrawing bromine atom disfavors such an unimolecular ring-opening reaction. Therefore, the cyclic dimer cannot be formed by a mechanism similar to that described for the cyclodimerization of **47a**.

3.2 Polymerization of Bicyclic Lactones Having a Bicyclo[2.2.2]octane Skeleton

2,6-Dioxybicyclo[2.2.2]octan-3-one (**49**), an isomer of **41**, is also a bicyclic acetal-ester, but behaves differently in its ring-opening polymerization. [51] This monomer was synthesized from acrolein and dimethyl malonate as starting materials. Its X-ray analysis revealed a significant shortening of the C(1)−O(6) bond (1.390 Å) compared with the C(5)−O(6) bond (1.439 Å) and a stretching of the C(1)−O(2) bond (1.458 Å) (cf. the ordinary C−O single bond, 1.43 Å). Thus, during its cationic polymerization, the bond cleavage of **49** is very likely to occur at the C(1)−O(2) bond, whereas the C(1)−O(6) bond is intact, according to the theory of Kirby et al. [52]

49   **50**

Bicyclic oxalactone (**49**) is readily polymerized with boron trifluoride etherate at or below −60 °C to give a high molecular weight polyester (**50**) (M_n 1×10^5) in a high yield. The bicyclic lactone is also polymerizable with anionic initiators, although higher temperature and longer reaction time are required. The polyester **50** is soluble in chloroform, dichloromethane, γ-butyrolactone, and pyridine, and swells in tetrahydrofuran and 1,4-dioxane. Transparent flexible film can be obtained by casting its chloroform solution, but it decomposes at about 150 °C (DSC).

The polyester **50** consisted of the *cis*- and *trans*-2,5-linked tetrahydropyran rings (hereafter referred to as *cis*- and *trans*-units, respectively). The proportions of the *cis*- and *trans*-units varied from 11/89 to 76/24, depending on the reaction conditions. The polymer having the lowest *cis/trans* ratio of 11/89 was obtained when the polymerization at −90 °C was terminated at a low conversion. Even at −90 °C, the *cis/trans* ratio increased to 30/70 with increasing conversion and reaction time. This ratio also tended to increase as the polymerization temperature was raised.

The propagation reaction in this polyester formation proceeds primarily through the S_N2 type alkyl-oxygen scission of the ester linkage of the growing oxonium ion at low temperature, and yields a product polyester consisting of *trans*-units. However, transacetalization inevitably occurs, converting the *trans* unit to the energetically more stable *cis* counterpart (anomeric effect). Thus, the structure of the polymer becomes less regular after the polymerization for a relatively long time or at higher temperature. Concurrent S_N1 type propagation by the oxacarbenium ion giving both *cis* and *trans* units cannot be ruled out at such temperature.

The anionic polymerization of **49** initiated with lithium benzophenone ketyl gave the polymer **50** having a *cis/trans* ratio of 42/58. Probably it proceeded through the acyl-oxygen scission of the ester linkage of the monomer. The formation of both *cis* and *trans* units can be interpreted in terms of the epimerization at the growing chain end. It is possible to convert the *cis* alkoxide anion (51c) producing the *cis* unit in the polymer chain to the *trans*-alkoxide anion (51t) through a ring-chain equilibrium. Thus, one *trans* unit is introduced by addition of a monomer to 51t. Epimerization induced by hydrogen abstraction of the methine proton adjacent to the carbonyl group seems to be negligible, if inferred from the anionic polymerization of 2,5-dioxabicyclo[2.2.2]octan-3-one (54) under similar conditions or even at higher temperatures, which will be discussed later.

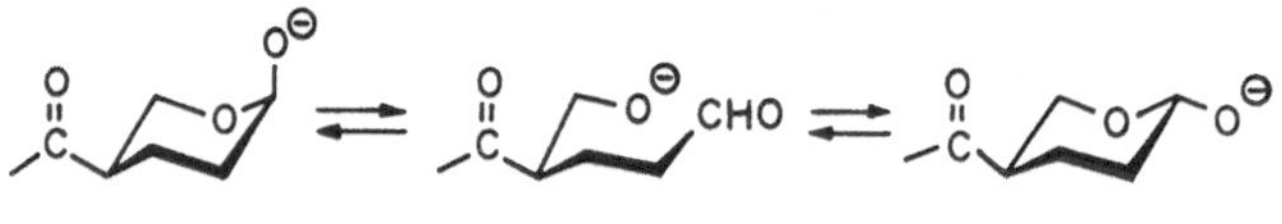

51c   **51t**

4-Methoxycarbonyl-2,6-dioxabicyclo[2.2.2]octan-3-one (52) readily undergoes cationic polymerization at $-90\,^\circ$C to give a high molecular weight polyester (53) having tetrahydropyran rings in the backbone chain and methoxycarbonyl groups as side chains. [53] Introduction of a methoxycarbonyl group at the bridge head carbon significantly reduces the reactivity as can be inferred from the apparent monomer reactivity ratios in the copolymerization of 49 with 52 (r_1(49) $= 17 \pm 4$, r_2(52) $= 0.3 \pm 0.1$).

52   **53**

Compared with 49, 2,5-dioxabicyclo[2.2.2]octan-3-one (54) prepared from sodium 3,4-dihydro-2*H*-pyran-2-carboxylate has a much low polymerization reactivity: [54] Lewis acids such as antimony pentachloride, phosphorus pentafluoride, and boron trifluoride etherate were not effective at all to initiate the polymerization of 54. Trifluoromethanesulfonic acid induced the polymerization of 54, but the yield and molecular weight of the polymer were low. Bicyclic lactone 54 was allowed to polymerize with anionic and coordination initiators such as butyllithium, lithiumbenzophenone ketyl, and tetraisopropyl titanate. However, the

54   **55**

yields and number-average molecular weights of the polymers were low. A fractionated polymer having a number-average molecular weight of 1.4×10^4 melted at 281–297 °C (DSC).

This polymer consists entirely of cis-2,5-disubstituted tetrahydropyran rings (cis-unit) (55). From the estimation of the conformational energies of its two conformers, it has been concluded that the dominant conformation of the repeating unit of the polyester 55 has the carbonyl carbon atom in the equatorial position and the ester oxygen atom in the axial position of the tetrahydropyran ring. The formation of 55 gives definite evidence for the occurrence of the acyl-oxygen scission of the ester linkage in the polymerization of 54 with anionic initiators. Thus, the anionic polymerization of 54 propagates through an alkoxide anion without side reactions, which alter the configuration of the asymmetric carbons in the repeating units.

Anionic polymerization behavior of 2-oxabicyclo[2.2.2]octan-3-one, an analogue of 54, is different from that described above. [55] Thus, the polymerization initiated with n-butyllithium propagates by alkoxide growing species, whereas the polymerization initiated by sodium tert-butoxide and sodium-potassium alloy propagates by both alkoxide and carboxylate growing species. The latter yields a polyester consisting of cis- and trans-1,4-linked cyclohexane rings.

One of the reasons for the lower reactivity of 54 in its anionic and coordination polymerizations is the poor solubility of the polymer, which tends to occlude the growing species in the precipitated polymer as the polymerization proceeds. Furthermore, there are at least two other factors: According to the quantum chemical calculation by Burgi et al., [56] a nucleophile approaches a carbonyl carbon atom along the path for which the angle α between the line joining the nucleophile with the carbonyl carbon and the direction of the C=O bond is in general 100–110 °C (Scheme 11 [54]). The anionic polymerization of 54, therefore, must overcome a severe steric hindrance, regardless of whether the alkoxide anion attacks the carbonyl carbon atom (56) from the upper or lower side of the plane containing the ester linkage. Moreover, the addition of an alkoxide anion to the carbonyl carbon converts the sp^2 carbon atom to the sp^3 carbon, thus giving rise to synperiplanar interactions of the polar C–O- and C–OR bonds with the electron pairs of the adjacent O(2) atom (57). [57] An additional synperiplanar interaction arises between one of the polar C–O bonds and the electron pair of the O(5) atom. This factor also makes the formation of the adduct 57 energetically unfavorable.

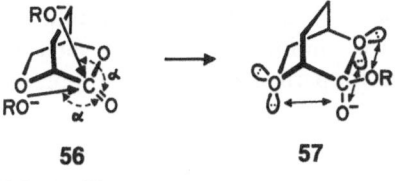

56   **57**

Scheme 11

A carboxylate growing anion should be excluded from the propagating species, because the polyester obtained contains cis-units only. This means that the strain energy of the bicyclic lactone **54** is not so large as to allow the formation of a carboxylate anion by the $C(1)-O(2)$ bond scission. In fact, the strain energy of 2-oxabicyclo[2.2.2]octan-3-one, an analogue of **49** and **54**, was evaluated to be 41.5 kJ/mol. [58] This is much smaller than the strain energies of β-propiolactone (94.2 kJ/mol) [59] and pivalolactone (85.8 kJ/mol), [60] which undergo anionic polymerization not only through the acyl-oxygen scission but also through the alkyl-oxygen scission.

The poor cationic polymerizability of **54** can be explained as follows: Addition of cationic species is most likely to take place on the carbonyl oxygen atom having the highest electron density in the monomer, thus leading to the alkyl-oxygen scission of the ester linkage. [61] However, such bond cleavage is less likely to occur, because of the relatively small strain energy of the bicyclic structure of **54**. In fact, the polyester obtained with trifluoromethanesulfonic acid entirely consisted of *cis*-2,5-linked tetrahydropyran rings. This implies that the polymerization proceeded through the acyl-oxygen scission, in sharp contrast to the anionic polymerization of **49**, which occurs through the alkyl-oxygen scission assisted by the presence of the adjacent endocyclic oxygen atom. Therefore, when a cationic species adds onto the ester oxygen atom $O(2)$ of **54**, the resulting oxonium ion would be cleaved at the acyl-oxygen bond by the nucleophilic addition of the monomer to the electropositive carbonyl carbon atom. However, such nucleophilic addition of the bulky monomer is severely impeded for the reasons mentioned above, thus lowering the cationic polymerizability of **54**.

4 Ring-Opening Polymerization of Bicyclic Lactams

4.1 Polymerization of Bicyclic Lactams Having a Bicyclo[3.2.1]octane Skeleton

Ring-opening polymerization of 8-oxa-6-azabicyclo[3.2.1]octan-7-one (**58**) has been most extensively investigated among bicyclic lactams. It is readily synthesized from sodium 3,4-dihydro-2*H*-pyran-2-carboxylate and polymerizes anionically at room temperature to give an amphiphilic polyamide (**59**) consisting of *cis*-2,6-linked tetrahydropyran rings. Membranes of **59** and its block copolymers were found to have excellent permeability for water, and permselectivity for alkali metal ions and solutes of various sizes in aqueous solution. [62, 63] Of particular interest is the unique phenomenon that when aqueous solutions of a protein such as myoglobin and cytochrome C were allowed to permeate through a porous

58   **59**

membrane of an ABA block copolymer whose A and B segments were **59** and polyethylene glycol, respectively, they were concentrated. [64, 65] This behavior will be discussed in detail elsewhere. [2]

Anionic ring-opening polymerization of lactams by the "activated monomer" mechanism always involved acyllactams in propagation as well as in initiation. In order to clarify the structure-reactivity relationship of acyllactams, aminolysis of a variety of N-acyllactams with n-butylamine was investigated kinetically in dimethylformamide at 25 °C. [66]. The amine reacts with both exocyclic and endocyclic carbonyl groups (path 1 and path 2 in Scheme 12) in N-benzoyl compounds of bicyclic oxalactam **58**. However, it reacts only with the exo-cyclic carbonyl group in monocyclic N-benzoyl-2-pyrrolidone and N-benzoyl-ε-caprolactam (path 1). As shown in Table 4, the electron-withdrawing substituent in the N-benzoyl derivative of **58** enhances the reactivities of both *exo-* and *endo-*carbonyl groups, while the electron-releasing one in this derivative reduces them. Therefore, N-benzoyllactams having an electron-withdrawing substituent should act more effectively as activators in the anionic polymerization of lactams.

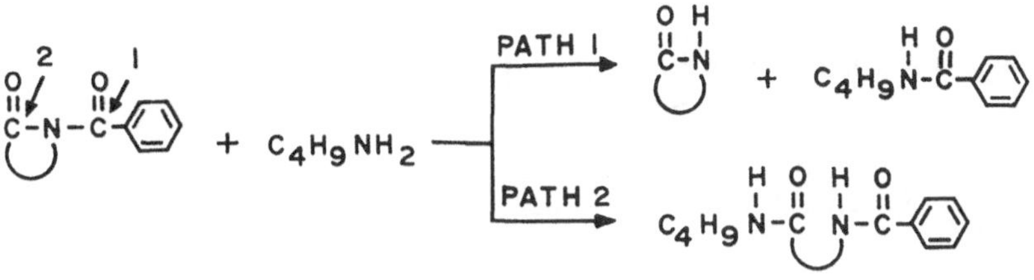

Scheme 12

The anionic polymerization of **58** shows typical equilibrium polymerization behavior. From the temperature dependence of the equilibrium monomer concentration, the thermodynamic parameters for the polymerization of **58** in dimethyl sulfoxide were evaluated to be $\Delta H_{ss} = -23.8 \pm 1.5$ kJ/mol and $\Delta S_{ss} = -71.5 \pm 4.2$ kJ/mol · deg (subscript ss refers to a solution state). [67] The ceiling temperature for 1 mol/L solution is about 60 °C. The enthalpy change in the polymerization of **58** is considerably larger than those for monocyclic lactams, pyrrolidone and piperidone, but no quantitative comparison of these data can be made, because the reported data refer to different experimental conditions. The significant entropy decrease in the polymerization is ascribable to the presence of the six-membered tetrahydropyran ring in the repeating unit.

Anionic polymerization of **58** activated by N-benzoyl lactam proceeds without side reactions. Since side reactions in the anionic polymerization of lactams are mainly caused by protonabstraction, the pKa value for the bridge-head methine proton adjacent to the lactam-carbonyl group in **58** must be higher than that for the α-methylene protons in 2-pyrrolidone. This is because the former monomer has a rigid bicyclic structure.

A monodisperse polyamide was prepared by the anionic polymerization of **58** using its N-benzoyl derivative and potassium pyrroridonate as the activator and

Table 4. Estimation of relative reactivity of N-(p-substituted benzoyl) derivatives of **58** in the aminolysis with n-butylamine in dimethylformamide at 25 °C [66]

Subst.	σ^a	exo-Carbonyl group $K_1(k_2 + k_3) \times 10^{3,b}$ $l^2\,mol^{-2}\,s^{-1}$	endo-Carbonyl group $K_1(k_2' + k_3') \times 10^{3,b}$ $l^2\,mol^{-2}\,s^{-1}$
NH$_2$	−0.66	0.2	0.4
OCH$_3$	−0.27	2.4	3.3
H	0	5.0	2.9
CN	0.66	102	108
NO$_2$	0.78	110	98

[a] σ-Values for p-substituents in the Hammett equation
[b] K_1, k_2, and k_3 are the equilibrium constant and rate constants of the following reactions. k_2' and k_3' are the rate constants of the corresponding reactions at the endo-carbonyl group

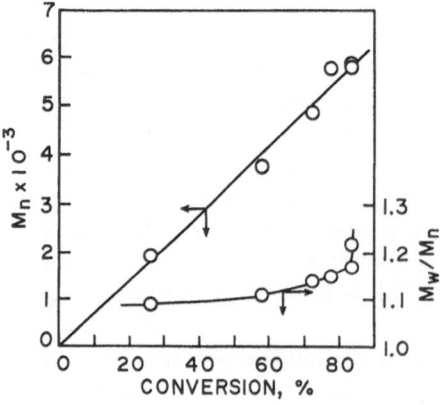

the catalyst, respectively, in dimethyl sulfoxide at 25 °C. [68] The number-average molecular weight of the polyamide **59** increased in direct proportion to the monomer conversion, and was consistent with the value calculated from the amounts of the consumed monomer and activator. Figure 2 shows that the

Fig. 2. Number-average molecular weight and molecular weight distribution of the polyamide obtained by anionic polymerization of bicyclic lactam **58** as a function of conversion. Polymerization conditions: [M]$_0$, 2.11 mol l^{-1} potassium pyrrolidonate, 0.5 mol%; activator, N-benzoyl derivative of **58**, 2 mol%; solvent, Me$_2$SO; temp., 25 °C

molecular weight distribution of **59** remains narrow ($M_w/M_n = 1.1$) until the polymerization reaches the middle stage (monomer conversion $<60\%$). The molecular weight distribution becomes broader in the later stage of the poly- merization, especially after the monomer conversion has reached a constant value (equilibrium conversion). The broadening can be attributed to the redistribu- tion of the molecular weight of **59**, due not only to the polymerization-depoly- merization equilibrium, but also to the transamidation between the polymer chains.

Even at room temperature, the acyllactam-type chain end is attacked by the amide anion in the polymer chain as well as by the lactamate anion in the later stage of the polymerization (Scheme 13 [68]). Not only the acyllactam group at the chain end but also the amide groups inside the polymer chain are attacked by these anions. Hence, scrambling of the monodisperse polymer chains formed in the initial stage should occur, thus broadening the molecular weight distribution. The total amount of the reactive acyllactam groups in the polyamide chain undergoes no change by the transamidation. Consequently, the number-average molecular weight will change little after the polymer- monomer equilibrium is reached, though the molecular weight distribution is broadened.

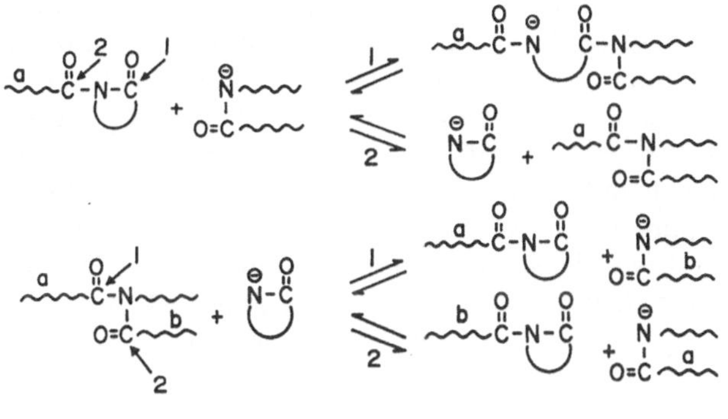

Scheme 13

In the presence of a cationic initiator such as trifluoromethanesulfonic acid and boron trifluoride etherate, **58**, gave a mixture of dimers and higher oligomers. [69] On the basis of structural analyses of the dimers prepared from the racemic and optically active monomers, it was concluded that the cationic oligomerization involved protonation onto the amide nitrogen of **58** followed by selective scission of the C(5)−N(6) bond, while the C(5)−O(8) bond remained intact.

Anionic polymerization of a sodium salt (**61**) of a bromosubstituted bicyclic lactam (**60**) is very unusual. It proceeds in dimethylformamide at 0 and 25 °C with elimination of sodium bromide and gives oligomers **62** having a bicyclic oxalactam ring in each repeating unit. [70]

From a nearly equimolar mixture of **60** and **61**, two diastereomeric dimers (major dimer **63**: minor dimer **64** = 87:13) were produced as the major products, along with a small quantity of higher oligomers. Structural analysis of the isolated

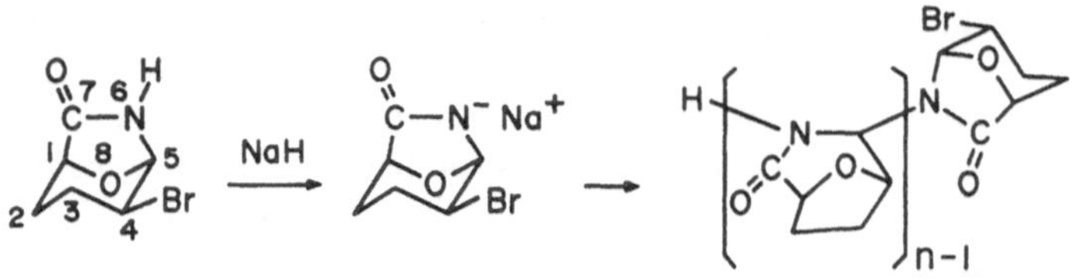

60   **61**   **62**

dimers indicated that the oligomerization had proceeded with high enantiomeric selectivity and completely retained the configuration of the bridge-head methine carbon adjacent to the nitrogen atom.

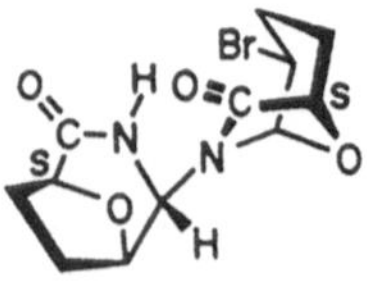

63 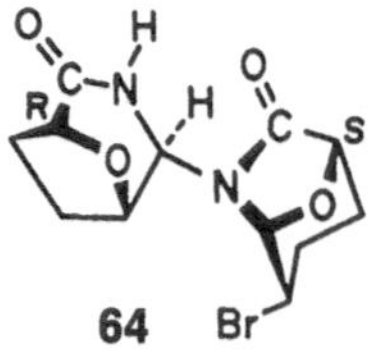   **64**   Br

Scheme 14 [70] illustrates a mechanism with which the major dimer **63** is formed. Nucleophilic attack of the anion **61** on the methine carbon adjacent to the nitrogen atom in **60** of the same configuration cleaves the C(5)−O(8) bond in the latter to form a seven-membered lactam ring. The alkoxide anion of this intermediate reacts intramolecularly with the bromo-substituted methine carbon. As a result, the bromine ion is eliminated and a bicyclic skeleton with a six-membered oxalactam ring is formed. Epimerization of the methine group between the two nitrogen atoms occurs and relieves the steric hindrance around the C−N bond joining the two units. This mechanism leads to the major dimer **63**. A similar reaction of the anion **61** with **60** having the opposite configuration gives the minor dimer **64**.

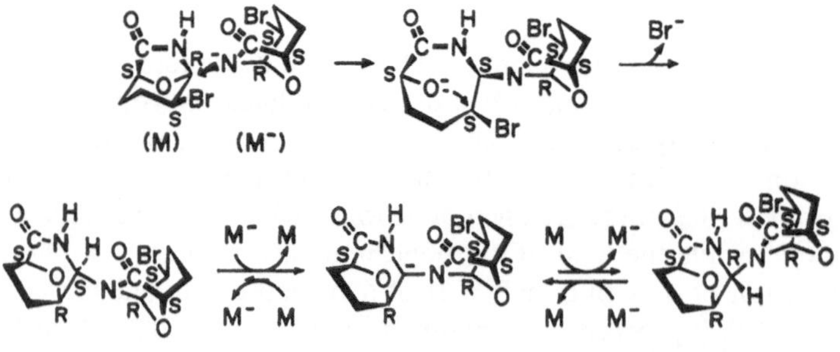

63

Scheme 14

8-Oxa-3-azabicyclo[3.2.1]octan-2-one (**65**) derived from methyl furoate was allowed to polymerize in bulk with its *N*-acetyl derivative as the initiator, and sodium (or potassium) tetraisobutyl aluminate as the catalyst. [71] The bicyclic lactam **65** was also allowed to polymerize hydrolytically in bulk. The resulting polyamide **66** having an intrinsic viscosity of 0.3 dL/g showed the glass transition at 118 °C. No melting was observed below 350 °C at which decomposition began.

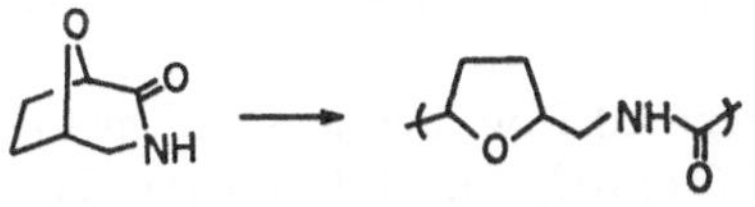

65   **66**

4.2 Polymerization of Bicyclic Lactams Having a Bicyclo[2.2.2]octane Skeleton

Anionic ring-opening polymerization of 2-oxa-5-azabicyclo[2.2.2]octan-6-one (**67**), which was derived from sodium 3,4-dihydro-2*H*-pyran-2-carboxylate, gave a polyamide (**68**) containing *cis*- and *trans*-2,5-linked tetrahydropyran rings in the main chain. [72] The *cis/trans* ratio varied from 70/30 to 95/5, depending on the reaction conditions. The polyamides having number average molecular weights of 3000–8000 were soluble in various solvents including methanol, ethanol, dimethyl sulfoxide, dimethylformamide, chloroform, and dichloromethane. They began to decompose at 280–320 °C, depending on the composition of *cis*- and *trans*-2,5-linked structural units. The *cis* unit content decreased with increasing temperature and also with increasing molar ratio of the catalyst to the activator.

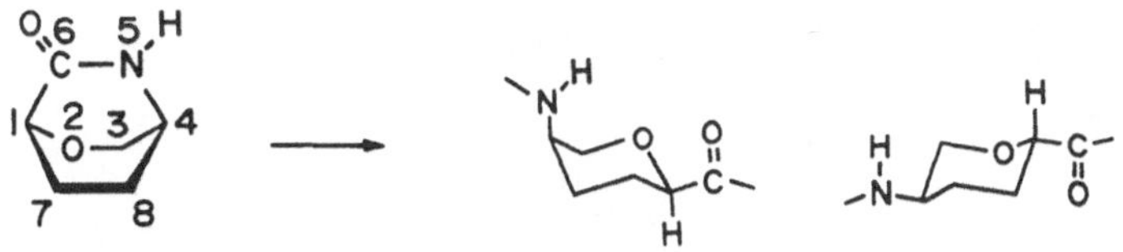

67   *cis*   **68**   *trans*

Anionic polymerization of **67** by the "activated monomer" mechanism should occur with the selective cleavage of the CO−NH bond of the monomer to give a polyamide composed of kinetically controlled *cis* units (**68c**). However, the *cis* units isomerize to the thermodynamically more stable *trans* units (**68t**) through the proton abstraction from the methine group adjacent to the carbonyl group. This was ascertained by the isomerization experiment in which a polymer consisting of 92% *cis* unit and 8% *trans* unit was converted to one containing 40% *cis* unit and 60% *trans* unit when heated in dimethyl sulfoxide at 80 °C for 6 hours in the presence of 15 mol% potassium pyrrolidonate.

The anionic polymerizability of **67** is much lower than that of **58**. This difference is due not only to the lower ring-strain of **67** and but also to a greater steric hindrance that the lactam anion of **67** receives when it comes closer to the carbonyl carbon of the **67** unit at the chain end. This hindrance has been discussed in relation to a relatively low anionic ring-opening polymerizability of the corresponding bicyclic oxalactone **54**. As to the ring strain, it is relevant to point out that **67** has the amide linkages C(1)—C(6)—N(5) of 109.9° and C(4)—N(5)—C(6) of 115.4°. These bond angles are larger than the corresponding ones for **58**, 104.8° and 110.1°, [73] and therefore indicate that the ring strain of **67** is smaller than that of **58**.

2-Oxa-6-azabicyclo[2.2.2]octan-5-one (**69**) prepared from acrolein and methyl malonate has the amide linkages C(4)—C(5)—N(6) of 109.9° and O(2)—C(1)—N(6) of 109.0°. [74] These bond angles are comparable to the corresponding ones for **67** but are larger than the corresponding ones for **58**. This difference means that the angle strain of the lactam ring moiety of **69** is smaller than that of **58**. In marked contrast to **58** and **67**, anionic polymerization of **69** gives no polyamide, but only a small amount of a dimeric adduct (**70**).

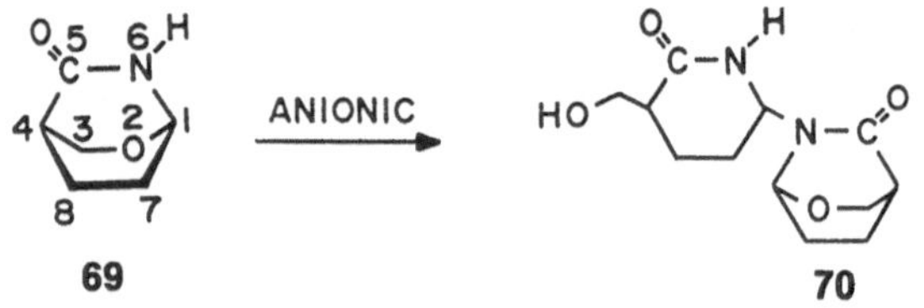

This adduct is presumably formed by the mechanism illustrated in Scheme 15 [74]. As in the conventional anionic polymerization of other lactams, **69** reacts with potassium pyrrolidonate to give a lactam anion **71**. However, as soon as **71** is

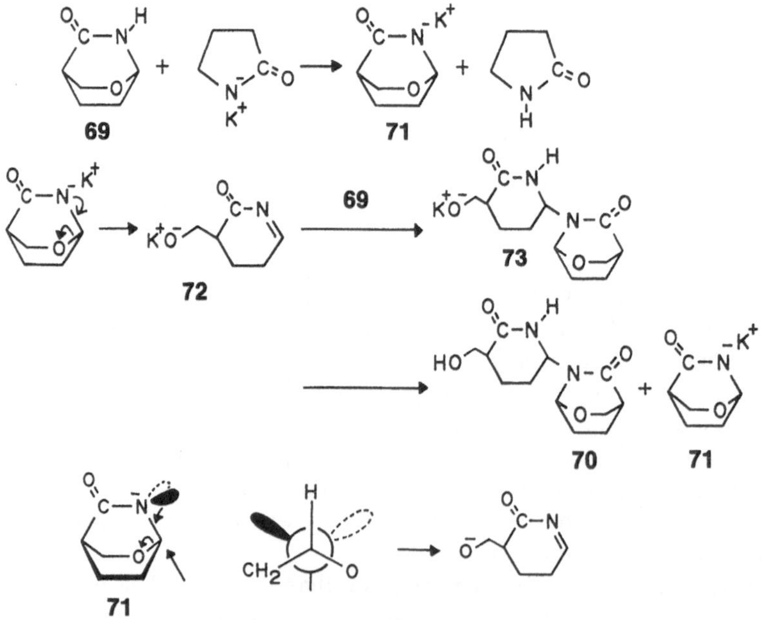

Scheme 15

formed, it must be ring-opened at the C(1)—O(2) bond to form an alkoxide anion 72 before attacking the activator molecule. The nucleophilic addition of 69 occurs on the reactive carbon atom of the C=N bond in 72 and produces a dimer anion 73. Proton abstraction from 69 by 73 would yield 70 and regenerate 71. However, this step should be slow, because 73 is of lower basicity than 71, thus explaining why the yield of 70 was low.

The lack of anionic polymerizability of 69 can be ascribed to the instability of 71 due to a stereoelectronic effect. [9] One of the lone pair orbitals on the nitrogen atom of 71 is perfectly antiperiplanar to the C(1)—O(2) bond, which causes this bond to cleave facily. Thus, as soon as 71 is formed, it is transformed to 72, so that anionic polymerization by the "activated monomer" mechanism does not take place.

According to X-ray analysis, the C(1)—O(2) bond length of 69 is 1.43 Å and the tetrahydropyran ring is nearly symmetrical. This is in sharp contrast to the shortness of the corresponding C—O bond (1.39 Å) and therefore to the asymmetrical molecular dimension of the tetrahydropyran ring in the bicyclic lactone 49. The C(1)—O(2) bond of this bicyclic oxalactone should undergo no cleavage in the cationic polymerization, according to Kirby's theoretical prediction. [52] However, the C(1)—O(2) bond of 69 is readily cleaved. Cationic polymerization of 69 proceeded relatively easily at or above −20 °C and yielded an oligomeric polyether (74) containing six-membered lactam rings in the main chain, not a polyamide containing tetrahydropyran rings. This finding gives definite evidence for the selective cleavage of the C(1)—O(2) bond in the cationic polymerization of 69.

4.3 Polymerization of Bicyclic Lactams Having a Bicyclo[2.2.1]heptane Skeleton

2-Azabicyclo[2.2.1]heptan-3-one (75) is more strained and therefore should be more reactive than the bicyclic lactams described in the foregoing sections. The monomer 75 was polymerized in bulk or in solutions in dimethyl sulfoxide and tetrahydrofuran at temperatures between 40 and 120 °C. [75] The resulting polyamide (76) having cyclopentane rings in the main chain has good thermal stability at temperatures up to 300 °C. Its melting point and decomposition temperature are about 307 and 335 °C, respectively.

An unsaturated bicyclic lactam, 2-azabicyclo[2.2.1]hept-5-en-3-one (77), which was synthesized by the Diels-Alder reaction of tosyl cyanide with cyclopentadiene

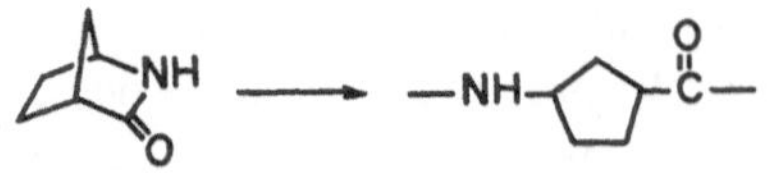

75   76

followed by hydrolysis, was polymerized by using metathesis catalysts tungsten chloride-organometallic compounds. [76] The best result (34% conversion and 0.18 dL/g inherent viscosity) was obtained when the molar ratio of **77** to tungsten chloride as the catalyst was 200 and that of tungsten chloride to triethylaluminum as the cocatalyst was 4. Spectroscopic analyses of the products showed that the ring-opening polymerization of **77** occurred to yield poly(2-pyrrolidone-3,5-diylvinylene) (**78**). This polymer was amorphous and showed only a second-order transition at 100 °C.

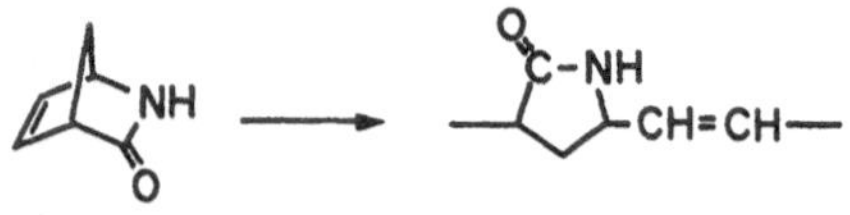

77   78

5 Polymerization of Bicyclic Orthoesters

In general, bicyclic acetals are highly reactive toward acid-catalyzed hydrolysis. For example, the hydrolysis of 2,7-dioxabicyclo[2.2.1]heptane (**2**) in aqueous acetone containing dichloroacetic acid was 2.5×10^4 times faster than that of an acyclic reference compound, dimethyl acetal. [77] 2,6-Dioxabicyclo[2.2.1]heptane (**79**) is more reactive, and hydrolyzed 6.9×10^5 times faster than dimethyl acetal. [78] These rate accelerations for bicyclic acetals arise from a partial liberation of the ring strain on hydrolysis, and have been correlated to the reactivities of bicyclic acetals in the cationic ring-opening polymerization. [5]

79   80

In marked contrast, bicyclic orthoesters are less reactive than another acyclic reference compound, trimethyl orthoformate. [79] For instance, the rate of hydrolysis of 2,6,7-trioxabicyclo[2.2.1]heptane (**80**) was slower by about 50% than that of trimethyl orthoformate. In addition, there are relatively small differences among the rate constants for the bicyclic orthoesters of [2.2.1]heptane, [3.3.1]no-nane, [3.2.1]octane and [4.2.1]nonane systems. This unexpected behavior was rationalized in terms of an early transition state of hydrolysis which involves very little ring-opening and hence little strain relief. [80] Acceleration of hydrolysis, though weak, by the phenyl substitution at the pro-acyl carbon atom in **80** indicates some carbon-oxygen bond breaking taking place before this transition state is reached. Differing from the bicyclic acetals, no correlation is found for the bicyclic orthoesters between the hydrolytic reactivity and the reactivity toward cationic initiators. The following order of polymerization reactivity has ben proposed: [2.2.1] > [3.2.1] > [3.3.1]. [81, 82] This order can be expected from the ring strains as in the polymerization of bicyclic acetals.

Polymerization of **80** at −78 °C gave a polymer consisting mostly of five-membered rings (**81**), which indicates that dominant cleavage takes place at the C(1)−O(2) bond. [81] 1-Methyl-2,6,7-trioxabicyclo[2.2.2]heptane also gave a polymer, which was found to consist of five membered rings from ^{13}C NMR spectra, contrary to an earlier report. [83]

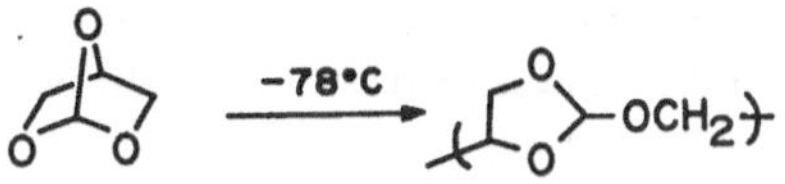

80   **81**

The selective bond cleavage of the C(1)−O(2) bond is kinetically controlled. The O(2) and O(6) atoms have a higher p-character than the O(7) atom, because the bond angles of the first two oxygen atoms are close to 100°, whereas the bond angle of the last oxygen atom is 90°. Therefore, the O(2) and O(6) atoms have higher nucleophilicity than the O(7) atom. Moreover, the C(1)−O(2) bond in the oxonium ion (**82**) produced by the nucleophilic addition of the O(2) oxygen is antiperiplanar to one of the orbitals of the unshared electrons on the O(6) and O(7) atoms. Consequently, it should be readily cleaved, according to Deslongchamps' theory. [9]

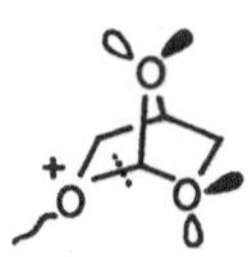

82

On the basis of kinetic studies on the polymerization of **80** with 1,3-dioxolan-2-ylium ion, an A_c2 mechanism (bimolecular addition on a carbenium ion) has been proposed for this polymerization. This mechanism assumes the bimolecular reaction of the cyclic dioxacarbenium ion with the monomer as the rate-determining step and predicts the preparation step to be nonstereospecific, that is, the formation of a polymer consisting of cis- and trans-2,4-linked 1,3-dioxolane rings.

At temperatures higher than 80 °C, polyethers (**83**) having pendant formate moieties were produced by double ring-opening in the polymerization of **80**.

[84, 85] This reaction is thermodynamically controlled, because the polymer containing 1,3-dioxolane rings converts itself to a polyether when allowed to stand at room temperature for several days or heated at 80 °C for a few hours in the presence of an acid catalyst. Similar double ring-opening polymerizations were observed for 2,6,7-trioxabicyclo[2.2.2]octane and its derivatives [86, 87] and for spiro orthoesters and spiro orthocarbonates as well (see Sects. 6 and 7).

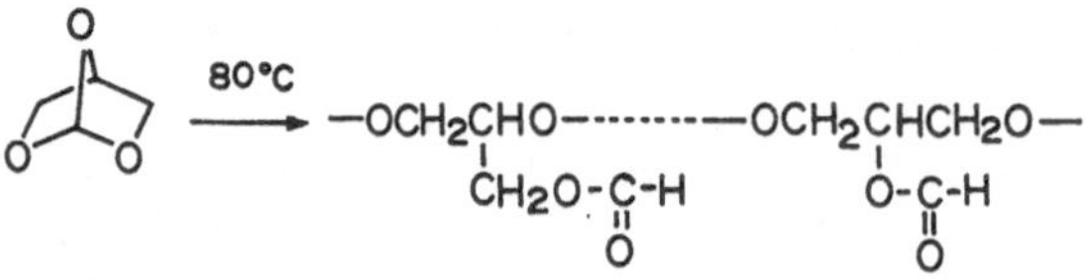

In the cationic polymerization of 2,7,8-trioxabicyclo[3.2.1]octane (84), most cleavage occurred at the C(1)−O(2) bond to yield a polymer consisting chiefly of five-membered rings (85). Six- and seven-membered rings (86 and 87) were also formed by the C(1)−O(7) and C(1)−O(8) bond cleavages, respectively, although their total content was about 15% [88].

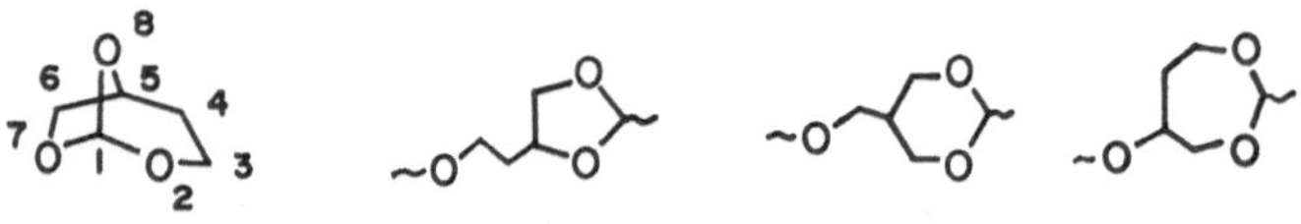

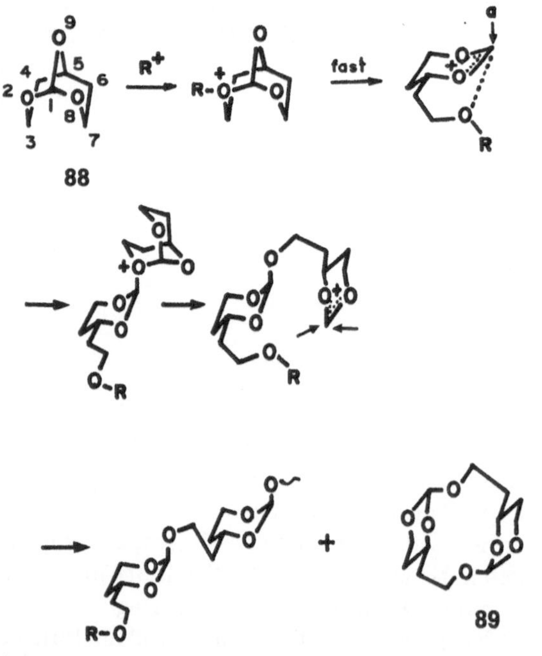

Scheme 16

2,7,8-Trioxabicyclo[3.3.1]nonane (**88**), though consisting of two-fused six-membered rings in the chair form, underwent cationic oligomerization that mostly yielded cyclic dimers (**89**). [88] This fact clearly indicates the existence of a small but significant ring strain, which probably is due to the repulsion between the two endo-protons at the C(3) and C(7) atoms. According to an NMR analysis of the dimer, only one isomer of the dimer is formed, and presumably, this fact is ascribable to a weak dipole-dipole interaction between the dioxacarbenium ion and the first oxygen atom in the chain, which forces the incoming monomer to react with the dioxacarbenium ion in a stereoselective fashion (Scheme 16). The formation of the dimer may be explained in terms of back-biting of the growing chain end.

6 Ring-Opening Polymerization of Spiro Orthoesters

Spiro orthoesters and spiro orthocarbonates as well as bicyclic orthoesters often undergo cationic polymerization involving double ring-opening. Because of the chemical transformation of compact bicyclic monomers to linear polymers, most of these monomers show little or no volume shrinkage on polymerization. [90]

On the basis of NMR studies, Matyjaszewski [91] suggested that the cationic polymerization of a spiro orthoester, 1,4,6-trioxaspiro[4.4]nonane (**90**), involved a fast single ring-opening reaction followed by slow intramolecular isomerization to give poly(ether-ester) (**91**).

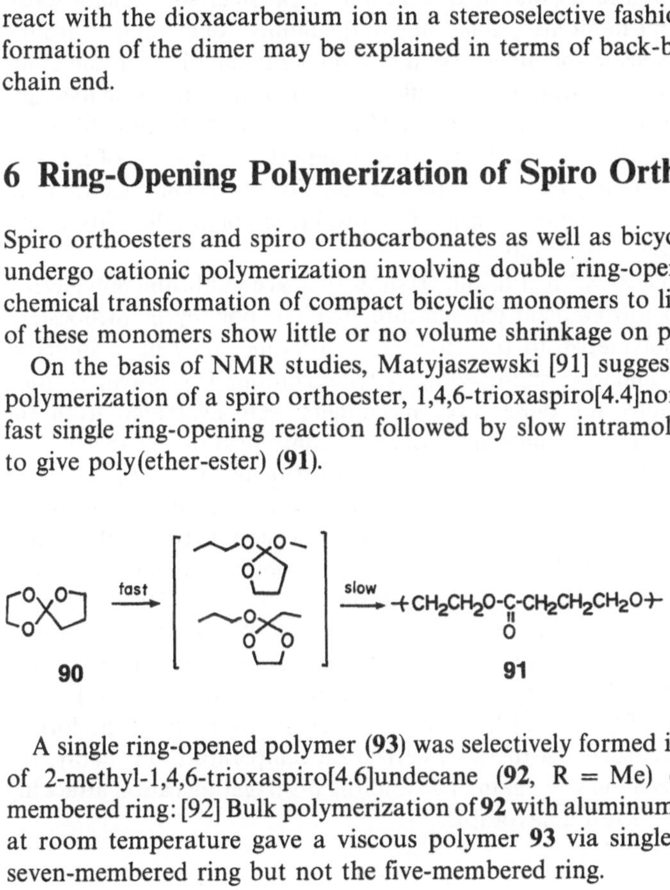

A single ring-opened polymer (**93**) was selectively formed in the polymerization of 2-methyl-1,4,6-trioxaspiro[4.6]undecane (**92**, R = Me) containing a seven-membered ring: [92] Bulk polymerization of **92** with aluminum (III) acetylacetonate at room temperature gave a viscous polymer **93** via single ring-opening of the seven-membered ring but not the five-membered ring.

Interestingly, neither 2-methyl-1,4,6-trioxaspiro[4.5]decane (**94**) containing a six-membered ring nor 2-methyl-1,4,6-trioxaspiro[4.4]nonane (**95**) containing a five-membered ring gave a polymer under the same conditions.

Me—O₄O   structures **94** and **95** (chemical structures, image-like)

94   **95**

According to an MM2 calculation, the strain energy of **92** is 84 kJ/mol, which is 21–25 kJ/mol larger than those of **94** and **95**. The difference is large enough to explain the experimental result that neither **94** nor **95** polymerized at all, whereas **92** did. Since these three monomers differ only in the number of methylene groups, it is reasonable to associate the difference in strain energy mainly with the strain of the ether ring alone. The strain energy difference between monocyclic five- and seven-membered rings being less than several kJ/mol, the 21–25 kJ/mol difference mentioned above appears to result from a concomitant suppression of freedom of the seven-membered ring of **92** by spiro cyclization. [91] The fact that the opening of only the seven-membered ring of **92** produced poly(cyclic orthoesters) **93** supports Matyjaszewski's proposal.

Spiro orthoesters (**92**, R = Me, Ph, and H) show typical equilibrium polymerization behavior at or below ambient temperature. [92] The poly(cyclic orthoester)s derived from **92** depolymerize to the monomers, although they have sufficient strains to be able to undergo ring-opening polymerization. The polymerization enthalpies and entropies for these three monomers were evaluated from the temperature dependence of equilibrium monomer concentrations (Table 5). The enthalpy became less negative as the size of the substituent at the 2-position in **92** was increased: H < Me < Ph. This behavior can be explained in terms of the polymer state being made less stable by steric repulsion between the bulky substituents and/or between the substituent and the polymer main chain. The entropy also changed in a similar manner with the size of the substituents.

Spiro orthoester containing a perfluoroalkyl group (**96**) underwent double ring-opening polymerization to give a poly(ether-ester) consisting of two kinds of structural units (**97** and **98**), which are formed by ring-opening at the O(4)—C(5) bond (path a) and the O(1)—C(5) bond (path b), respectively. [93] The former unit **97** dominated regardless of the polymerization temperature. Probably, the steric hindrance between the side chain of **96** and the propagating cation attacking the O(1) atom makes path b unfavorable.

Table 5. Thermodynamic parameters $\Delta H°$ and $\Delta S°$ for polymerization of spiro orthoester **92** [92]

	$\Delta H°$, kJ mol^{-1}	$\Delta S°$, J mol^{-1} deg^{-1}°C	T_c,[a]
92 (R = Me)	− 8.0	−30.7	254
92 (R = Ph)	− 6.7	−25.7	258
92 (R = H)	−10.6	−39.6	172

[a] Ceiling temperature for bulk polymerization obtained by extrapolation

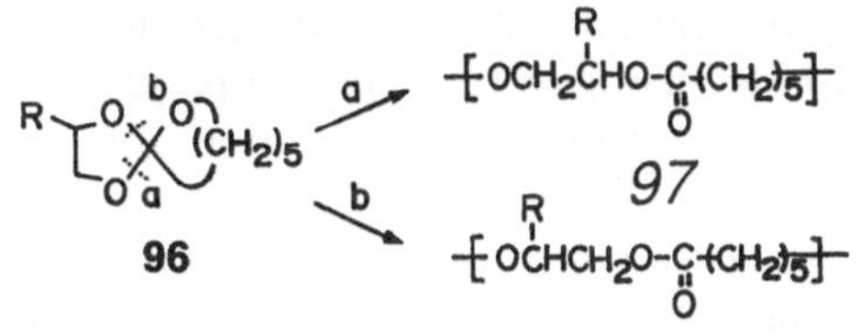

$$R = CF_3(CF_2)_7CH_2CH_2OCH_2$$

Polymerization of spiro orthoester **96** accompanied a 2.7% volume expansion. This behavior is similar to that observed in the polymerization of nonfluorinated spiro orthoester **99** (0.9% expansion). Thus, it appears that the perfluoroalkyl group does not significantly affect the feature of zero shrinkage during the polymerization of sipro orthoesters.

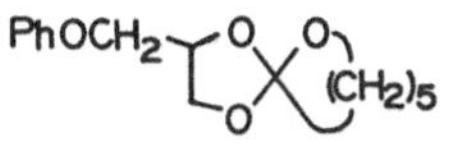

99

Unsaturated spiro orthoesters (**100**) having one methylene group in the acetal ring undergo radical double ring-opening polymerization to yield poly(keto-ester)s (**101**). [94] They also copolymerize with commercially available vinyl monomers such as ethylene and styrene to give polymers containing keto and ester groups in the main chains. The resulting copolymers, therefore, are expected to have photo- and biodegradabilities.

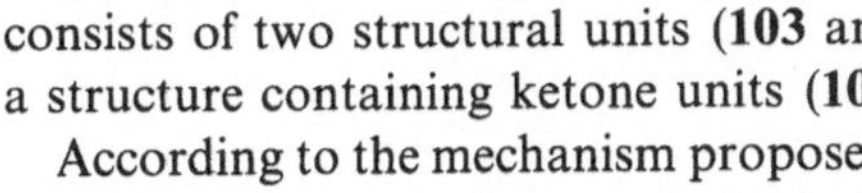

Polymerization behavior of 4'-methylenespiro[2-benzofuran-2,2'-(1,3-dioxola-ne)] (**102**) is rather complicated. Han et al. [95] claimed that the resulting polymer consists of two structural units (**103** and **104**), but Pan et al. [96] later proposed a structure containing ketone units (**105**) instead of keto-ester units **103**.

According to the mechanism proposed by Pan et al. (Scheme 17 [96]), the radical **106** undergoes ring-opening to form the radical **107**, which is a tertiary benzyl radical stabilized by the adjacent oxygen. The radical **107** is stabler than the primary benzyl radical **109** which is a product of double ring-opening polymerization. This stability explains why double ring-opening polymerization does not occur with **107**. If **107** adds to the monomer **102**, **106** is regenerated, which is much less stable than **107**. Thus, reaction (4) cannot compete with reactions (1) and (2), so that the structural unit **110** would not be formed. Reaction (2) is driven by the formation of a very stable phthalide and the radical **108** stabilized by the adjacent carbonyl group. Therefore, only reactions (1) and (2) occur during the polymerization.

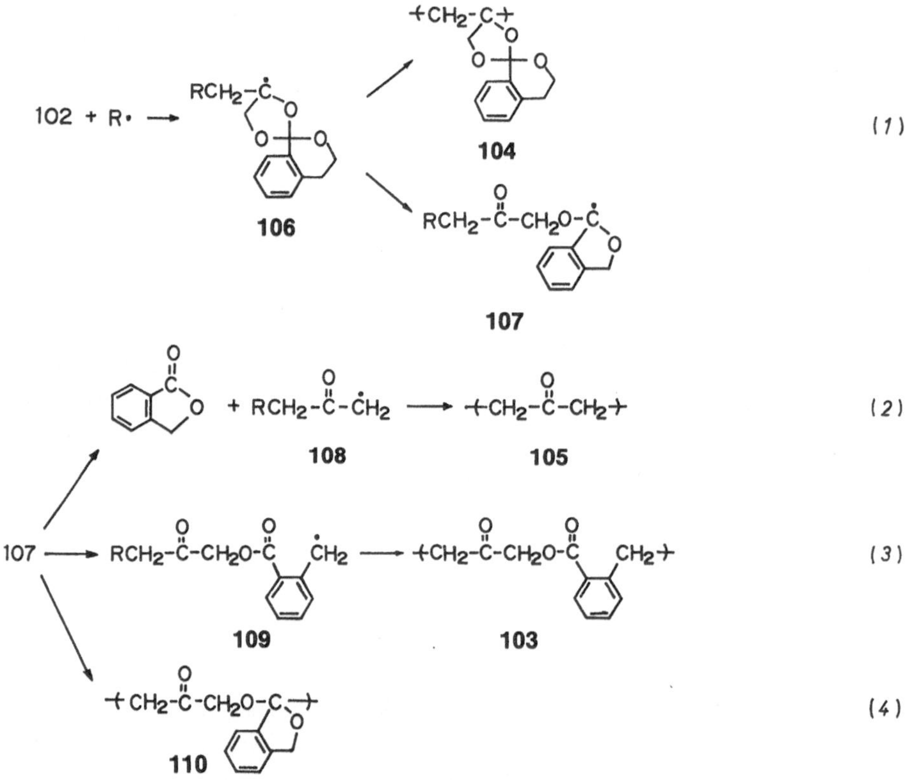

Scheme 17

2-Methyl-7-methylene-1,4,6-trioxaspiro[4,4]nonane (**111**) undergoes radical ring-opening polymerization in bulk to give a polymer containing ketone and cyclic acetal moieties (**112**), which are the products of a single ring-opening reaction. [97]

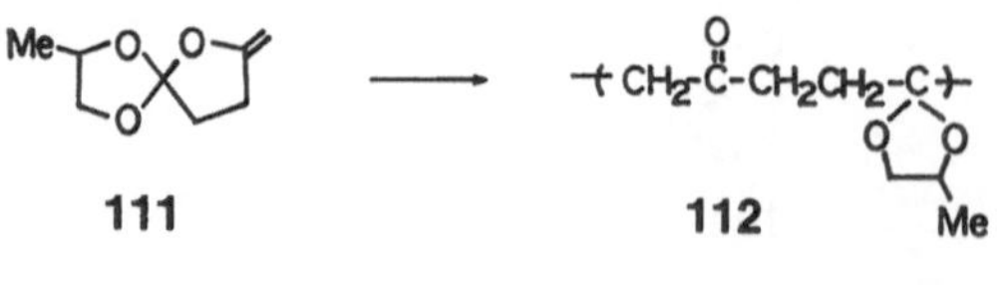

7 Ring-Opening Polymerization of Spiro Orthocarbonates

Sakai et al. [98] was the first to report the cationic ring-opening polymerization of three types of spiro orthocarbonates (**113**, **114**, and **115**) with boron trifluoride etherate as the initiator. As shown in Scheme 18, 1,4,6,9-tetraoxaspiro[4.4]nonane (**113**) having two five-membered rings mainly gives poly(ethylene oxide) (**116**) with the loss of ethylene carbonate. 1,5,7,11-Tetraoxaspiro[5.5]undecane (**114**) having two six-membered rings gives the poly(ether-carbonate) alternating copolymer **117** (n = 3). 1,6,8,13-Tetraoxaspiro[6.6]tridecane (**115**) having two seven-membered rings yields polycarbonate (**118**) with the complete elimination of tetrahydrofuran.

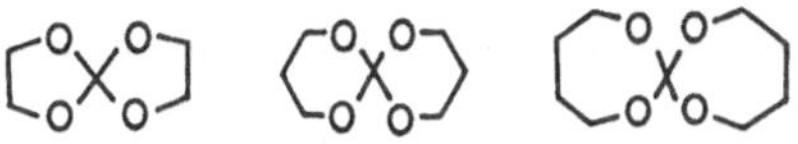

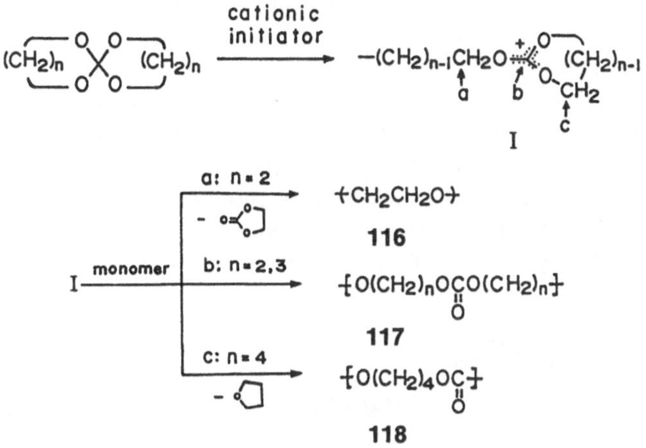

Scheme 18

Takata et al. [99] found for two spiro orthocarbonates (**119** and **121**) having 1,4,6,9-tetraoxaspiro[6.6]tridecane skeletons different cationic ring-opening polymerization behavior: Irrespective of solvent, initiator, and temperature (from room temperature to 150 °C), **119** gave polycarbonate (**120**), whereas **121** gave poly(ether-carbonate) (**122**).

Concerning the three possible sites attacked by the monomer illustrated in Scheme 18, the attack along path b produces polymer **120**, while the attack along path c gives polymer **122**. This marked difference in polymerization behavior indicates that the propagating species of the benzo derivative **121** is more reactive, i.e. more electrophilic, than that of **119**.

For some seven-membered spiro orthocarbonates (**115** and **119**), the attack along path b predominates over that along path c, whereas for **114**, its derivatives, and **121**, path c is strongly favored. The stability of the leaving moiety appears to play a decisive role in determining the reaction course. In the case of **114** and its derivatives, four-membered oxetane derivatives are so highly strained that they are not formed at all. As for **121**, dihydrobenzofuran structure is unstable owing to its antiaromatic 4π system and therefore, its formation is not favored.

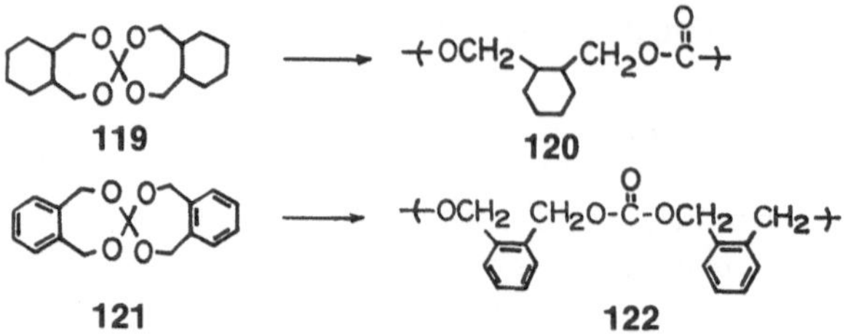

Cationic polymerization of spiro orthocarbonates leading to a poly(ether-carbonate) alternating copolymer proceeds via the trialkoxycarbenium ion as a possible propagating species. Since this type of carbenium ion was found to be quite stable, [100] the growing chain end is expected to remain living after the completion of the polymerization. [101]

Cationic polymerization of 3,9-dibenzyl-1,5,7,11-tetraoxaspiro[5.5]undecane (**123**) with tri-phenylmethylcarbenium tetrafluoroborate or boron trifluoride etherate proceeded at room temperature. After the complete consumption of the monomer, the polymerization took place again when the monomer was added to the reaction system. This fact strongly indicates the presence of a stable and long-lived propagating chain end (**124**) in the polymerization of **123**.

The number-average molecular weights of the polymer samples obtained in the second stage increased roughly in proportion to the monomer conversion. Their polydispersity indices (M_w/M_n) were approximately constant (1.45–1.61) during the polymerization. [101] Therefore, the polymerization of spiro orthocarbonate **123** is not ideally living, although it exhibits some characteristics of living polymerization. Most interesting is the great stability of the growing end species. In fact, ^{1}H, ^{13}C, and ^{19}F NMR analyses showed that the active chain end was the trialkoxycarbenium ion as expected.

Cationic polymerization of norbornene spiro orthocarbonate (**125**) differs from that of other spiro orthocarbonates in that it gives a structurally complicated polymer. [102] When polymerized with boron trifluoride etherate at 100 °C in chlorobenzene for 24 h, the ^{1}H NMR of a polymer of **125** showed very reduced olefinic proton signals (22% of the original monomer). The monomer was

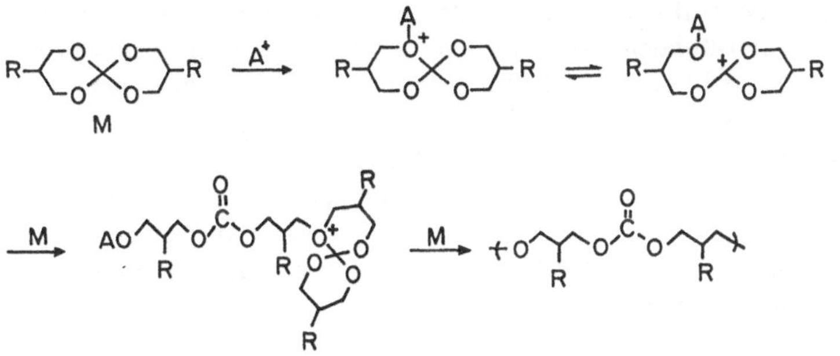

Scheme 19

completely consumed in only 5 min, but no polymer was formed. Chromatographic separation yielded a spirocyclic carbonate (**126**, 54%), a dimer of an oxetane (**127**, 25%), and an oligomeric mixture (21% yield, 72% olefin content). Since the yield of **127** was sufficiently high (95% of the theoretical yield), most of the monomer did not undergo polymerization by the same mechanism as that illustrated in Scheme 19 [102].

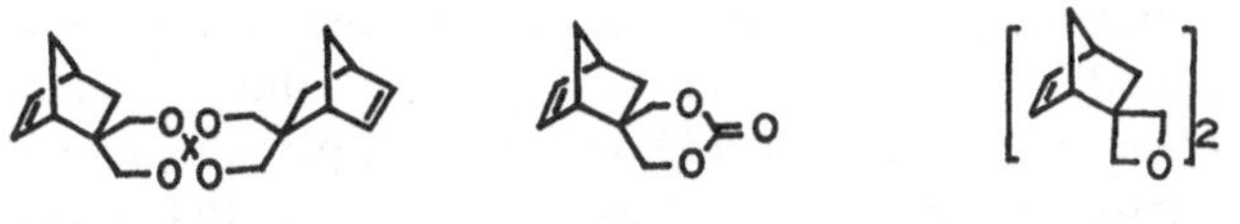

125	**126**	**127**

The olefin seems to disappear mainly in the second stage after the decomposition of **125**. Probably, a nonclassical type of carbenium ion intermediate (**128**) is formed with the help of the olefin group located at an intramolecularly favorable position, and this intermediate obviously makes the polymer structure complicated by a possible cyclization-skeleton rearrangement prior to the nucleophilic attack of the monomer.

125 $\xrightarrow{\text{A}^+}$

128

Referring to the cationic polymerization of spiro tetrathioorthocarbonates, a five-membered spiro compound, 1,4,6,9-tetrathiaspiro[4.4]nonane (**129**) quantitatively gave poly(ethylene sulfide) with the complete extrusion of ethylene trithiocarbonate as found in the corresponding oxygen counterpart. [103] A six-membered spiro compound, 1,5,7,11-tetrathiaspiro[5.5]undecane (**130**), did not polymerize at all under the conditions examined. A seven-membered spiro compound, 1,6,8,13-tetrathiaspiro[6.6]tridecane (**131**), underwent ring-opening polymerization giving polymers which were soluble in hot p-chlorophenol and insoluble in common organic solvents. The polymers consisted of three structural units **132**, **133**, and **134**, whose proportions depended on the polymerization conditions. Thus, the polymers obtained at room temperature contained **134** and **132** as the major and minor constituents respectively, with a very small content of **133**. On the other hand, the polymer prepared at 100 °C contained **133** (21–55%) and **134** (27–57%) as the major components.

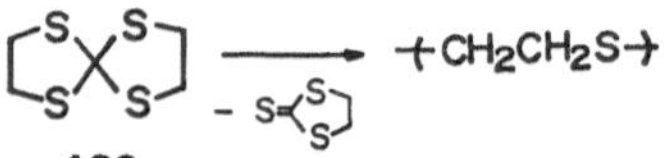

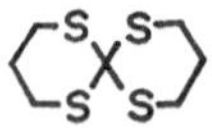

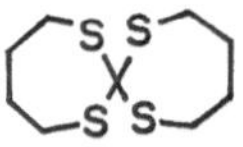

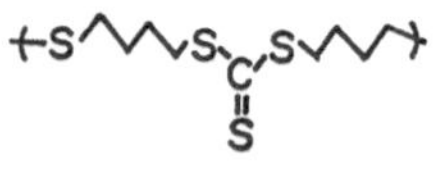

Scheme 20 [103] shows that the polymerization of **131** proceeds via a stable tris(alkylthio)carbenium ion **135**. This species can be attacked in three different ways by the second monomer, producing **134**, **133**, and **132** via paths a, a′, and b, respectively. Kinetically, path a should dominate over the others. Path a′ is one of the possible pathways, but presumably it is of minor importance. Although the conversion of **134** to **132** cannot be ruled out, **132** is more likely to be formed via path b, as is the case with six-membered spiro orthocarbonates. At high

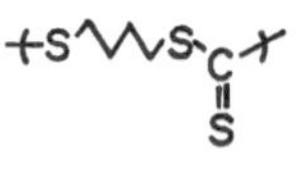

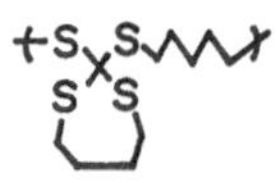

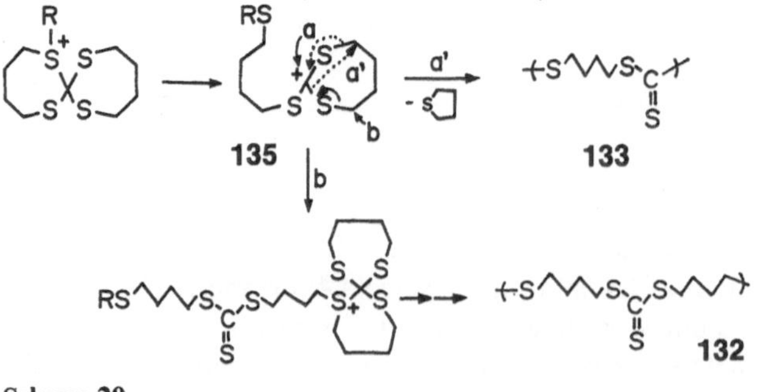

Scheme 20

temperature, path c becomes important. Elimination of tetrahydrothiophene from **134** through path c causes the polymer main chain to cleave and **134** to transform to **133**.

It has been established that 4-methylene-1,3-dioxolanes (**136**) having one or two radical stabilizing groups at the 2-position undergo radical ring-opening polymerization with the C(2) – O(3) bond cleavage. [104] In some cases, the polymerization is accompanied by the elimination of the corresponding carbonyl compounds such as benzaldehyde, phthalide, benzophenone, and methyl benzoate to give polyketone (**137**).

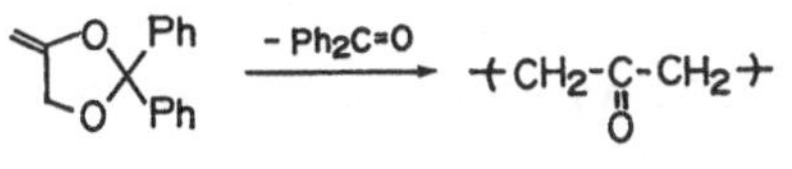

In contrast, radical polymerization of 2,3-benzo-7-methylene-1,4,6,9-tetraoxa-spiro[4.4]nona-2-ene (**138**) gave only a vinyl polymer (**139**) with no occurrence of ring-opening isomerization, even in the solution polymerization at 165 °C. [105] It appears that the propagating radical having a spiro orthocarbonate structure is too stable to liberate the corresponding carbonyl compound, phenylene carbonate.

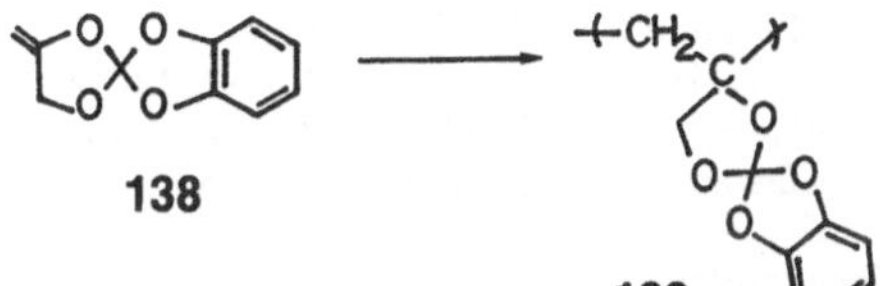

8 Concluding Remarks

Sophisticated functions displayed by polysaccharides, proteins, and nucleic acids basically originate from the controlled molecular structure (molecular weight, molecular weight distribution, stereoregularity, functional groups, branchings, conformations etc.). In order to chemically synthesize speciality polymers having a specific function comparable to that of naturally occurring polymers, we must develop more elaborate techniques which allow us to control completely the polymerization processes and thereby to synthesize the polymers having a precisely designed molecular structure. As we have discussed in this review, the reactivities of heterobicyclic monomers vary over a wide range, and the mechanisms of ring-opening polymerizations are often quite complicated. Our techniques have advanced remarkably recently, but they are not yet mature enough to produce desired polymers at will. For instance, the method of living polymerization is obviously indispensable for synthesizing polymers having a controlled molecular weight and a narrow molecular weight distribution, but only a few bicyclic monomers have so far been reported to be living in their ring-opening polymerization. Therefore, a prerequisite is to discover new catalysts for living polymerization that are widely applicable to ring-opening polymerization of various types of bicyclic monomers in order to make an epoch-making breakthrough in this field of polymer synthesis.

9 References

1. Hall HKJr (1958) J Am Chem Soc 80: 6412; (1960) J Am Chem Soc 82: 1209
2. Okada M (to be published) Prog Polym Sci
3. Sumitomo H, Okada M (1978) Adv Polym Sci 28: 47
4. Penczek S, Kubisa P, Matyjaszewski K (1980) Adv Polym Sci 37: 1
5. Yokoyama Y, Hall HKJr (1982) Adv Polym Sci 42: 107
6. Sumitomo H, Okada M (1984) In: Ivin K, Saegusa T (eds) Ring-opening polymerization, Elsevier, London, vol I, p 229
7. Penczek S, Kubisa P, Matyjaszewski K (1985) Adv Polym Sci 68/69: 1
8. Hall HKJr, Fr DeBlauwe, Carr LJ, Rao VS, Reddy GS (1976) J Polym Sci Symposia No 56: 101
9. Deslongchamps P (1975) Tetrahedron 31: 2463
10. Okada M, Sumitomo H, Sumi A, Sugimoto T (1984) Macromolecules 17: 2451
11. Okada M, Sumitomo H, Sumi A (1985) Carbohydr Res 143, 275
12. Hirasawa T, Okada M, Sumitomo H (1988) Macromolecules 21: 1566
13. Hirasawa T, Okada M, Sumitomo H (1988) Polym J 20: 587
14. Okada M, Hirasawa T, Sumitomo H (1989) Makromol. Chem. 190: 1289
15. Uryu T, Sakamoto Y, Hatanaka K, Matsuzaki K (1984) Macromolecules 17, 1307
16. Ponomarenko VA, Berman EL, Sakharov AM, Nysenko ZN (1984) In: Goethals EJ (ed) Cationic polymerization and related processes, Academic, London, p 155
17. Matyjaszewsky K (1983) Eur Polym J 19: 787
18. Okada M, Sumitomo H, Sumi A (1982) Macromolecules 15: 1238
19. Kobayashi K, Ichikawa H, Sumitomo H, Schuerch C (1988) Macromolecules 21: 542
20. Ichikawa H, Kobayashi K, Sumitomo H, Schuerch C (1988) Carbohydr Res 179: 315
21. Okada M, Sumitomo H, Komada H (1979) Macromolecules 12: 395
22. Okada M, Hirasawa T, Sumitomo H (1988) Makromol Chem Rapid Commun 9: 641
23. Okada M, Sumitomo H, Hirasawa T (1985) Macromolecules 18: 2345
24. Sumitomo H, Okada M (1987) In: Ottenbrite RM, Utracki LA, Inoue S (eds) Current topics in polymer science, Hanser, Munich, vol 1, p 15
25. Korshak VV, Golova OP, Sergeev VA, Merlis NA, Schneer RY (1961) Vysokomol Soedin 3: 477
26. Schuerch C (1981) Adv Carbohydr Chem Biochem 39: 157
27. Schuerch C (1985) Encyclopedia of polymer science and engineering, 2nd edn, John Wiley & Sons, New York, vol 13, p 162
28. Uryu T (1989) In: Penczek S (ed) Models of Biopolymers by Ring-Opening Polymerization, CRC Press, Boca Raton, FL, p 133
29. Micheel F, Broddle OE (1974) Liebig Ann Chem 702
30. Uryu T, Yamaguchi C, Morikawa K, Terui K, Kanai T, Matsuzaki K (1985) Macromolecules 18: 599
31. Yoshida T, Arai T, Mukai Y, Uryu T (1988) Carbohydr Res 177: 69
32. Uryu T, Yamanouchi J, Kato T, Higashi S, Matsuzaki K (1983) J Am Chem Soc 105: 6865
33. Hagino A, Yoshida S, Shinpuku T, Matsuzaki K, Uryu T (1986) Macromolecules 19: 1

34. Koyama Y, Harima K, Matsuzaki K, Uryu T (1985) J Polym Sci, Polym Chem Ed 23: 2989
35. Ichikawa H, Kobayashi K, Sumitomo H (1990) Macromolecules 23: 1884
36. a) Veruovic B, Schuerch C (1970) Carbohydr Res 14: 199
    b) Masura U, Schuerch C (1970) Carbohydr Res 15: 65
37. Uryu T, Yamanaka M, Hemmi M, Hatanaka K (1986) Carbohydr Res 157: 157
38. a) Goethals EJ (1977) Adv Polym Sci 23: 103 b) Penczek S, Slomkowski S (1989) In: Eastmond GC, Ledwith A, Russo S, Sigwalt P (eds) Comprehensive polymer science. Pergamon, Oxford, vol 3, p 725
39. Ito K, Hashizuka Y, Yamashita Y (1977) Macromolecules 10: 821
40. Jacobson H, Stockmayer WH (1950) J Chem Phys 18: 1600
41. Okada M, Sumitomo H (1985) Makromol Chem, Suppl 14: 29
42. Okada M, Sumitomo H, Atsumi M (1984) Macromolecules 17: 1840
43. Okada M, Sumitomo M, Atsumi M (1983) Makromol Chem, Rapid Commun 4: 253
44. a) Tajima I, Okada M, Sumitomo H (1981) Macromolecules 14: 1180 b) Okada M, Tajima I, Sumitomo H (1984) In: Bailey WJ, Tsuruta T (eds) Contemporary topics in polymer science, Vol 4, Plenum, New York, p 415
45. Tanaka I, Tajima I, Hayakawa Y, Okada M, Bitoh M, Ashida T, Sumitomo H (1980) J Am Chem Soc 102: 7873
46. Imaeda M, Tanaka I, Ashida T, Tajima I, Okada M, Sumitomo H (1982) Polym J 14: 197
47. Sakuragi I, Tanaka I, Ashida T, Tajima I, Okada M, Sumitomo H (1982) 104: 6035
48. Tajima I, Okada M, Sumitomo H (1981) J Am Chem Soc 103: 4096
49. Okada M, Sumitomo H, Atsumi M (1984) J Am Chem Soc 106: 2101
50. Okada M, Sumitomo H, Ito K, Goto S, Atsumi M (1988) Polym J 20: 55
51. Okada M, Sumitomo H, Atsumi M, Hall HKJr, Ortega RB (1986) Macromolecules 19: 503
52. Jones PG, Kirby AJ (1984) J Am Chem Soc 106: 6207
53. Okada M, Sumitomo H, Atsumi M, Hall HK Jr (1987) Macromolecules 20: 1199
54. Okada M, Sumitomo H, Yamada S, Atsumi M, Hall HK Jr, Chan RJH, Ortega RB (1986) Macromolecules 19: 953
55. Ceccarelli G, Andruzzi F, Paci M (1979) Polymer 20: 625
56. Burgi HB, Dunitz JD, Lehn JM, Wipff G (1974) Tetrahedron 30: 1563
57. Balackburn GM, Dodds HLH (1977) J Chem Soc, Perkin Trans 2 1977: 377
58. Andruzzi F, Pilcher G, Hacking JM, Cavell S (1980) Makromol Chem 181: 923
59. Boyesso B, Nakase Y, Sunner S (1966) Acta Chem Scand 20: 803
60. Hall HK Jr (1969) Macromolecules 2: 488
61. Hofman A, Szymanski R, Slomkowski S, Penczek S (1984) Makromol Chem 185: 655
62. Sumitomo H, Okada M (1978) Adv Polym Sci 28: 47
63. Sumitomo H, Hashimoto K (1985) Adv Polym Sci 64: 63
64. Hashimoto K, Sumitomo H, Yamamori H (1987) Polym J 19: 249
65. Hashimoto K, Sumitomo H, Yamamori H (1987) Polym J 19: 1139
66. Hashimoto K, Sumitomo H, Washio A (1988) Polym J 20: 615
67. Hashimoto K, Sumitomo H, Shinoda H, Washio A (1989) J Polym Sci Part C, Polym Lett 27: 307
68. Hashimoto K, Sumitomo H, Washio A (1989) J Polym Sci, Part A, Polym Chem 27: 1915
69. Hashimoto K, Sumitomo H (1984) J Polym Sci, Polym Chem Ed 22: 1733
70. Hashimoto K, Sumitomo H, Suzuki M (1987) Macromolecules 20: 2797
71. Moore JA, Pertain EM III (1982) J Polym Sci, Polym Lett 20: 521
72. Okada M, Sumitomo H, Mori H, Hall HK Jr, Chan RJH, Bruck M (1990) J Polym Sci, Part A, Polym Chem 28: 3251
73. Gu Y-X, Yamane T, Ashida T, Hashimoto K, Sumitomo H (1986) Bull Chem Soc Jpn 59: 2085
74. Okada M, Sumitomo H, Sassa T, Takai M, Hall HK Jr, Bruck M (1990) Macromolecules 23: 2427

75. Cho H, Choi K, Choi S (1985) J Polym Sci Polym Chem Ed 23: 623
76. Cho H, Choi S (1985) J Polym Sci Polym Chem. Ed 23: 1469
77. Hall HK Jr, De Blauwe F, Carr LJ, Rao VS (1976) J Polym Sci, Polym Symp 56: 101
78. Hall HK Jr, De Blauwe F (1974) J Am Chem Soc 96: 7265
79. Hall HK Jr, De Blauwe FG, Pyriadi T (1975) J Am Chem Soc 97: 3854
80. Burt RA, Chiang Y, Hall HK Jr, Kresge AJ (1982) J Am Chem Soc 104: 3687
81. Yokoyama Y, Padias AB, De Blauwe F, Hall HK, Jr (1980) Macromolecules 13: 252
82. Yokoyama Y, Padias AB, Bratoeff EA, Hall HK Jr (1982) Macromolecules 15: 11
83. Padias AB, Szymanski R, Hall HK, Jr (1985) In: McGrath JE (ed) Ring-opening polymerization kinetics, mechanisms and synthesis. ACS Symp Ser 286 p 313
84. Endo T, Saigo K, Bailey WJ (1980) J Polym Sci, Polym Lett Ed 18: 457
85. Hall HK Jr, Yokoyama Y (1980) Polym. Bull. 2: 281
86. Syzmanski R, Hall HK Jr (1983) J Polym Sci, Polym Lett. Ed 21: 177
87. Uno H, Endo T, Okawara M (1985) J Polym Sci, Polym Chem Ed 23: 63
88. Yokoyama Y, Hall HK Jr (1980) J Polym Sci Polym Chem Ed 18: 3133
89. a) Bailey WJ (1975) J Macromol Sci Chem A9: 849 b) Bailey WJ, Sun RL, Katsuki H, Endo T, Iwama H, Tsushima R, Saigo K, Bitritto MM (1977) In: Saegusa T, Goethals EJ (ed) Ring-opening polymerization, ACS Symp. Ser. 59: 38
90. Matyjaszewski K (1984) J Polym Sci, Polym Chem Ed 22: 29
91. Chikaoka S, Takata T, Endo T (1990) J Polym Sci, Part A, Polym Chem 28: 3101
92. Chikaoka S, Takata, Endo T (1990) Macromolecules 24: 331
93. Yokozawa T, Sato M, Endo T (1990) J Polym Sci, Part A, Polym Chem 28: 1841
94. Endo T, Okawara M, Yamazaki N, Bailey WJ (1981) J Polym Sci, Polym Chem Ed 19: 1283
95. Han YK, Choi SK (1983) J Polym Sci, Polym Chem Ed 21: 353
96. Pan C, Lu S, Bailey WJ (1987) Makromol Chem 188: 1651
97. Pan C, Wang Y (1986) Makromol Chem, Rapid Commun 7: 627
98. Sakai S, Fujinami T, Sakurai S (1973) J Polym Sci, Polym Lett Ed 11: 631
99. Takata T, Endo T (1988) Macromolecules 21: 900
100. Olah GA, Olah JA, Svoboda JJ (1973) Synthesis 490
101. Endo T, Sato H, Takata T (1988) Macromolecules 21: 1187
102. Takata T, Amachi K, Kitazawa K, Endo T (1989) Macromolecules 22: 3188
103. Takata T, Endo T (1988) Macromolecules 21: 2314
104. a) Hiraguri Y, Endo T (1987) J Am Chem Soc 109: 3779. b) Gong MS, Chang S-I, Cho I (1989) Makromol Chem, Rapid Commun 10: 201
105. Sugiyama J, Yokozawa T, Endo T (1990) J Polym Sci, Part A, Polym Chem Ed 28: 3529

Editor: H. Fujita
Received May 17, 1991

The Development of Well-Defined Catalysts for Ring-Opening Olefin Metathesis Polymerizations (ROMP)

B. M. Novak[1], W. Risse[2] and R. H. Grubbs*
Division of Chemistry and Chemical Engineering, California Institute
of Technology, Pasadena, California 91125, USA

The present article reviews the development of well-defined transition metal compounds
which catalyze the ring-opening polymerization of strained cyclic olefins by the olefin
metathesis reaction (ROMP). Several titana- and tantalacyclobutane compounds and
tungsten- and molybdenum carbenes polymerize norbornene and norbornene derivatives
by a "living" mechanisms, thus producing polymers of controlled molecular weights with
narrow molecular weight distributions. Some of these systems have been successfully applied
to the synthesis of block and graft copolymers. Studies on Ru-based catalysts resulted in the
development of olefin metathesis polymerizations in aqueous solutions.

1 History of Olefin Metathesis 48
2 Olefin Metathesis with Organotitanium Catalysts 50
  2.1 Living Polymerizations 52
  2.2 Di- and Tetrafunctional Initiators 56
  2.3 Conversion of Titanacyclobutane Based Metathesis Catalysts into
      Ziegler Natta Catalysts 58
  2.4 Living Polymerizations Catalyzed by Group VB and Group VIB
      Complexes 59
      2.4.1 Tantalacyclobutane Catalysts 59
      2.4.2 Tungsten and Molybdenum Catalysts 60
  2.5 Utilization of ROMP Techniques in the Synthesis of Speciality
      Polymers 64
      2.5.1 Synthesis of Polyacetylenes 64
      2.5.2 ROMP of Functionalized Monomers in Protic Solvents . . . 67
3 Conclusions 69
4 References 69

---

[1] present address: University of California, Berkeley, Department of Chemistry, Berkeley, CA 94720.
[2] present address: Philipps Universität Marburg, Fb. Physikalische Chemie, Polymere, Hans Meerweinstr., West Germany.
* to whom correspondence should be addressed

1 History of Olefin Metathesis

A patent by A. W. Anderson et al. (Du Pont) disclosed the first transition metal catalyzed polymerization of a cyclic olefin in 1955 [1]. Subsequent to the discovery of Ziegler-Natta polymerization [2] norbornene was found to polymerize in the presence of the catalyst systems $TiCl_4$/EtMgBr. Eleuterio [3a] and Truett et al. [3b] obtained polynorbornene by using transition metal catalysts based on Mo and Ti, respectively. IR-studies and ozonolysis of the polymer revealed the presence of carbon-carbon double bonds, indicating that polymerization had occured by unexpected ring opening Eq. 1 [3].

$$n \quad \underset{}{\triangle\!\!\!\!\triangle} \quad \xrightarrow{TiCl_4/LiAl(n\text{-}C_7H_{15})_4} \quad \left(\!\!\!\!\!\bigcirc\!\!\!\!\!\diagup\!\!\!\!\!\diagdown\right)_n \qquad (1)$$

In the following years, a number of other monocyclic, bicyclic and polycyclic olefins including cyclobutene, [4] cyclopentene [3a, 5] and cyclooctene [6] were found to undergo ring-opening polymerization.

Catalysts for the reaction can either be homogeneous or supported. Compounds of the transition metals Ti, Zr, Hf, V, Nb, Ta, Cr, Mo, W, Re, Ru, Os, Rh, Ir have been used [7]. The most active catalysts are based on Mo, W and Re. Classic non-supported catalysts were prepared from a transition metal halide complex and an organometallic or Lewis acid cocatalyst. Some typical catalyst systems are WCl_6/$EtAlCl_2$/EtOH [8]; $MoCl_2(NO)_2L_2$/Me_3Al (L = phosphine or py) [9]; $WOCl_4$/Me_4Sn [10]; $ReCl_5$/Et_3Al/O_2 [11]; WCl_6/Ph_4Sn [12]. Some examples among a number of supported catalysts are MoO_3/CoO/Al_2O_3 [13]; WO_3/SiO_2 [14] and Re_2O_7/Al_2O_3 [15].

Independently from the results of ring-opening polymerization, Banks and Bailey [16] reported a disproportionation reaction of acyclic olefins, the formation of ethylene and 2-butene from propylene Eq. (2).

$$\underset{CH_3}{=\!\!\diagdown} \quad \xrightarrow[160\ °C]{MoO_3/Al_2O_3} \quad CH_2{=}CH_2 \quad + \quad CH_3CH{=}CHCH_3 \qquad (2)$$

Calderon et al. [17] demonstrated that disproportionation, Eq. (2), and ringopening polymerization Eq. (1) are one and the same chemical reaction. The catalyst system WCl_6/$EtAlCl_2$/EtOH was successfully applied for the polymerization of cyclooctene and cycloocta-1,5-diene and also for the disproportionation of 2-pentene. The cross reaction of 2-butene and 2-butene-d_8 gave only one product, 2-butene-d_4, thus demonstrating that the reaction proceeds by breaking double bonds and exchanging alkylidene units. Accordingly, this type of reaction was named olefin metathesis.

A pairwise mechanism involving a "quasi-cyclobutane" structure was suggested first, Eq. (3) [18], but cross metathesis between cyclopentene and 2-pentene produced a statistical ratio of cross-products 1–3 (**1:2:3** = 1:2:1), Eq. (4) [19]

even at very low conversions. The formation of only **2** would have been expected from the simple pairwise mechanism, Eq. (3). Based on these results, Hérrison and Chauvin [19] proposed a new mechanism involving metal carbenes Eq. (5).

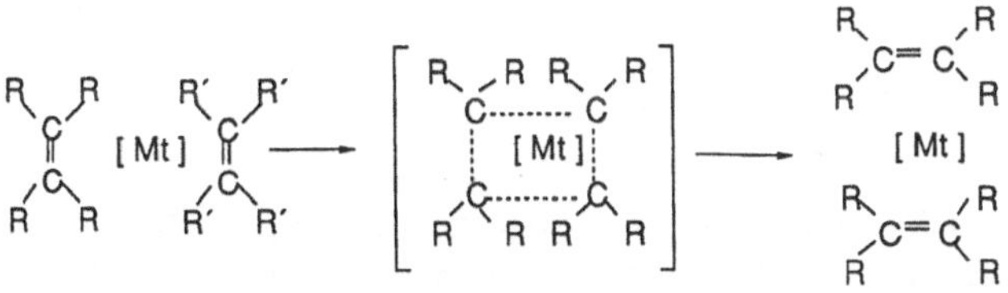

$$\tag{3}$$

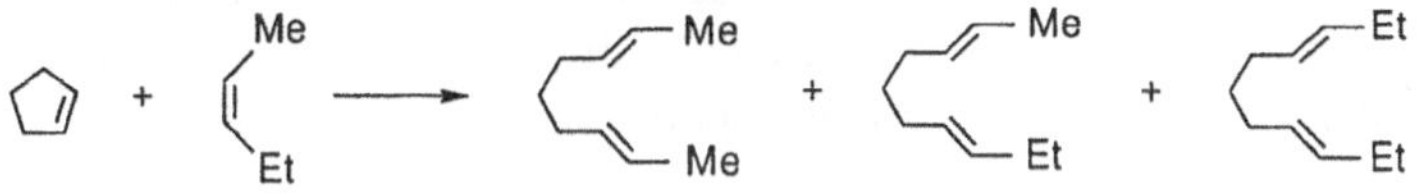

$$\tag{4}$$

$$(1 \;:\; 2 \;:\; 1)$$

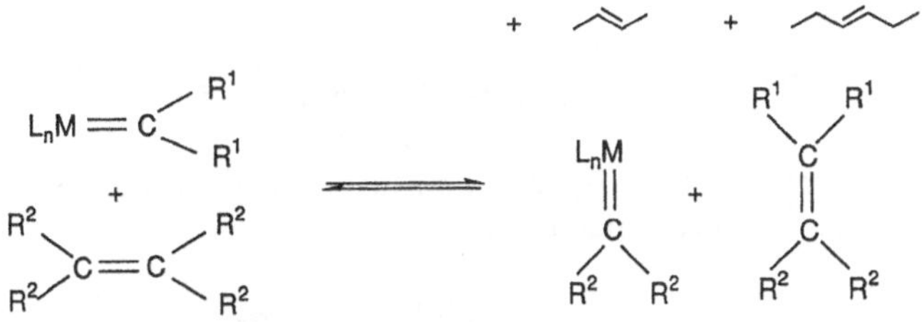

$$\tag{5}$$

Isotope labeling studies by Grubbs [20] and Katz [21] presented further proof for the metal carbene mechanism.

The observation that the metal carbene complex, $(CO)_5W=C(Ph)_2$ [22], catalyzed the polymerization of cyclic olefins to ring opened polymers containing the diphenylmethylene unit of the catalyst provided additional evidence that carbenes were involved in the catalytic cycle. The formation of the initiating metal carbenes in the classic systems that consist of transition metal halides and alkylating agents was proposed to involve metal alkylation followed by α-hydrogen loss, Eq. (6). Methane and propene were detected in the early stages of these reactions [23].

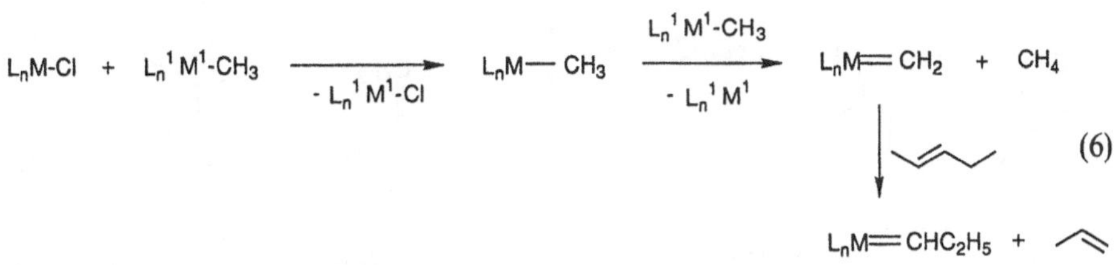

$$\tag{6}$$

Ring-opening Olefin Metathesis Polymerization (ROMP) gives a mixture of linear high molecular weight polymer and a series of cyclic oligomers for a number of monomer/catalyst systems (monomers include cyclopentene, cyclooctene and norbornene) [6, 24]. The formation of cyclic oligomers and monomer from high molecular weight polymer is caused by an intramolecular backbiting reation, Eq. (7). Secondary cis/trans isomerization [25] also occurs as the reaction proceeds.

$$L_nM = \text{(structure)} \longrightarrow L_nM{=}CHP_{n-a} \; + \; \text{(structure)} \tag{7}$$
$$P_{n-a} =$$

Metallacyclobutanes, fomed upon addition of olefin to the metal carbene, were proposed as intermediates or transition states for olefin metathesis. Metal-acyclobutanes are also known to undergo reductive and β-hydride elimination [27] to produce cyclopropanes, Eq. (8a), and olefins, Eq. (8b), respectively. These reactions are termination reactions for olefin metathesis.

$$L_nM\text{(structure)} \quad \begin{matrix} a \nearrow \quad L_nM \;+\; \triangle \\[1em] b \searrow \quad \underset{L_nM\text{---}}{\overset{H}{}} \longrightarrow L_nM \;+\; \text{(structure)} \end{matrix} \tag{8}$$

2 Olefin Metathesis with Organotitanium Catalysts

The Tebbe reagent 4 [28], prepared from Cp_2TiCl_2 and two equivalents of Me_3Al, Eq. (9) is of great significance for olefin metathesis.

$$Cp_2Ti\overset{Cl}{\underset{Cl}{\diagup}} \;+\; 2\,Me_3Al \;\xrightarrow[-\,Me_2AlCl]{}\; Cp_2Ti\underset{Cl}{\overset{CH_2}{\diagdown}}Al\underset{Me}{\overset{Me}{\diagup}} \tag{9}$$

4

Complex 4 can be regarded as titanium carbene $Cp_2Ti{=}CH_2$ coordinated by Me_2AlCl. This is the first example of a well-defined metal carbene that catalyzes olefin metathesis and is re-isolated in high yield at the end of the reaction, Eq. (10). In a comparatively slow degenerative metathesis reaction (near equilibrium after 47 h at 51 °C). The ^{13}C label of isobutylene was shown to grow into the methylene group of methylene cyclohexane [29].

$$\underset{CH_3\;CH_3}{\overset{^{13}CH_2}{\diagdown}} \;+\; \underset{}{\overset{CH_2}{\diagdown}}\;+\;\text{(structure)} \;\;\underset{\xrightleftharpoons}{Cp_2Ti\underset{Cl}{\overset{CH_2}{\diagdown}}Al\underset{Me}{\overset{Me}{\diagup}}}\;\; \underset{CH_3\;CH_3}{\overset{CH_2}{\diagdown}} \;+\; \overset{^{13}CH_2}{\text{(structure)}} \tag{10}$$

$$Cp_2Ti\underset{Cl}{\overset{CH_2}{\diagdown}}Al\underset{Me}{\overset{Me}{\diagup}} \;+\; \underset{CH_3\;CH_3}{\overset{^{13}CH_2}{\diagdown}} \;\xrightleftharpoons\; Cp_2Ti\underset{Cl}{\overset{^{13}CH_2}{\diagdown}}Al\underset{Me}{\overset{Me}{\diagup}} \;+\; \underset{CH_3\;CH_3}{\overset{CH_2}{\diagdown}}$$

Titanacyclobutane compound **5** [30] was isolated from the reaction of Tebbe reagent with d_1-neohexene, Eq. (11). On addition of dimethylaluminium chloride, the Tebbe reagent (partially deuterated) was recovered, Eq. (12).

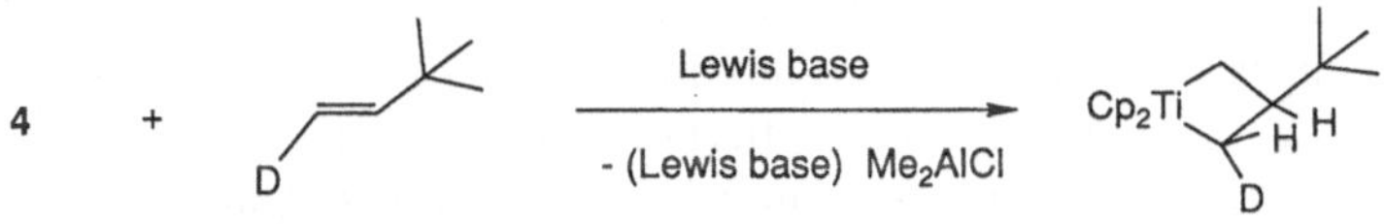

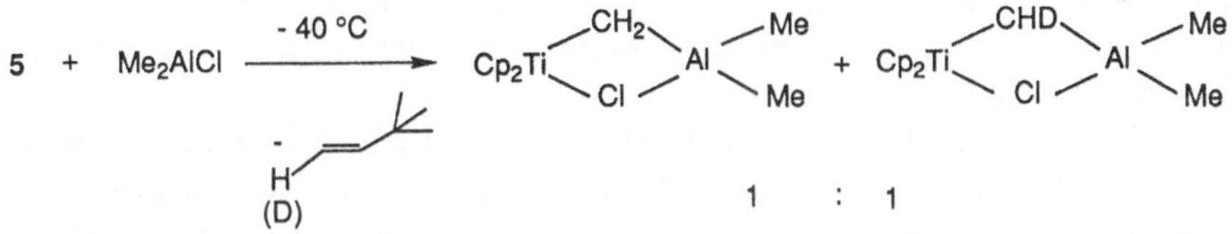

This reaction sequence demonstrates the presence of titanacyclobutanes as intermediates in the olefin metathesis reaction. Thus, the fundamental steps in olefin metathesis involve the interconversion of metal carbenes and metallacyclobutanes, Eq. (13).

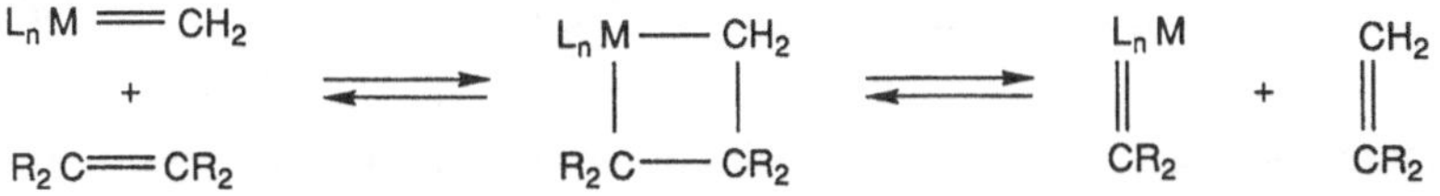

Titanacyclobutanes were found to provide aluminium-free sources of "$Cp_2Ti=CH_2$" that catalyze olefin metathesis, Eq. (14) [31].

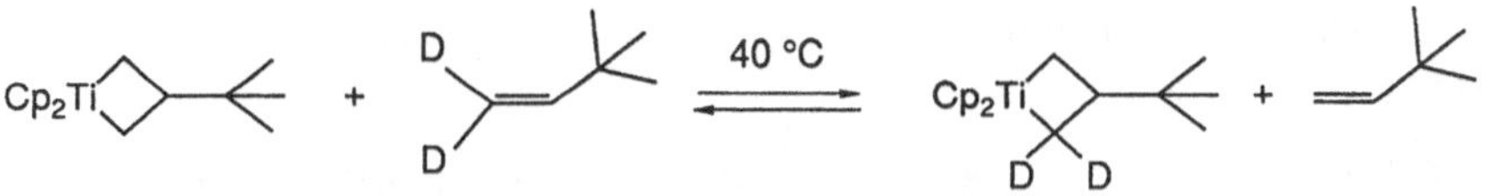

Trapping of **6** with diphenylacetylene yielded a titanacyclobutene derivative, Eq. (15) [31]. The reaction was found to be first order in **5a** and zeroth order in the substrate diphenylacetylene. Thus, ring-opening (k_1) of the titanacycle is the rate determining step ($k_2 \gg k_1$).

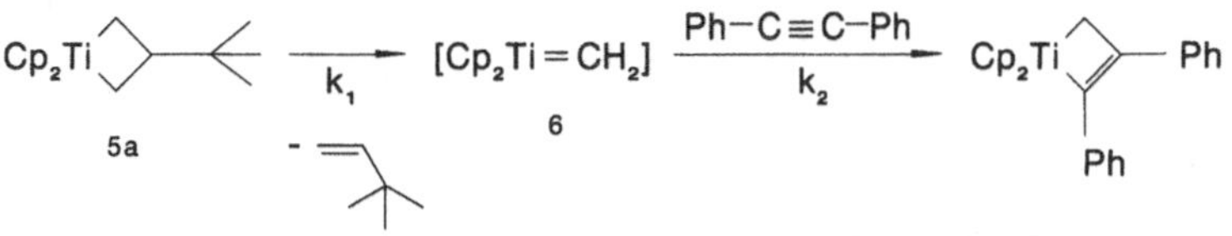

As will be seen later, the lack of possible side reactions such as cyclopropanation or β-hydride elimination is key to the use of titanacyclobutanes as initiators for living ROMP polymerization.

2.1 Living Polymerizations

As early as 1928, K. Ziegler [32] recognized the principles of living polymerizations, when he studied the reactions of alkyl potassium compounds with stilbene and styrene. Polymer chain growth proceeds by continuous monomer addition to an active propagating species without chain termination and chain transfer. The active center of the polymerization remains intact on the polymer chain after consumption of all monomer molecules. After renewed monomer addition, polymer growth continues. When the rate of initiation is comparable to or faster than the rate of propagation, a narrow molecular weight distribution is obtained [33]. Anionic polymerization of vinyl monomers was first demonstrated by Szwarc [34] to provide the control required to produce polymer samples with a polydispersity ($\bar{M}_w/\bar{M}_w$) less than 1.1.

The relatively low reactivity of titanacyclobutanes for olefin metathesis and the large ring strain in norbornene and norbornene derivatives are favorable properties for developing living polymerization by ring-opening olefin metathesis. Gilliom and Grubbs [35] discovered that bis(n^5-cyclopentadienyl)titanacyclobutane (7), derived from norbornene, initiates the polymerization of norbornene upon heating to temperatures above 65 °C, Eq. (16).

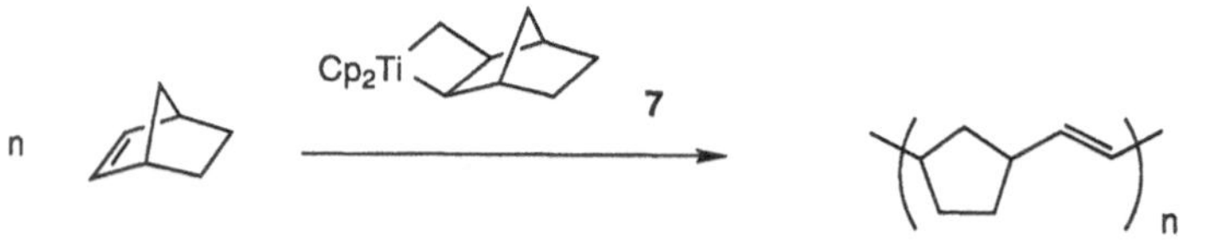

$$\text{(16)}$$

The polymerization proceeds without termination or chain transfer to give polynorbornene with a narrow molecular weight distribution [35]. After an induction period, the rate of monomer consumption (rate of polymerization) becomes constant, indicating a zero-order dependence on the monomer concentration. The induction period is caused by part of titanacycle 7 undergoing non-productive, but rapidly reversible, cleavage to norbornene and the titanium methylene complex, Eq. (17a).

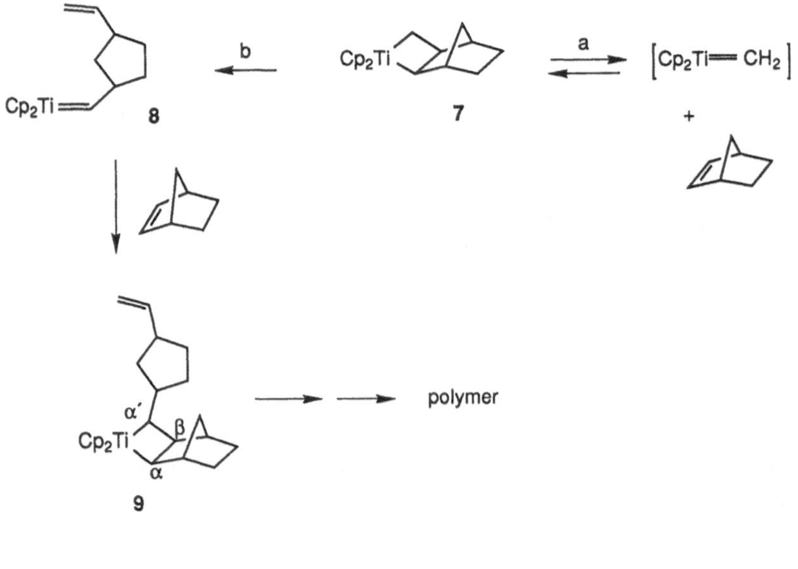

$$\text{(17)}$$

Productive cleavage, Eq. (17b), gives the unstable carbene structure **8** that is trapped by excess norbornene to form α,β,α'-trisubstituted titanacyclobutane structure **9**. a,b,a'-Trisubstituted titanacycle **9** cleaves exclusively in a productive fashion, Eq. (18). The zero-order dependence indicates that the ring-opening of the titanacycle is the rate determining step, similar to the reaction of titanacycle **5** with diphenylacetylene, Eq. (15).

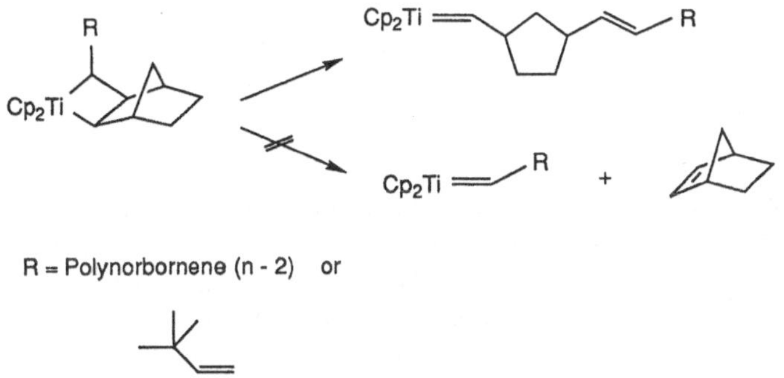

$$\text{(18)}$$

R = Polynorbornene (n - 2)   or

The ring-strain in 3,3-dimethylcyclopropene causes titanacyclobutane **10** to open exclusively in a productive fashion at temperatures above 23 °C. The absence of an equilibrium as shown in Eq. (17a) and the lower reaction temperature for ring-opening provide a polymerization reaction with initiation faster than chain propagation, Eq. (19).

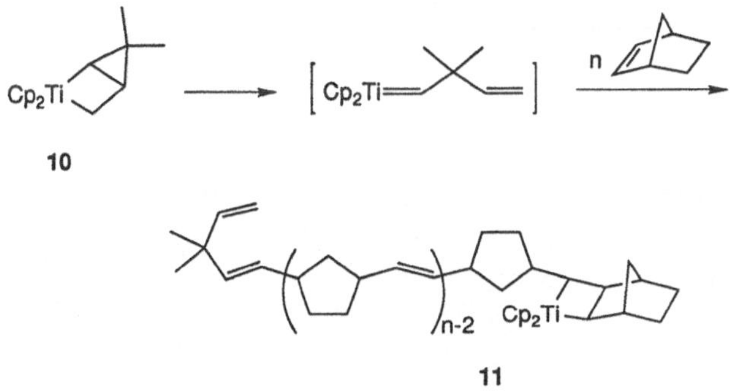

10

11

$$\text{(19)}$$

Thus, **10** is better than **7** as an initiator for the polymerization of norborne. When **10** was used to polymerize norbornene, no induction period was observed, there was a linear dependence of molecular weight on monomer consumption and polymers with high molecular weights (M_n of $10^4–10^5$) and polydispersities lower than 1.1 were prepared [35]. Similar to living anionic polymerization, quantitative monomer consumption (preferably limited to 95–98% of monomer conversion in order to avoid any side reactions) resulted in a polymer with a molecular weight determined by the molar ratio of monomer to initiator. In addition, zero-order kinetics allowed polymerization to any desired molecular weight by adjusting the reaction time. Backbiting and secondary isomerization were not observed.

Polymer chain **11** contains one titanacyclobutane end group. The number of end groups was determined by trapping with benzophenone in a Wittig-type reaction, Eq. (20) [37]. UV- and ^{1}H NMR spectroscopy revealed one diphenylethenyl end group, [12].

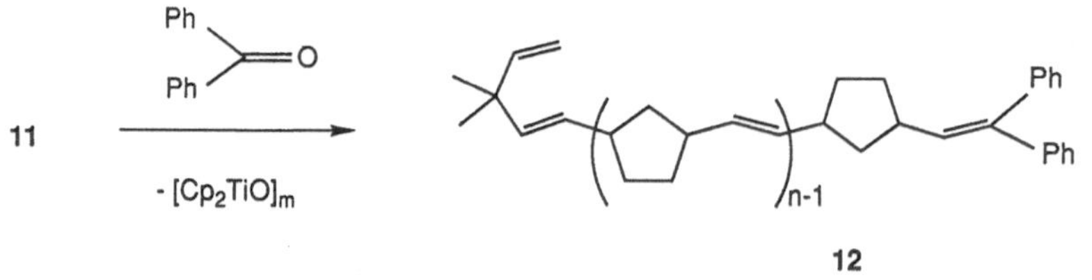

11 $\xrightarrow{\quad -[Cp_2TiO]_m \quad}$ (20)

12

The polymerization of norbornene, Eq. (19), is stopped by cooling the reaction mixture to room temperature. The active polymer **11** can be stored for long periods of time. Heating **11** to temperatures above 65 °C in the presence of monomer causes renewed chain propagation. The subsequent addition of different cyclic olefins, such as *endo-* and *exo*-dicyclopentadiene, benzonorbornadiene and 6-methylbenzonorbornadiene resulted in the formation of well-defined AB- and ABA-type block copolymers, Eq. (21) [38]. Triblock copolymers **13** with narrow molecular weight distributions (polydispersity = 1.14) were prepared. Thus, the living character enables the preparation of new uniform block copolymers of predictable composition, microstructure and molecular weight.

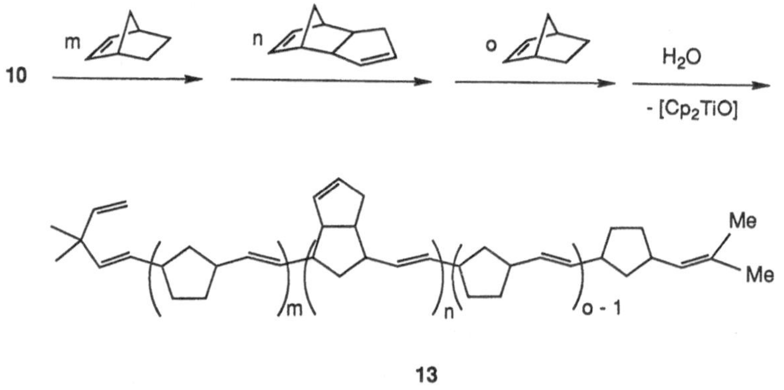

10 $\xrightarrow{m} \xrightarrow{n} \xrightarrow{o} \xrightarrow[-[Cp_2TiO]]{H_2O}$ (21)

13

A second route to block copolymers with low polydispersities applies the transformation of the metathesis active end group into a structure that provides for polymer growth by a different mechanism. The reaction of the polymeric titanacyclobutane **11** with a 9-fold molar excess of terephthalaldehyde gives polynorbornene with one *p*-formylstyrene end group **14**, Eq. (22) [39]. The aldehyde group of **14** is a polymeric initiator for the Aldol group transfer polymerization of *t*-butyldimethylsilylvinyl ether, Eq. (23). Polynorbornene-*block*-poly(silylvinyl ether) **15** and poly(*exo*-dicyclopentadiene)-*block*-poly(silylvinyl ether) with low polydispersities were prepared. These block copolymers were transformed into hydrophobic-hydrophilic block copolymers by cleaving off the silyl groups, Eq. (24)

[39]. Poly(vinyl alcohol) is the resulting hydrophilic block.

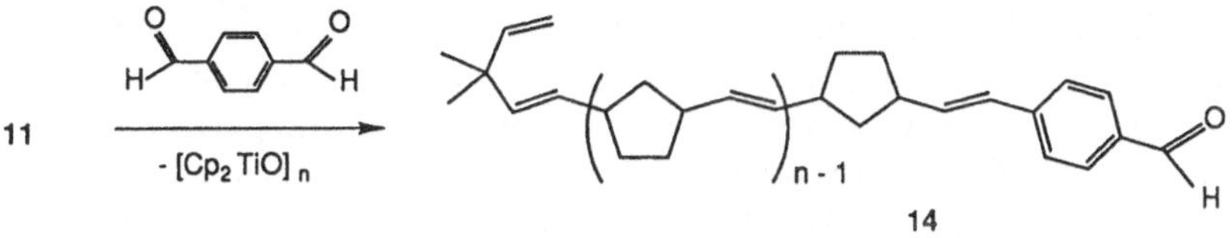

(22)

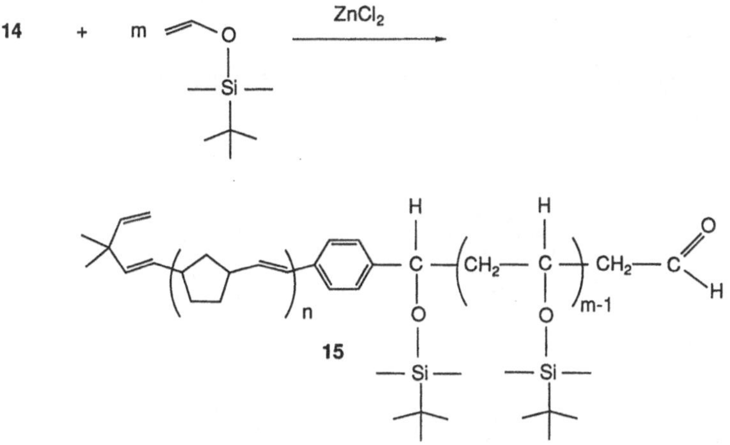

(23)

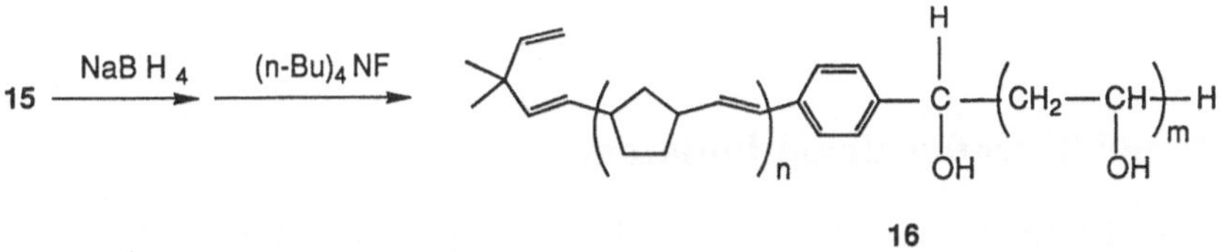

(24)

In addition to block copolymer synthesis by subsequent polymer growth along one polymer chain, Eq. (21), and the reaction sequence of Eqs. (19, 22–24), preformed polymer blocks can be linked via reactive end groups. Polynorbornene with one titanacyclobutane end group was reacted in a Wittig-type reaction with

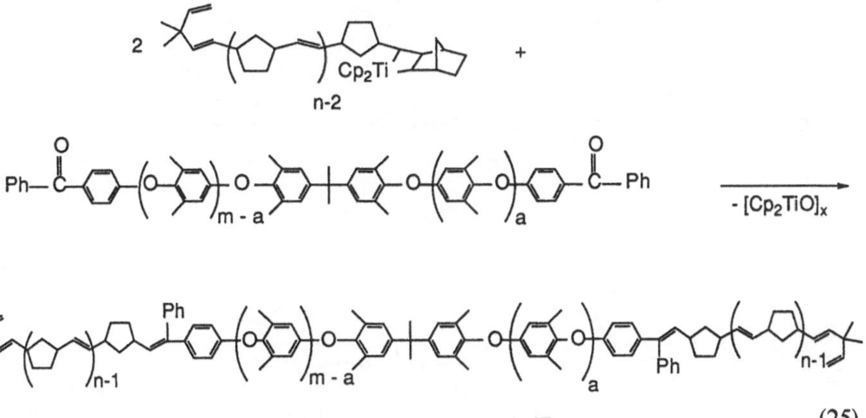

(25)

polymers containing ketone end groups. ABA-triblock copolymers containing polynorbornene and polyphenylene ether segments, **17**, were prepared by the reaction of titanacyclobutane ended polynorbornene **11** with poly(oxy-2,6-dimethyl-1,4-phenylene) containing two ketone end groups, Eq. (25) [40]. The same type of reaction was used for the synthesis of graft copolymers with polynorbornene side chains **19**, Eq. (26). The precursor for the main chain is polyether ketone **18** derived from isopropylidenediphenol and 4,4'-difluorobenzophenone.

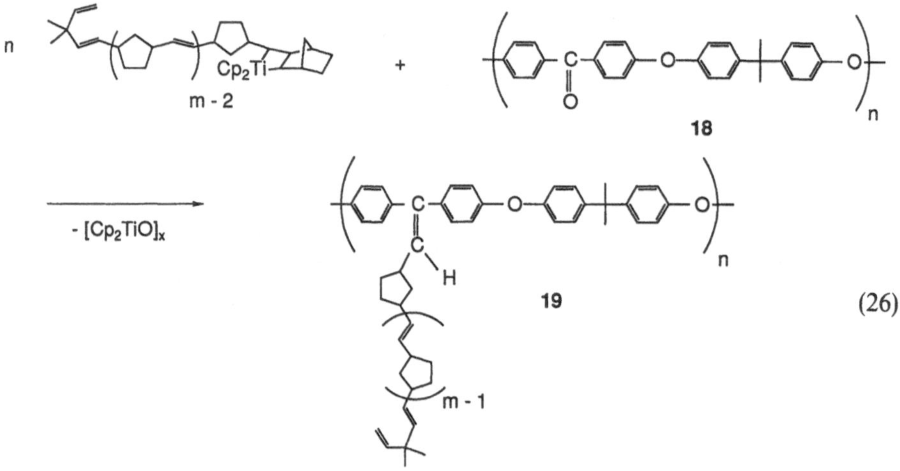

(26)

2.2 Di- and Tetrafunctional Initiators

Titanacycle **10** ring-opens photochemically at temperatures well below 0 °C, Eq. (27) [41]. The unstable titanium carbene **20** can be trapped by *exo-trans-exo*-diene **21**, derived from (2 + 2) cycloaddition of norbornadiene, to produce a mixture of two isomeric bis(titanacyclobutanes), **22a, b** [42]. Bis(titanacycle) **22a, b** could not be synthesized cleanly by thermolysis of **10** because formation of **22a, b** competes with ring-opening and the addition of a second equivalent of diolefin.

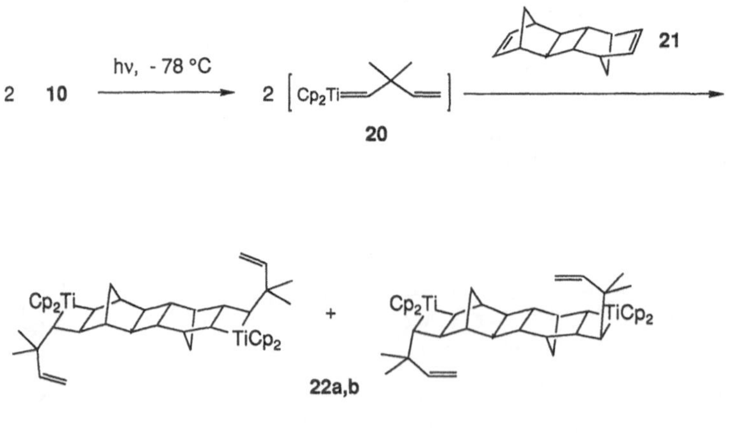

(27)

Bis(titanacyclobutane) **22a, b** allows chain growth in two directions, Eq. (28). Polymers with polydispersities of 1.15 were produced.

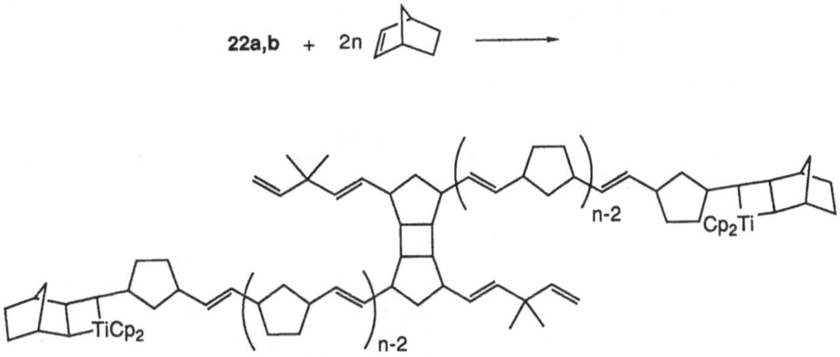

$$\textbf{22a,b} \quad + \quad 2n \quad \text{[norbornene]} \quad \longrightarrow \tag{28}$$

The monomer consumption rate was twice as large as for the polymerization initiated by monometallacycle **10** (50 Eq./h at 75 °C with **22a, b** compared to 23 Eq./h at 73 °C with 10). Compounds **24** and **25** are siloxane based di- and tetra-functional initiators for ring-opening polymerization of cyclic olefins. Tetrakis-(titanacyclobutane) **25** initiates polymerization in four directions resulting instar-shaped polymers [42a]. The molecular weight distributions of the polymers prepared from the siloxanes based initiators are slightly broadened, because the presence of small amounts of nortricyclene units reduces the functionality of these initiators.

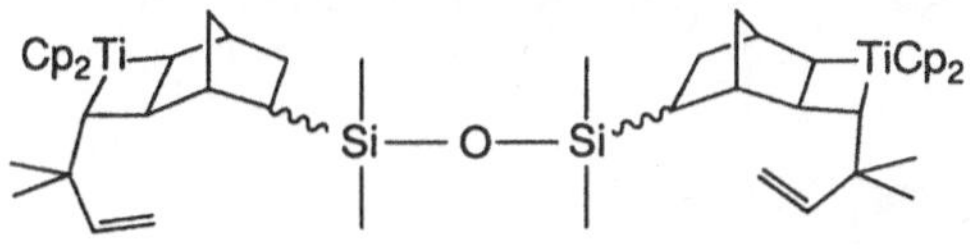

24

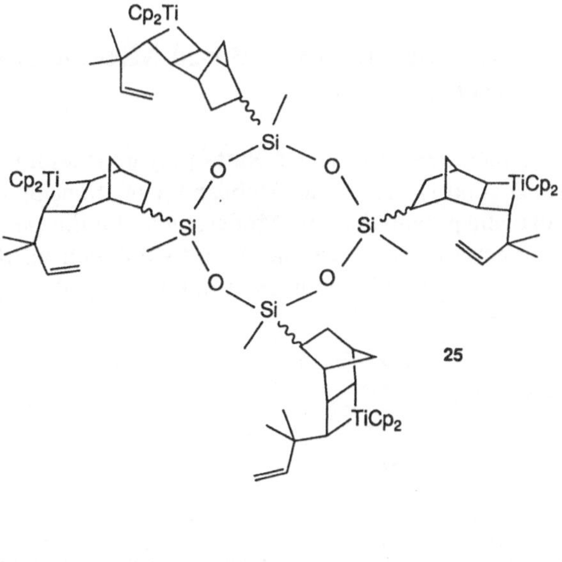

25

Two ended living polynorbornene **23** was prepared with difunctional initiator **22a, b**, and *anti*-7,8-benzotricyclo[4.2.2.0^{2,5}]deca-3,7,9-triene **26** was added as a second monomer to produce a triblock copolymer with a poly(norbornene) middle segment, Eq. (29) [42b]. Naphthalene cleavage occured upon heating to temperatures above 120 °C. The product of thermolysis is polyacetylene/polynorbornene/-polyacetylene triblock copolymer **28**.

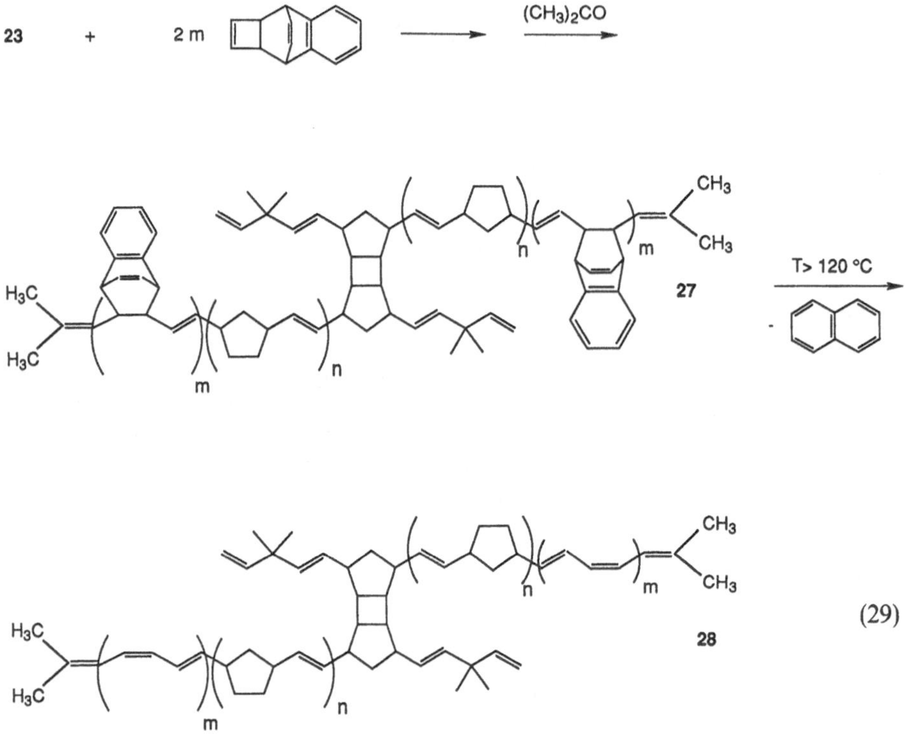

2.3 Conversion of Titanacyclobutane Based Metathesis Catalysts into Ziegler Natta Catalysts

Switching from olefin metathesis to Ziegler Natta polymerization is of interest in order to prepare block copolymers and to establish the relationship between these two related modes of olefin polymerization. Model studies for this purpose included the conversion of titancyclobutanes **10** and **7a** into titanium alkyl compounds Eq. (30a and b) by the addition of 1 equivalent of ethanol [43].

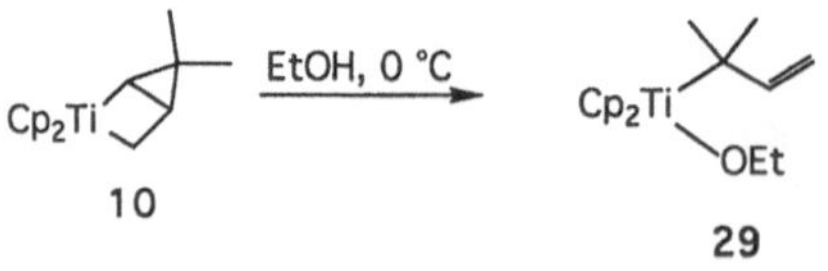

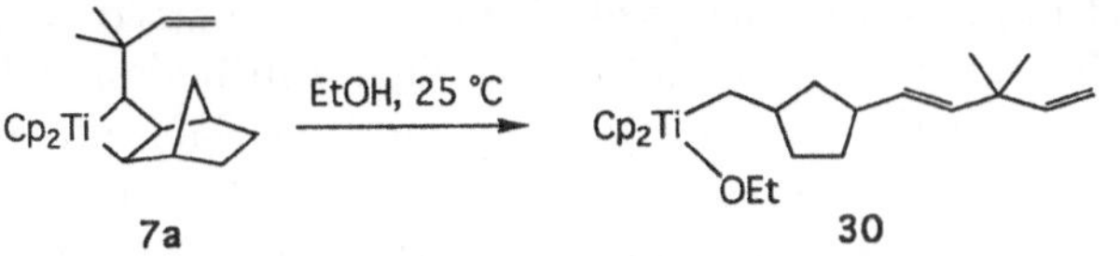

(30b)

7a   **30**

The titanium complexes **29** and **30** were found to polymerize ethylene after the addition of the Lewis acid $AlEtCl_2$. The reaction of 10 with methyl alumoxane resulted in the *m*-methyl bridged bimetallic compound **31**, Eq. (31), which was able to polymerize ethylene in the absence of a Lewis acid cocatalyst, when dichloromethane was used as solvent.

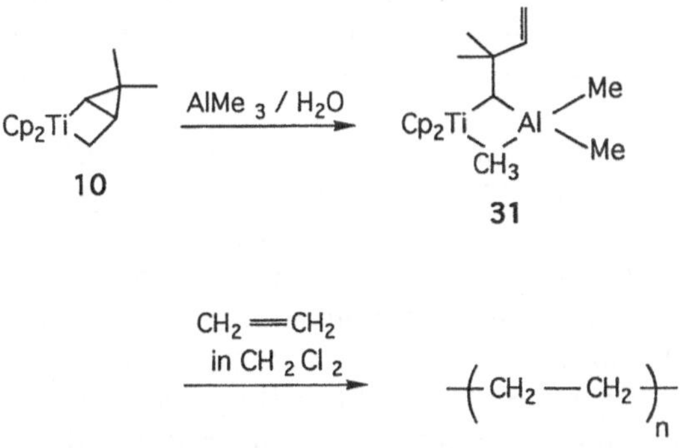

10   **31**

$$\xrightarrow[\text{in } CH_2Cl_2]{CH_2=CH_2} \quad \left(CH_2-CH_2 \right)_n$$ (31)

2.4 Living Polymerizations Catalyzed by Group VB and VIB Complexes

The living polymerization principles developed using the titanacene metallacyclobutane catalysts have been extended to ROMP reactions catalyzed by other transition-metal complexes. These new systems include tantalacyclobutane complexes and preformed tungsten(VI) carbene complexes.

2.4.1 Tantalacyclobutane Catalysts

Metallacycle complexes analogous to the titanium system can be prepared from the reaction of $Ta(CHt-Bu)(OR)_3(THF)$[44] complexes (OR = 2,6-diisopropylphenoxide (DIPP), **32**, or 2,6-dimethylphenoxide (DMP), **33**) and norbornene at $-30\,°C$ Eq. (32) [45, 46].

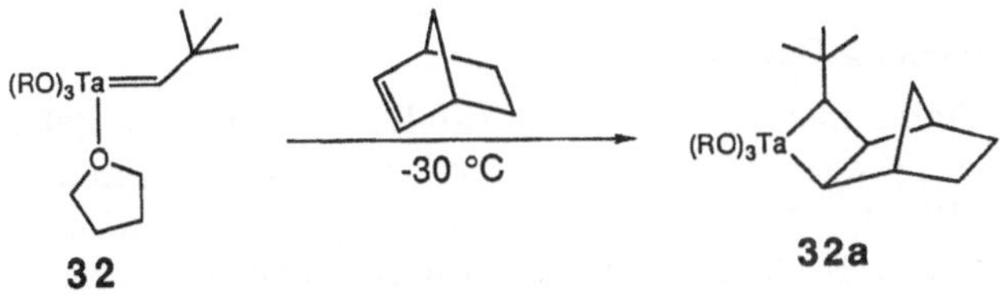

(32)

32   **32a**

The DIPP tantalacyclobutane complex, **32a**, has been shown to polymerize excess norbornene at $50\,°C$ to yield polynorbornene with a molecular weight proportional to the amount of monomer employed. The polymer obtained after 75% monomer conversion has a very narrow molecular weight distribution with $M_w/M_n = 1.04$ for the low molecular weight fraction comprising 94% of the sample. The DMP tantalacycle, 33, is an active polymerization catalyst as well. It is, however, more reactive, and as a result, not as well-behaved as the DIPP derivatives. The polydispersity of the polymer prepared using 100 equivalents of norbornene was measured to be 2.71.

The tantalum and titanium systems are remarkably similar. For the DIPP tantalacyclobutane system, the rate of polymerization (after an initiation period) is zero order in monomer and first order in catalyst concentration. The kinetic parameters calculated for these two systems are: $\Delta H^+ = 24.9$ kcal/mol, $\Delta S^+ = 7.5$ eu, and $\Delta G^+_{338} = 22.4$ kcal/mol for the tantalacycle polymerization, and $\Delta H^+ = 27.1$ kcal/mole, $\Delta S^+ = 9$ eu, and $\Delta G^+_{338} = 24$ kcal/mol for the titanacycle polymerization [35].

If the phenoxide ligands of 32 and 33 are replaced by 2,4,6-triisopropylbenzene-thiolate ligands, TIPT, to form $Ta(CHt\text{-}Bu)(TIPT)_3(THF)$ [46], 34, the reactivity of the catalyst towards acyclic olefins can be dramatically reduced. Complex 34 will not react with simple olefins such as cis-2-hexene or 2-pentene. However, 34 reacted rapidly with norbornene at 25 °C to give polynorbornene (ca. 70% cis-double bonds). Using ^1H NMR, no intermediate metallacyclobutane species can be detected during this reaction. The molecular weights of the polynorbornene samples increased linearly with the number of monomer equivalents employed and the molecular weight distributions were very narrow (PDI < 1.1). Importantly, no broadening of the distribution was observed when the reaction was allowed to run to completion before quenching with benzaldehyde.

As with the titanium metallacyclobutane complexes, these tantalum catalysts react with functional groups containing heteroatoms. Two competitive reactions are observed when the tantalacycles are allowed to react with organic carbonyl groups. A Wittig type reaction via an incipient tantalum alkylidene is observed, along with a competitive direct insertion of the carbonyl into the tantalacycle to form an oxytantalacyclohexane complex Eq. (33) [46].

$$(33)$$

2.4.2 Tungsten and Molybdenum Catalysts

As a complement to the stable metallacyclobutane catalysts, a series of stable alkylidene catalysts have been prepared and shown to be active living polymerization catalysts. The complex $W(CHt\text{-}Bu)(NAr)(Ot\text{-}Bu)_2$ [47, 48] (Ar = 2,6-diisopropylphenyl) (35) was reacted with 50–200 equivalents of norbornene in toluene at 25 °C, followed by end-capping with benzaldehyde, yielding polymers in which the major component has a molecular weight proportional to the number of equivalents of norbornene consumed, with dispersities of approximately 1.05,

Eq. (34). (A minor fraction of high molecular weight polymer was also produced.)

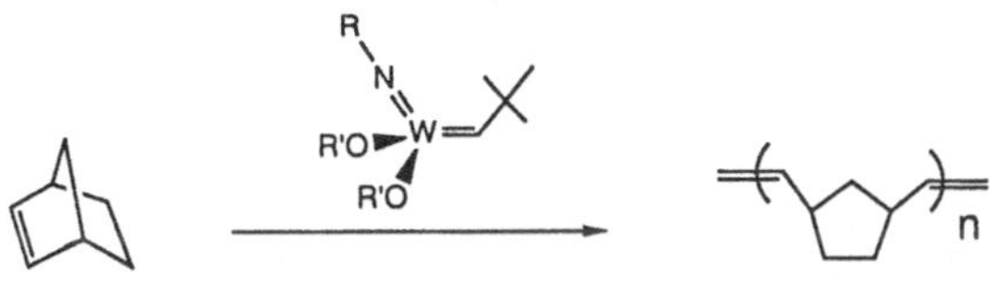

$$
(34)
$$

R = 2,6-diisopropylphenyl, R′ = tert.-butyl

The living nature of this catalyst is also attributed, in part, to its relative in-activity towards unactivated double bonds (i.e. the internal olefins present in the polymer backbone). More active yet less selective catalysts can be synthesized by modification of the alkoxide ligands on the catalyst. Substitution of the t-butoxide ligands by $-OC(CH_3)_2CF_3$ (36), and $-OCCH_3(CF_3)_2$ (37) [49a, b] leads to increasingly more active catalysts (i.e. polymerization of norbornene at tempera-tures as low as $-60\,°C$); however, the molecular weight of the polymer produced is independent of both the reaction time, and the amount of the monomer present [47]. This attenuation of the living behavior is attributed to both a slow initiation rate relative to propagation rate and to secondary metathesis of the double bonds along the polymer chain. Unlike the t-butoxide catalyst, 35, the catalysts containing the fluorinated alkoxides are very active acyclic olefin metathesis catalysts (i.e. 1000 turnovers per minute of cis-2-pentene) [49b].

The high reactivity of carbene complex 37 was successfully employed by K. B. Wagner for polymerizations of acyclic dienes, Eq. (35) [50]. Vinyl addition reactions could be avoided under these Lewis acid free reaction conditions.

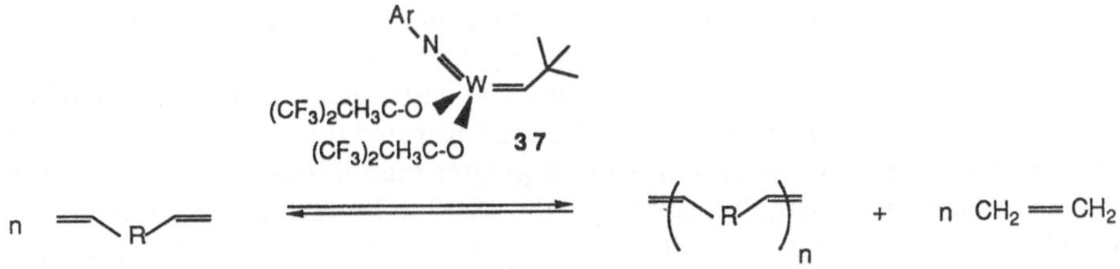

$$
(35)
$$

R = -CH$_2$CH$_2$Si(Me)$_2$-, -CH$_2$Si(Me)$_2$PhSi(Me)$_2$CH$_2$-
   -(CH$_2$)$_m$- (m = 2,6), -(CH$_2$)$_m$O(CH$_2$)$_m$- (m = 3,4),
   -((CH$_2$)$_8$C(O)OCH$_2$)$_2$-

The polymerization proceeds in a step growth fashion and forms ethylene. By removing the ethylene from the reaction mixture high molecular weight polymers ($\bar{X}_n > 10$) could be obtained. Polymers containing alkylene, silyl, ether, aromatic and ester groups have been prepared via this route [50].

Another attractive route to the synthesis of highly reactive tungsten carbene complexes involves alkylidene transfer from phosphoranes. Arylimido tungsten carbene 38 was obtained in a "one-pot" version by Johnson and Grubbs [51a] via the reaction sequence according to Eq. (36).

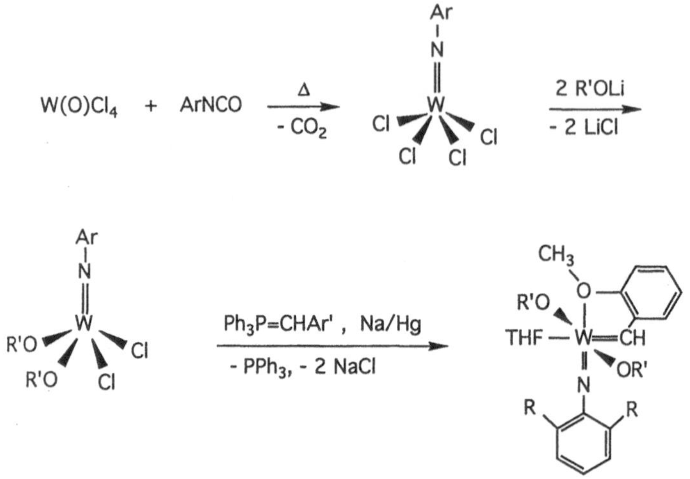

OR' = OCCH$_3$(CF$_3$)$_2$
Ar' = 2-CH$_3$O(C$_6$H$_4$)-
R = Me, H and *i*-Pr

                                                                                    (36)

38

The addition of 5 to 10 equivalents of trimethylphosphine or phenyldimethylphosphine to tungsten carbenes **37** and **38** resulted in catalyst systems with substantially reduced activity since most of the tungsten carbenes is converted into an unreactive phosphine adduct [51b]. A rapid equilibrium between the free carbene, which is present in very small amounts, and the corresponding carbene with the phosphine attached affords a highly-controlled polymerization system. for example, the polymerization of cyclobutene with **37** results in polybutadiene with a very broad polydispersity. It can be demonstrated that for this very reactive monomer the rate of propagation is much greater than initiation [51b]. Addition of 5 equivalents of trimethylphosphine to the reaction leads to polybutadiene with polydispersities less than 1.1. The decreased polydispersity results from the selective poisoning of the propagation rate relative to the initiation rate which results in a reversal of the relative initiation and propagation rates. As a result, cyclobutene can be polymerized to poly-1,4-butadiene (free of 1,2-butadiene units) with a very narrow molecular weight distribution (M_w/M_n as low as 1.02). Hydrogenation yields low-dispersity polyethylene without side chains [51b].

Further synthetic utility can be introduced by changing the metal in the above alkylidene complexes from tungsten to molybdenum [52]. Molybdenum which is less oxophilic than tungsten will tolerate monomers containing mildly reactive functionalities such as esters [52a] and nitriles [52b] without appreciable catalyst deactivation during the polymerization reaction. For example, the polymerization of *endo,endo*-5,6-dicarbomethoxynorbornene (DCNBE) with Mo(CHtBu)(NAr) (OtBu)$_2$ (Ar = 2,6-diisopropylphenyl) (35a) has been reported to give the corresponding ring-opened polymer with molecular weights proportional to the equivalents of monomer used, and polydispersities ranging from 1.11 to 1.22 [52a]. It was found that **35a** would tolerate approximately 100 equivalents of DCNBE before any signs of catalyst deactivation could be noticed [46, 52a]. The polymerization of 200 equivalents of 2,3-dicarbomethoxy-norbornadiene resulted in the formation of a polymer, Eq. (37), with a lower polydispersity: $M_w/M_n = 1.07$ [52b, c]. Additional work concerning the ROMP of functionalized olefins will be covered later.

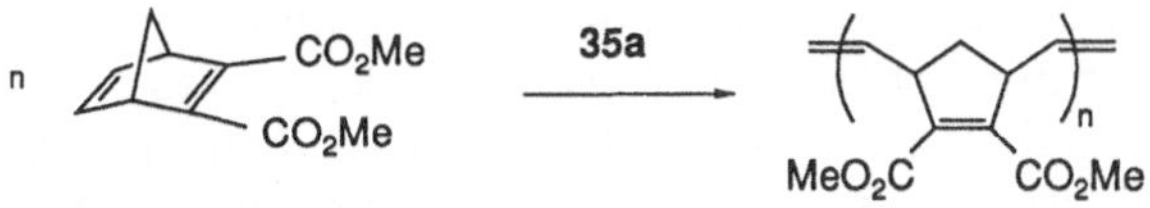

$$ \text{(37)} $$

Schrock, Gibson et al. [52d] found that styrene and 1,3-pentadiene could be used as chain transfer reagents for the living ring-opening olefin metathesis polymerization of norbornene with molybdenum based catalyst **35a**. Renewed norbornene addition to a polymerization mixture containing initiator **35a** and 30 equivalents of styrene resulted in the formation of polynorbornene with a low polydispersity and a molecular weight controlled by the number of norbornene equivalents in each of the individual monomer solutions, Eq. (38). This method allows a more efficient use of the catalyst.

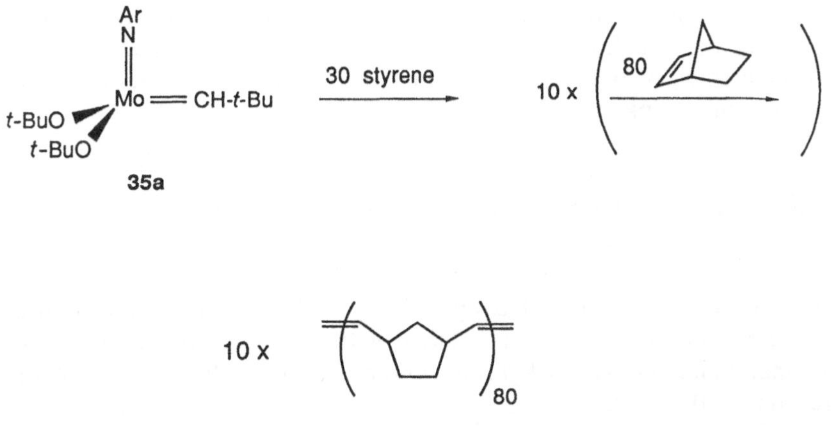

$$ \text{(38)} $$

A variety of isolated pentacoordinate tungsten-carbene complexes are known to be active metathesis catalysts [53]. At least one of these systems has been proposed to be living based primarily on ^{1}H NMR identification of the propagating alkylidene [53e]. To date, verification of the living nature of these catalysts through GPC determination of polydispersities are still pending. The solution NMR studies do confirm the mechanism of the metathesis reaction, but do not insure that all of the requisite factors for a living system are met.

W(CHtBu)(OCH$_2$tBu)$_2$ [53e] (39) is a mildly active ROMP catalyst [54] that can be further activated by the addition of Lewis acids such as GaBr$_3$ to form highly active complexed (39a), and cationic (39b) metathesis species, Eq. (39) [55].

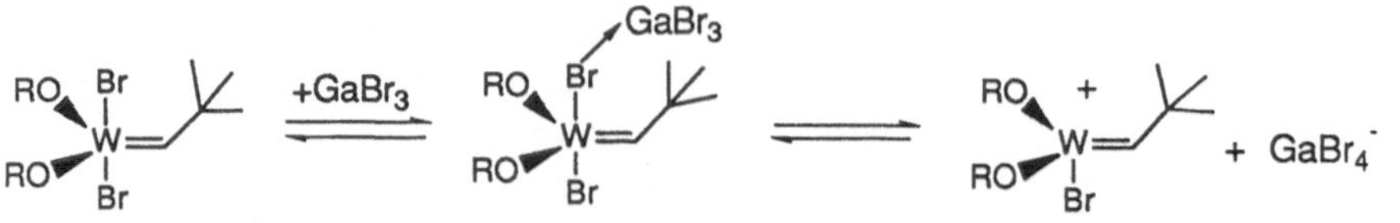

As shown by ^{1}H NMR, the activated catalyst mixture reacts with norbornene (or a series of methyl-substituted norbornenes) to be *partially* converted to a new carbene species [53]. From the ratio of product carbene and residual initiator carbene concentrations, it was estimated that the rate constant for propagation is at least 3 times that for initiation. The three species present in the equilibrium situation of Eq. (39) may all possess their own intrinsic activities, resulting in a more complex polymerization behavior. In addition, a substantial amount of secondary metathesis occurs, as shown by changes in both the *cis* content of the polymer and head/tail ratio of the substituted carbenes when the catalyst was left in solution (120 min at 20 °C) after consumption of the monomer.

In an important extension of the **39**/GaBr$_3$ work [56a, b], the conversion of the initiating metal-carbene complex into the initial metallacyclobutane complex by the addition of a first equivalent of norbornene was directly observed at low temperature by ^{1}H NMR. Subsequent orthogonal cleavage of the metallacyclobutane and its re-formation by the addition of further equivalents of norbornene were also observed. The direct observation of these primary steps may lead to a better understanding of both the kinetic and thermodynamic factors operative during these propagating steps.

2.5 Utilization of ROMP Techniques in the Synthesis of Speciality Polymers

The availability of new catalysts has opened the way for the polymerization of new monomers and the synthesis of materials with new properties. Some recent examples that build on the background outlined above will demonstrate the new directions in this area.

2.5.1 Synthesis of Polyacetylenes

Ring-opening metathesis polymerizations are unique in that all of the unsaturation present in the monomers is conserved in the polymeric product. This feature makes ROMP techniques very attractive for the preparation of highly unsaturated, and fully conjugated materials. It is therefore anticipated that controlled ROMP methodologies could have a large impact in the area of conducting polymers. Using ROMP chemistry, one can envision that the most direct route to polyacetylene would be through the simple metathesis of one of the double bonds of polycyclooctatetraene (COT), Eq. (40).

$$\text{(structure)} \xrightarrow{\mathbf{37}} \text{(polymer structure)}_n \tag{40}$$

The preparation of polyacetylene via the ROMP of COT provides an illustration of the enhanced activity and control which can be realized by using preformed tungsten alkylidene complexes. In 1985, Höcker et al. reported the polymerization of COT with the "classic" metathesis catalyst system $WCl_6/AlEt_2Cl$ [57]. Dilute solution conditions resulted in low yields of insoluble powder, and vapor deposition of COT onto a catalyst slurry over a period of days produced films containing chlorine and saturation defects. If, however, the $WCl_6/AlEt_2Cl$ catalyst system is replaced by the preformed alkylidene catalysts 37 or 38, the condensed-phase polymerization of COT proceeds smoothly, forming high-quality polyacetylene films [58]. Properties of these poly-COT films are nearly identical to those of polyacetylene produced using the Shirakawa method [59].

Extending the metathesis polymerization methodology to other cyclooctate-traene derivatives, provides a convenient route to a variety of substituted polyacetylene derivatives. For example, soluble conjugated polyacetylene der-ivatives can be prepared through the ROMP of trimethylsilylcyclooctatetraene (40) Eq. (41) [60a–d].

$$\underset{\mathbf{40}}{\underset{\text{SiMe}_3}{\text{(structure)}}} \xrightarrow{\mathbf{37}} \underset{\text{SiMe}_3}{\text{(polymer structure)}_n} \tag{41}$$

The polymer of 40, with one TMS substituents for every eight carbons, is readily soluble in CCl_4, benzene and THF. The polymer as prepared having a high cis-double bond content can be converted to the fully trans-polyacetylene derivative by photolysis. Upon isomerization to the all trans form, the UV maximum absorbance shifts from 380 nm to 512 nm, which indicates an increase in the average conjugation length to greater than 15 double bonds [60a]. Several substituted cyclooctatetraenes have been polymerized [60a–e] by the use of highly reactive tungsten carbene catalysts 37 and 38. These experimental results indicate that secondary or tertiary substituents immediately adjacent to the main chain result in solubility for both the predominantly cis and trans form of the substituted poly-COT [60b–e]. Alkyl substituents containing a methylene group next to the polymer main chain only result in partial solubility of the poly-COT-derivative. The substituents are thought to impart twisting of the polymer backbone around single bonds adjacent to the substituents. This twist inhibits close packing of the polymer chains and increases the conformational flexibility of the backbone, thus improving polymer solubility. The spacing between the side groups allows the polymer to maintain conjugated conformations. The twisting required for solubility decreses the conjugation length and conductivity. For example, trans-Poly-40 is oxidized upon exposure to iodine vapor, resulting in a conductivity of 0.2 S/cm.

The ability to cast thin films of conducting polymers with modest conductivities has been exploited in solar cell applications [60d]. Soluble and highly conjugated polycyclooctatetraene derivatives containing chiral substituents [60e] have been prepared from chiral monosubstituted cyclooctatetraenes. The backbone p → p* transition of the corresponding polymers with the stereogenic center a to the main chain shows substantial circular dichroism. The magnitude of the CD effect is characteristic of a disymmetric chromophore indicating that the chiral side group twists the main chain in predominantly one sense rather than just electronically perturbing the chromophore.

Unsubstituted polyacetylene, like many other conductive polymers, is an intractable material and thus its processing into useful shapes and morphologies is limited. One solution to the processing problems has been the use of soluble precursor polymers that can be transformed into conductive polymers. Application of ROMP in the formation of soluble polyacetylene precursors was elegantly pioneered by Feast and coworkers [61]. Using this approach, a precursor polymer is synthesized by the ROMP of a cyclobutene derivative. Once synthesized, the precursor polymer can undergo a thermally promoted, retro-Diels Alder reaction to split off an aromatic fragment and produce polyacetylene, Eq. (42).

$$(42)$$

One drawback to the existing precursor routes is that they generally rely on the cleavage of molecular fragments. These cleaved fragments may comprise a substantial fraction of the total mass and/or have a disruptive effect on the polymer morphology. In pursuit of an alternative precursor route that does not rely on the splitting off of small molecules, the ROMP of benzvalene, **41**, was investigated [62]. The selective metathesis polymerization of **41** produces a polymer containing a bicyclobutane moiety. Effecting the bicyclobutane to butadiene rearrangement would complete the conjugation, forming polyacetylene, Eq. (43).

$$(43)$$

The titancyclobutane catalysts proved ineffective in the polymerization of **41**. If, however, monomer **41** is allowed to react with either of the preformed tungsten carbenes **37** or **38**, the desired metathesis polymer is obtained. It was found that the bicyclobutane rings of poly-**41** could be rearranged to 1,3-dienes by treating thin films of poly-**41** with mild Lewis acids such as $HgCl_2$. Films of polyacetylene prepared in this fashion are strong, flexible and of much lower crystallinity than polyacetylene prepared using other methods [63].

2.5.2 ROMP of Functionalized Monomers in Protic Solvents

Second only to the issues of activity and selectivity, an underlying concern in the development of new catalysts is the degree of stability of the active species to ubiquitous impurities such as air and water [64]. In addition to affecting the usefulness of a catalyst in a practical sense, sensitivity to oxygen containing impurities often foreshadows a catalyst's intolerance to substrates possessing polar functional groups [65]. This becomes an important issue in the design of new speciality materials. The properties of polymers can be dramatically altered by the systematic incorporation of functional groups. As a vehicle for this study, efforts were focused on the ROMP of 7-oxanorbornene derivatives [67]. Although the 1,4-bridging epoxide unit is normally highly deactivating toward the early transition-metal catalysts [68], it has been found that these monomers are polymerized in good yields, Eq. (44) [54], using certain group VIII complexes (a number of different ruthenium and osmium complexes in oxidation states II or III), which were first reported as ROMP catalysts in the early 1960s [69].

$$\underset{\textbf{42}}{}\quad\xrightarrow[C_6H_6/C_2H_5OH]{RuCl_3}\quad \underset{ca.\,70\%}{}\qquad 26\ h \qquad\qquad (44)$$

Polymerizations employing these group VIII complexes are typically preceded by a lengthy induction period (hours, or even days, depending on the system) [70]. During attempts to minimize the induction period of the $RuCl_3$ catalyzed polymerization of **42** by employing more stringent (anaerobic and anhydrous) reaction conditions, it was surprisingly discovered that water acts as a cocatalyst for these ROMP reactions [71]. An important distinction between this system and the Ziegler-Natta [72] and ROMP [17, 73] polymerizations that are cocatalyzed by alcohol, or water, is that these ruthenium reactions are not deactivated by the presence of excess water (i.e. no decrease in initiation rates is observed, even when the water concentration approaches 25%). This observation resulted in the finding that the polymerization of the 7-oxanorbornene monomers proceeds rapidly in water alone to produce the desired ROMP polymer in nearly quantitative yields, Eq. (45) [74].

$$\underset{\textbf{43}}{}\quad\xrightarrow[H_2O]{RuCl_3}\quad \underset{100\%}{}\qquad 35\ min. \qquad\qquad (45)$$

The fact that very high molecular weight materials form under these aqueous conditions indicates that if a termination reaction involving the hydrolysis of the carbon-metal bonds is occuring in either the metallacycle or metal carbene intermediates, it has a much slower rate (by several orders of magnitude) than the rate of polymer propagation.

An important issue in any type of catalysis reaction is the recyclability of the catalyst. This is particularly true of transition-metal catalyzed reactions in which a considerable amount of time and money may be expended in just the catalyst preparation alone. Ruthenium catalysts were demonstrated to be recyclable when used for the polymerization of norbornene in methanol. $Ru(H_2O)_6(tos)_2$ (tos = p-toluenesulfonate) [76], **44**, catalyzes the polymerization of norbornene with an induction period of 55 ± 5 s at 50 °C. The same catalyst solution can be recycled without any detectable loss of activity [77].

Using $RuCl_3$ in water, a 1.0 M solution of **43** typically displays an initial initiation time of 37.5 min at 55 °C. Subsequent polymerizations using the resulting aqueous catalyst solution show substantially increasing initiation rates. By the third polymerization, the initiation time for **43** is reduced from over 35 min to only 10–12 s. "Used" $RuCl_3$ solutions have been recycled up to 14 consecutive times without a detectable change in the initiation rate. Significant progress in uncovering the active species in these recycled solutions was made using $Ru(H_2O)_6(tos)_2$ (**44**) as the catalyst. When n equivalents of monomer **43** are allowed to react with **44** in D_2O, ^1H NMR reveals that (n − 1) equivalents of monomer polymerize and the remaining equivalent cleanly forms a 1:1 Ru^{2+}-olefin complex, **45**, (less than 1% is converted to active catalyst) Eq. (46) [71, 77].

$$ (46) $$

When the olefin adduct **45** is allowed to react with excess monomer, polymerization is observed to occur in just under 10 s at 55 °C, Eq. (47).

$$ (47) $$

These and other observations suggest that a Ru^{2+} olefin complex is the catalyst precursor. In the absence of a reducing agent, the formation of Ru^{2+} from Ru^{3+} salts under the reaction conditions has been shown to occur through a disproportionation process. The increased activity of these recycled Ru^{2+} and Ru^{3+} catalysts is attributed to: 1) The reduction of the Ru^{3+} to Ru^{2+} (when applicable); and 2) the in situ formation of Ru^{2+}-olefin complexes.

These highly activated aqueous ROMP catalysts can be applied to the polymerization of monomers hitherto reluctant to polymerize in aqueous solution. This can be illustrated with the following example. When either 7-oxanorbornene-2,3-dicarboxylic acid, **46**, or its dipotassium salt, which posses both the 1,4-bridging epoxide and the dicarboxylate moieties, is allowed to react with a wide range of metathesis catalysts, only catalyst deactivation is observed, Eq. (48) [79].

No Reaction

$$ (48) $$

Likewise, attempts at polymerizing the relatively deactivated anhydride, **47**, in dry organic solvents (to prevent hydrolysis of the anhydride), resulted in catalyst deactivation as well. It was found, however, that monomer **47** could be readily polymerized using aqueous dipotassium ruthenium pentachloride or complex **45**, Eq. (49).

$$
\begin{array}{c}
\text{(structure of 47)} \xrightarrow[\ \ H_2O\ \ ]{\ \ 45\ \ } \text{(polymer structure } m \text{)}
\end{array}
\tag{49}
$$

That polymerization proceeds in aqueous solution is quite surprising given that hydrolysis of the anhydride moiety occurs simultaneously, producing **46**, the known catalyst poison, as well as the desired polyacid material. There is a competition in these aqueous systems between the rate of chain initiation and the rate of hydrolysis of the anhydride to give the catalyst poison, **46**. Under the polymerization conditions (i.e.) neutral water at 57 °C), there is a 35% conversion of **47** into **46** during a typical initiation period (15 min) for the aqueous Ru^{3+} system. It is thought that during reactions in which the initiation process is delayed, the hydrolysis of the anhydride moiety competes successfully, thereby deactivating the ruthenium catalysts.

3 Conclusions

The developments over the past few years in the understanding of organometallic mechanisms and structure has led to a new generation of catalysts for the polymerization of cyclic olefins by the ring opening metathesis mechanism. These catalysts provide the means to control molecular weight and molecular weight distribution and to introduce functionality into polyolefins.

4 References

1. Anderson AW, Merckling NG (Du Pont) (1955) US Pat 2,721,189; (1956) Chem Abstr 50: 3008i
2. (a) Ziegler (1952) Angew Chem 64: 323; (b) Natta G, Pino P, Mazzanti G, Giannini U, Mantica E, Peraldo (1957) M Chim Ind Milan 39; 19; (c) Natta G, Pino P, Mantica E, Danusso F, Mazzanti G, Peraldo M (1956) Chim Ind Milan 38: 124
3. (a) Eleuterio HS, (Du Pont) (1961) Ger Pat 1,072,811 (1960); Chem Abstr 55: 16005 ; (b) Truett WL, Johnson DR, Robinson IM, Montague, B A (1960) J Am Chem Soc 82, 2337
4. (a) Dall'Asta G, Mazzanti G, Natta G, Porri L (1962) Makromol Chem 56, 224; (b) Natta G, Dall'Asta G, Mazzanti G, Motroni G (1963) Makromol Chem 69, 163; (c) Kormer VA, Yufa TL, Poletaeva IA (1969) Dokl Akad Nauk SSSR 185, 873. (d) Dall'Asta G, Motroni G, Motta L (1972) J Polym Sci, A-1, 10: 1601
5. Natta G, Dall'Asta G, Mazzanti G (1964) Angew Chem, Int Ed 3: 723
6. (a) Alkema HJ, van Helden R (Shell) (1968) Brit Pat 1,117,968 Chem Abstr 69: 95906 (1968). (b) Höcker H, Musch R (1972) Makromol Chem 157: 201.

7. Reviews on Olefin Metathesis: (a) Grubbs RH (1982) In: Wilkinson G (ed) Comprehensive organometallic chemistry, Pergamon, Oxford, vol 8, p 499; (b) Dragutan V, Balaban AT, Dimonie M (1985) Olefin metathesis and ring-opening polymerization of cyclo-olefins, Wiley, Chichester; (c) Ivin KJ (1983) Olefin metathesis, Academic, London

8. Calderon N, Chen HY, Scott KW (1967) Tetrahedron Lett 3327

9. Hughes WB (1972) Organomet Chem Synth 1: 341

10. Hérisson JL, Chauvin Y, Phung NH, Lefebvre GCR (1969) Hebd Seances Acad Sci, Ser C 269: 661

11. Uchida Y, Hidai M, Tatsumi T (1972) Bull Chem Soc Jpn 45: 1158

12. Ivin KJ, O'Donnell JH, Rooney JJ, Steward CD (1979) Makromol Chem 180: 1975

13. Grubbs RH, Swetnick SJ (1980) J Mol Catal 8: 25

14. Mol JC, Moulijn JA (1975) Adv Catal 24: 131

15. Sata H, Tanaka Y, Taketomi T (1977) Makromol Chem 178: 1993

16. Banks RL, Bailey GC (1964) Ind Eng Chem, Prod Res Dev 3: 170

17. Calderon N, Ofstead EA, Ward JP, Judy WA, Scott KW (1968) J Am Chem Soc 90: 4133

18. Bradshaw CPC, Howman EJ, Turner L (1967) J Catal 7: 269

19. Hérisson JL, Chauvin Y (1971) Makromol Chem 141: 161

20. (a) Grubbs RH, Burk PI, Carr DD (1975) J Am Chem Soc 97, 3265; (b) Grubbs RH, Carr DD, Hoppin C, Burk PI (1976) J Am Chem Soc 98: 3478

21. Katz TJ, Rothchild R (1976) J Am Chem Soc 98: 2519

22. Katz TJ, Lee SJ, Acton, N (1976) Tetrahedron Lett 4247

23. (a) Soufflet J-P, Commereuc D, Chauvin YCR (1973) R Hebd Seances Acad Sci, Ser C 169; (b) Muetterties EL (1975) Inorg Chem 14: 951; (c) Grubbs RH, Hoppin CR (1977) J Chem Soc, Chem Commun 634

24. (a) Scott KW, Calderon N, Ofstead EA, Judy WA, Ward JP (1969) Am Chem Soc, Adv Chem Ser 91: 399; (b) Höcker H, Reimann W, Riebel K, Szentivanyi, Z(1976) Makromol Chem 177: 1707

25. (a) Tanaka Y, Sato H, Hatada K, Terawaki Y (1977) Makromol Chem 178: 1823; (b) Calderon N, Scott KW (Goodyear) (1981) Can Pat 1,095,646 Chem Abstr 95: 8034 (1981); (c) Syatkowsky AI, Denisova TT, Abramenko EL, Khatchaturov AS, Babitsky BD (1981) Polymer 22: 1554; (d) Ivin KJ, Rooney JJ, Bencze L, Hamilton JG, Lam LM, Lapienis G, Reddy BSR, Ho HT (1982) Pure Appl Chem 54: 447

26. (a) Jolly PW, Wilke G (1974) In: The organic chemistry of nickel, Academic Press: New York, Vol 2; (b) Maitlis PM (1971) In: The organic chemistry of palladium, Academic: New York

27. (a) Ephritikhine M, Green MLH, Mackenzie RE (1976) J Chem Soc, Chem Commun 619; (b) Adam GJA, Davies SG, Ford KA, Ephritikhine M, Todd PF, Green MLH (1980) J Mol Catal 8: 15

28. Tebbe FN, Parshall GW, Reddy GSJ (1978) J Am Chem Soc 100: 3611

29. Tebbe FN, Parshall GW, Ovenall DW (1979) J Am Chem Soc 101: 5074

30. Howard TR, Lee JB, Grubbs RH (1980) J Am Chem Soc 102: 6878

31. Lee JB, Ott KC, Grubbs RH (1982) J Am Chem Soc 104: 7491

32. (a) Ziegler K, Bähr K (1928) Chem Ber 61: 253; (b) Ziegler K, Dersch F, Wollthan H (1934) Justus Liebigs Ann Chem 511: 13

33. Flory PJ (1940) J Am Chem Soc 62: 1561

34. (a) Szwarc M, Levy M, Milkovich R (1956) J Am Chem Soc 78: 2656; (b) Szwarc M (1983) Adv Polym Sci 49: 1

35. Gilliom LR, Grubbs RH (1986) J Am Chem Soc 108: 733

36. Gilliom LR, Grubbs RH (1986) Organometallics 5: 721

37. Cannizzo LF, Grubbs RH (1987) Macromolecules 20: 1488

38. Cannizzo LF, Grubbs RH (1988) Macromolecules 21: 1961

39. Risse W, Grubbs RH (1989) Macromolecules 22: 1558

40. Risse W, Grubbs RH (1989) Macromolecules 22: 4462

41. Tumas W, Wheeler DR, Grubbs RH (1987) J Am Chem Soc 109: 6182

42. (a) Risse W, Wheeler DR, Cannizzo LF, Grubbs RH (1989) Macromolecules 22: 3205; (b) Stelzer F, Park JW, Risse W, Grubbs RH (unpublished results)

43. Tritto I, Grubbs RH (1990) Proceedings of the international symposium on recent developments in olefin polymerization catalysts; Kodansha, Tokyo pp 301–312
44. Wallace KC, Dewan JC, Schrock RR (1986) Organometallics 5: 2162
45. Wallace KC, Schrock RR (1987) Macromolecules 20: 450
46. Wallace KC, Liu AH, Dewan JC, Schrock RR (1988) J Am Chem Soc 110: 4964
47. Schrock RR, Feldman J, Cannizzo LF, Grubbs RH (1987) Macromolecules 20: 1169
48. Cannizzo LF (1988) PhD Thesis, California Institute of Technology
49. (a) Schaverien CJ, Dewan JC, Schrock RR (1986) J Am Chem Soc 108: 2771; (b) Schrock RR, DePue RT, Feldman J, Schaverien CJ, Dewan JC, Liu, AH (1988) J Am Chem Soc 110: 1423
50. (a) Bauch CG, Wagener KB, Boncella JM 377; (b) Wagener KB, Puts RD, 379; (c) Brzezinska K, Wagener KB, 381; (d) Wagener KB, Smith jr DW 373; (e) Wagener KB, Konzelmann J 375
51. (a) Johnson LK, Virgil SC, Grubbs RH (1990) J Am Chem Soc 112: 5384; (b) Wu Z, Grubbs submitted to J Am Chem Soc
52. (a) Murdzek JS, Schrock RR (1987) Macromolecules 20: 2640; (b) Schrock RR (1990) Acc Chem Res 23: 158; (c) Bazan GC, Khosravi E, Schrock RR, Feast WJ, Gibson VC, O'Regan MB, Thomas JK, Davies WM (1990) J Am Chem Soc 112: 8378; (d) Crowe WE, Mitchell JP, Gibson VC, Schrock RR (1990) Macromolecules 23: 3536
53. (a) Aguero A, Kress J, Osborn JA (1985) J Chem Soc Chem Commun 793; (b) Quignard F, Leconte M, Basset JM (1985) J Chem Soc, Chem Commun 1816; (c) Quignard F, Leconte M, Basset JM (1986) J Mol Catal 36: 13; (d) Quignard F, Leconte M, Basset JM (1985) J Mol Catal 28: 27; (e) Kress J, Osborn JA, Greene RME, Ivin KJ, Rooney JJ (1985) J Chem Soc, Chem Commun 874
54. Novak BM, Grubbs RH (1988) J Am Chem Soc 110: 960
55. Kress J, Osborn JA (1983) J Am Chem Soc 105: 6346
56. (a) Kress J, Osborn JA, Greene RME, Ivin KJ, Rooney JJ, (1987) J Am Chem Soc 109: 899; (b) Greene RME, Ivin KJ, Rooney JJ, Kress J, Osborn J (1988) Makromol Chem 189: 2797
57. Korshak YV, Korshak V, Kansichka G, Hocker H (1985) Makromol Chem, Rapid Commun 6: 685
58. Klavetter FL, Grubbs RH (1988) J Am Chem Soc 110: 7807
59. Shirakawa H, Ikeda S (1971) Polym J (Tokyo) 2: 231
60. (a) Ginsburg EJ, Gorman CB, Marder SR, Grubbs RH (1989) J Am Chem Soc 111: 7621; (b) Gorman CB, Ginsburg EJ, Marder SR, Grubbs RH (1990) Polym Prepr 31 (1): 386; (c) Ginsburg EJ, Gorman CB, Sailor MJ, Lewis NS, Grubbs RH (1990) In: Olefin metathesis and polymerization catalysts, Imamoglu Y (ed), Kluwer, Dordrecht, Netherlands p 537; (d) Sailor MJ, Ginsburg EJ, Gorman CB, Kumar A, Grubbs RH, Lewis NS (1990) Science 249: 1146; (e) Moore JS, Gorman CB, Grubbs RH (1991) J Am Chem Soc 113: 1704
61. Edwards JH, Feast WJ (1987) Polym J 28: 567
62. Swager TM, Grubbs RH, Doughtery DA (1988) J Am Chem Soc 110: 2973
63. Swager TM, Grubbs RH (1989) J Am Chem Soc 111: 4413
64. For a general overview of Ziegler-Natta polymerizations of functionalized substrates see: Boor J In: Ziegler-natta catalysts and polymerizations, Academic, New York: 1979
65. (a) Parshall GW In: Homogeneous Catalysis; Wiley, New York, 1980; (b) Vandenberg (1963) J Polym Sci C1, 207; (c) Kuntz, EG CHEMTECH 1987, 570; (d) Chung TC (1988) Macromolecule 21: 865
66. For references on the metathesis of polar substrates (both cyclic and acyclic) see: (a) Murdzek JS, Schrock RR (1987) Macromolecules 20: 2640; (b) van Dam PD, Mittelmijer MC, Boelhouwer C (1972) J Chem Soc, Chem Commun 1221; (c) Mol JC (1982) J Mol Catal 15: 35; (d) Matsumoto S, Komatsu K, Igarashi K (1977) Am Chem Soc, Polym Preprints 18: 110; (e) Matsumoto S, Komatsu, K, Igarashi K (1977) Am Chem Soc, Polym Preprints, 18: 110
67. Novak BM, Grubbs RH (1987) Proc Am Chem Soc Div PMSE 57: 651

68. (a) Saegusa T, Matsumoto S, Motoi M, Fuji H (1972) Macromolecules 5: 236; (b) Kops J, Spanggaard H (1972) Makromol Chem 151: 21

69. (a) Michelotti FW, Keaveney WP (1963) Am Chem Soc, Polym Preprints 4: 293; (b) Michelotti FW, Keaveney WP (1965) J Polym Sci A-3: 895; (c) Michelotti FW, Carter JH (1965) Am Chem Soc, Polym Preprints 6: 224; (d) Ho HT, Ivin KJ, Rooney JJ (1982) J Mol Catal 15: 245; (e) Porri L, Diversi P, Lucherini A, Rossi R (1975) Makromol Chem 176: 3131; (f) Porri L, Rossi R, Diversi P, Lucherini A (1974) Makromol Chem 175: 3097

70. The induction period is defined as the time elapsed from the initial mixing and heating of the reaction mixture until the initial polymer formation

71. Novak BM, Grubbs RH (1988) J Am Chem Soc 110: 7542

72. (a) Breslow DS, Newberg NR (1957) J Am Chem Soc 79: 5072; (b) Breslow DS, Newberg NR (1959) J Am Chem Soc 81: 81; (c) Anderson A, Cordes HG, Herwig J, Kaminsky W, Merck A, Mottweiler R, Pein J, Sinn H, Vollmer HJ (1976) Angew Chem 88: 689; (d) Kaminsky W, Miri M, Sinn H, Wold R (1983) Makromol Chem, Rapid Commun 4: 417; (e) Adema EH, Bartelink HJM, Smidt, J (1961) Rec. Trav Chim 80: 173

73. (a) Ivin KJ, Reddy BSR, Rooney JJ (1981) J Chem Soc, Chem Commun 1062; (b) Uchida Y, Hidai M, Tatsumi T (1972) Bull chem Soc Jpn 45: 1158

74. Early attempts at emulsion ROMP systems have been reported. These systems, however, either fail for many monomers or, at best, give low yields of polymer (typically less than 9%). See: Rinehart RH, Smith HP (1965) J Polym Sci, Polym Lett 3: 1049

75. Schrock RR, Yap KB, Yang DC, Sitzmann H, Sita LR, Bazan GC (1989) Macromolecules 22: 3191

76. Sullivan BP, Baumann JA, Meyer TJ, Salmon DJ, Lehmann H, Ludi A (1977) J Am Chem Soc 99: 7368

77. Novak BM, Grubbs RH J Am Chem Soc Submitted for Publication

78. Seddon KR, Seddon EA In: Chemistry of ruthenium, Elsevier, New York: 1984

79. Novak BM, Grubbs RH Macromolecules, Submitted for Publication

Editor H.-H. Kausch
Received September 5, 1991

Captodative Olefins in Polymer Chemistry

J. Penelle[1], A. B. Padias, H. K. Hall, Jr.
C. S. Marvel Laboratories, Department of Chemistry, The University of Arizona, Tucson, Arizona 85721, USA

H. Tanaka
Department of Chemical Science and Technology, Tokushima University, Mina-mijosanjima, Tokushima 770, Japan

The radical polymerization behavior of captodative olefins such as acrylonitriles, acrylates, and acrylamides α-substituted by an electron-donating substituent is reviewed, including the initiated and spontaneous radical homo- and copolymerizations and the radical polymerizations in the presence of Lewis acids. The formation of low-molecular weight products under some experimental conditions is also reviewed. The reactivity of these olefins is analyzed in the context of the captodative theory. In spite of the unusual stabilization of the captodative radical, the reactivity pattern of these olefins in polymerization does not differ significantly from the pattern observed for other 1,1-disubstituted olefins. Classical explanations such as steric effects and aggregation of monomers are sufficient to rationalize the observations described in the literature. The spontaneous polymerization of acrylates α-substituted by an ether, a thioether, or an acylamido group can be rationalized by the Bond-Forming Initiation theory.

1   Introduction 75

2   Captodative Radicals and Captodative Theory 75

3   Reactions of Captodative Olefins with Radicals Leading to Low
    Molecular Weight Products 76

4   Radical-Initiated Homopolymerization of Captodative Olefins 80

5   Radical-Initiated Copolymerization of Captodative Olefins 84

6   Spontaneous Polymerization of Captodative Olefins 87

7   Lewis Acid-Assisted Spontaneous Polymerization of Captodative
    Olefins 90
    7.1 Polymerization 90
    7.2 Structure of Lewis Acid-Coordinated Radicals 91

---

[1] Present address: Département de Chimie, Université Catholique de Louvain, Place L. Pasteur 1, B-1348 Louvain-la-Neuve, Belgium

8   The Bond-Forming Initiation Theory for Spontaneous Polymerizations   92
8.1 Background on Thermal (Spontaneous) Polymerizations of Vinyl
     Monomers   92
8.2 Proposed Theory: The Bond-Forming Initiation Theory   93
8.3 Generation of Tetra- and Trimethylene Diradicals from Cyclic
     Compounds   95

**9   Application of Bond-Forming Initiation Theory to the Spontaneous
     Polymerization of Captodative Olefins**   96

10   Conclusions and Perspectives   99

11   References   100

List of Abbreviations

AIBN	Azo-bis(isobutyronitrile)
CT	Charge transfer
DTBP	Di-*tert*-butyl peroxide
DTBPO	Di-*tert*-butyl peroxalate
EDA	Electron-donor-acceptor
ESR	Electron spin resonance (spectroscopy)
HPLC	High-pressure liquid chromatography
IBN	Isobutyronitrile radical
LALLS	Low angle laser light scattering
MA	Methyl acrylate
MMA	Methyl methacrylate
N.A.	Not available
NVCz	*N*-Vinylcarbazole
SEC	Size exclusion chromatography
St	Styrene
T	Tetramethylene
T_c	Ceiling temperature
VA	Vinyl acetate

1 Introduction

Captodative olefins are olefins geminally substituted with both an electron-withdrawing group c (Latin "captare", to take) and an electron-donating group d (Latin "dare", to give) [1–8]. These olefins often display a special reactivity in free radical chemistry due to the possibility of obtaining very stabilized captodative radicals, sometimes also named merostabilized or push-pull radicals.

It is the purpose of this paper to review the implications of this unusual stabilization for the radical polymerization of captodatively substituted olefins. Also, the possibility of producing 1,4-tetramethylene diradicals from these olefins opens an important new area for the spontaneous polymerization of olefins. The latter will be discussed in the context of the Bond-Forming Initiation Theory [9–10].

2 Captodative Radicals and Captodative Theory

In general, the stabilization afforded to a radical by stabilizing substituents is not additive. The origin of this non-additivity is related to the steric difficulty for di- or trisubstituted radicals to reach a geometry where all substituents are able to fully interact with the single electron.

The special stabilization of captodative radicals originates in the synergy of action that results when the radical is substituted by both an electron-withdrawing and an electron-donating group. This interaction can be easily visualized by a simple model. The only resonance structure that can be drawn for a radical substituted by just an electron donor implies a charge separation and the appearance of a negative charge on the carbon atom:

$$\cdot CH_2\text{-}D \quad \longleftrightarrow \quad :CH_2^-\text{-}D\cdot^+$$

$$A\text{-}CH\cdot\text{-}D \quad \longleftrightarrow \quad :\overset{\displaystyle |}{\underset{\displaystyle A}{C}}H^-\text{-}D\cdot^+ \quad \longleftrightarrow \quad A^- = CH\text{-}D\cdot^+$$

When in addition there is an electron acceptor present on the carbon, it allows the delocalization of that negative charge, resulting in extra stabilization.

However, this electronic effect can sometimes be counterbalanced by steric effects as discussed earlier, hindering the full benefit of the electronic effect. Therefore, the overall stabilization energy of a captodative radical will be the sum of the individual stabilization energies of each substituent, plus the synergistic electronic stabilization due to the captodative effect, minus the destabilizing effects due to geminal steric interactions. This means that not all captodative radicals will necessarily be very stabilized. If the individual stabilization energies are low, the final stabilization energy will be low even with the occurrence of a synergistic effect between the substituents. However, for normal polar substituents (esters, nitriles, amides, ethers, thioethers, amines, etc.), large total stabilization energies are expected.

In the literature, a controversy has recently developed on the existence of the captodative effect. The extra stabilization energy of various di- and trisubstituted radicals has been experimentally or theoretically determined [11–21]. In some cases a captodative effect was found [11–16], but other researches failed to find any such effect [17–21]. These discrepancies can be explained by the relatively large experimental or theoretical errors (2–3 kcal/mol) which all methods involved in the determination of quantities used in free radical thermochemistry generate [14, 22, 23]. Because the extra stabilization energy cannot be determined directly (it is a quantity composed of several experimental values), the statistical propagation of errors leads to experimental errors larger than the determined values and does not permit any firm conclusions on the existence of a captodative effect. However, when the substituents are carefully chosen to provide a very large synergistic effect, the extra stabilization becomes large enough compared to the errors to be detectable [15]. Another important cause of discrepancy is the lack of consensus in the literature on how to define the stabilization energy of a radical [24–29].

Investigators using experimental tools like ESR [2, 30–39] or kinetics methods [40–48], which are able to quantitatively investigate small stabilization effects more accurately than thermochemical methods, always find a synergistic effect in accordance with the captodative theory. Accordingly there is little doubt, at the present time, of the validity of the captodative theory.

3 Reactions of Captodative Olefins with Radicals Leading to Low Molecular Weight Products

Viehe and his collaborators investigated the addition of different radicals to various captodative olefins. They first studied the reaction of an equimolar amount of azo-bis-isobutyronitrile (AIBN) with captodative olefins in benzene at 80 °C [2]. The only isolated products were the bisadducts 1 of two isobutyronitrile radicals

bisadduct   adduct-dimer

1   2

(IBN) to the olefin, accompanied in some cases by the adduct-dimers 2 formed by dimerization of two IBN-olefin adduct radicals (Table 1). Most olefins gave only moderate yields (20–49%) of the bisadducts 1; however when substituted by bulky *tert*-butylthio or morpholino groups, higher yields were obtained. The

Table 1. Reaction of captodative olefins $CH_2=C(c)d$ with large initial amounts of AIBN[a]

c	d	Bisadduct 1 Yield (%)	Adduct-dimer 2 Yield (%)	Ref.
CN	StBu	88	—	[2]
CN	SEt	N.A.	43	[2]
CN	SMe	18	51	[2]
CN	Morpholino	68	—	[2]
CN	NMe$_2$	38	—	[2]
CN	OSiMe$_3$	33	12	[49]
COOMe	StBu	76	—	[2]
COOMe	SMe	35	—	[2]
COOMe	NEt$_2$	37	—	[2]
COOMe	OMe	49	—	[2]
CONMe$_2$	NMe$_2$	40	—	[2]
COOSiMe$_3$	OSiMe$_3$	20	—	[50]

[a] Ratio olefin/AIBN = 1, 80 °C, benzene

α-alkylthioacrylonitriles showed a substantial steric effect. α-*tert*-Butylthioacrylo-nitrile gave only the bisadduct **1a** (88% isolated yield), while α-methylthioacryloni-trile gave mainly the adduct-dimer **2b** (51%), with the bisadduct **1b** as a minor product (18%).

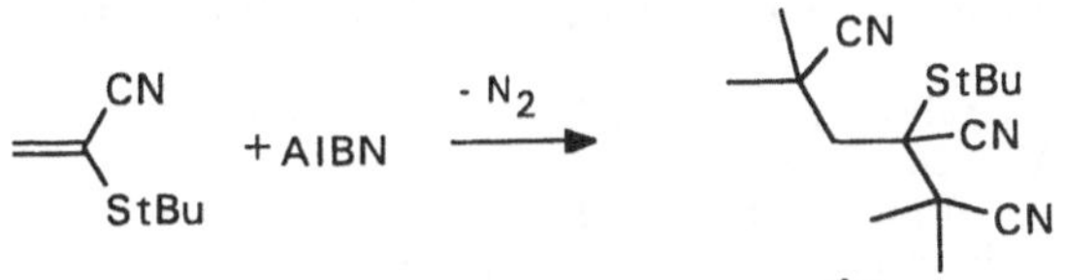

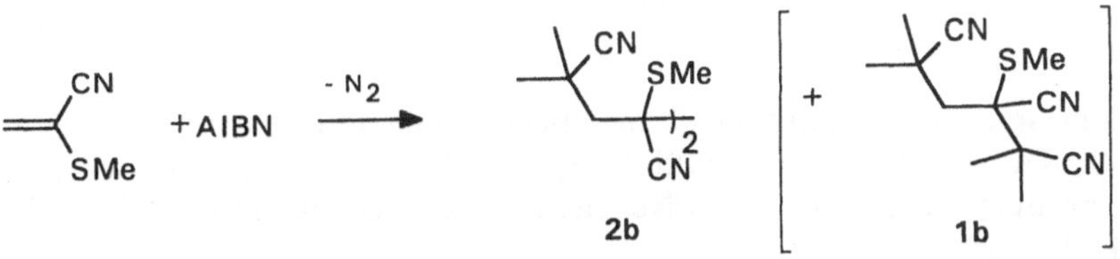

The reactions of α-*tert*-butylthioacrylonitrile with various radicals were tho-roughly studied by Viehe and his coworkers [51–56]. Abstraction of a hydrogen atom from various substrates (aldehydes, thiols, . . .) was induced by di-*t*-butyl peroxide (DTBP) or di-*t*-butyl peroxalate (DTBPO). The results are described in Table 2. With the exception of the IBN-radical addition described above, all radicals gave adduct-dimers **3** in substantial yields.

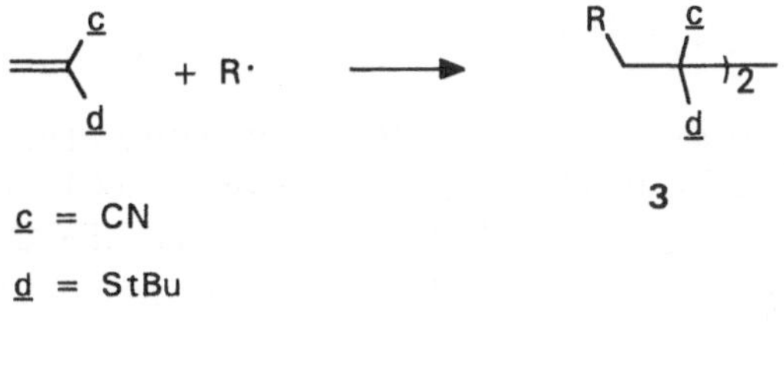

$\underline{c}$ = CN

$\underline{d}$ = StBu

Table 2. Addition of various radicals to α-*tert*-butylthioacrylonitrile

Radical	Method to produce the radical[a]	Adduct-dimers **3** Yield (%)[b]	Ref.
R^1—CO—C—R^2R^3	A	38–63	[51, 56]
R—C=O	A	52.5–70	[51, 54, 56]
R　(R = cycloalkyl)	A	66–79	[56]
Ph—CH_2	A	51	[56]
Ph—CH—Me	A	59	[52]
Ph—C—Me_2	A	0	[52]
Bicyclo [1.2.2] heptane-2-yl	A	58	[52]
R^1O—CH—R^2	A, B	46–72	[52, 53, 56]
R^1S—CH—R^2	A	48–68.5	[52]
R^1CO—NR^2—CH—R^3	A	46–74	[52, 54, 56]
R^1R^2N—CH_2	C, A	33–70	[52, 54, 56]
R^1R^2C—OH	A	34–62	[52]
$(RO)_2$CH	B	59–64	[52]
$(RS)_2$CH	B	50.5–57	[52]
RS	A	61–71	[54]
Me_3C—O	D	–	[55]
	E	11	[55]
	F	26	[55]

[a] Method A: H-abstraction by DTBPO; substrate/olefin/DTBPO = 20/2/1; 6 h, 60 °C.
Method B: H-abstraction by U.V.-irradiation of DTBP; substrate/olefin/DTBP = 20/2/1; benzene, room temperature.
Method C: H-abstraction by DTBP; substrate/olefin/DTBP = 20/2/1; 12 h, 130 °C.
Method D: DTBP, benzene, 140 °C; olefin/DTBP = 2/1.
Method E: DTBPO, benzene, 60 °C; olefin/DTBP = 2/1.
Method F: DTBP, U.V.-irradiation, benzene, 20 °C; olefin/DTBP = 2/1.
[b] Calculated based on the olefin; c = StBu, d = CN.

Neumann studied the addition of metal-containing radicals $Ph_2C^•$—MMe_3 (M = C, Si, Ge, Sn) to α-*tert*-butylthioacrylonitrile resulting in the 1,4-C,N-addition products **4** due to excessive steric hindrance of these very bulky radicals [57].

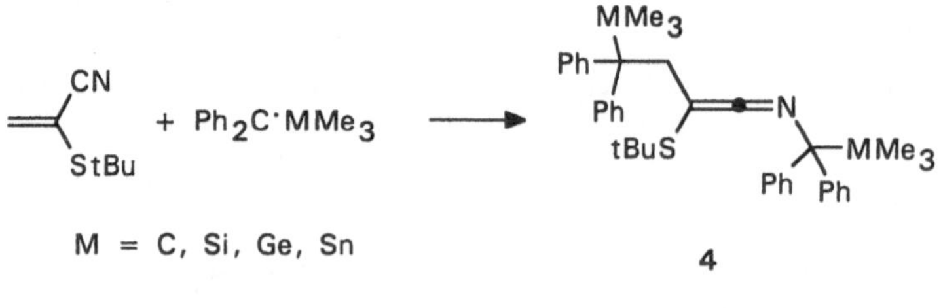

M = C, Si, Ge, Sn

4

The generalization of these reactions to other captodative olefins was often successful from a preparative point of view, as demonstrated by the reported results on the addition of tBuO$^•$, MeCO$^•$ [55], RS$^•$ [54], and MeCONMe—$CH_2^•$ [54] radicals to various captodative olefins (Table 3). Additions of (*N*-methyl-*N*-

phenylamino)methyl radical produced three different types of product **5**, **6**, and **7**, depending on the structure of the initial olefin [54] (Table 4).

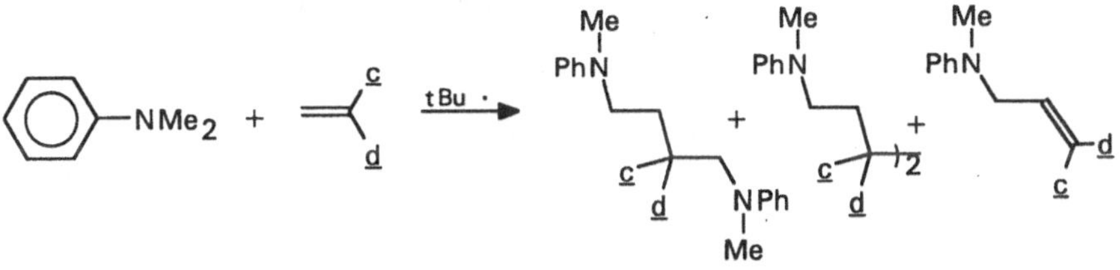

Table 3. Addition of various radicals to captodative olefins $CH_2=C(c)d$

Radical R	Method to produce the radical[a]	c	d	Adduct-dimers 2 Yield (%)	Ref.
Me$_3$CO	A	CN	StBu	0[b]	[55]
	B	CN	StBu	11[c]	[55]
	B	CN	SPh	21[d]	[55]
	B	Ph	SPh	74	[55]
	C	CN	StBu	26	[55]
	C	CN	SPh	22	[55]
	C	CN	OMe	21[e]	[55]
	C	CN	NMe$_2$	59	[55]
	C	CN	morpholino	73	[55]
	C	CN	piperidino	64	[55]
CH$_3$C=O	D	CN	StBu	67	[54]
	D	CN	SEt	63	[54]
	D	CN	SMe	67	[54]
	D	CN	SPh	71	[54]
	D	CN	OMe	71	[54]
	D	COOMe	StBu	64	[54]
	D	COOMe	OMe	61	[54]
PhS	D	CN	StBu	71	[54]
	D	CN	SEt	55	[54]
	D	CN	SMe	69	[54]
tBuS	D	CN	StBu	61	[54]
	D	CN	SEt	53	[54]
	D	COOMe	StBu	64	[54]
	D	COOMe	OMe	71	[54]
Ac-NMe−CH$_2$	D	CN	StBu	74	[54]
	D	CN	OMe	51	[54]
	D	COOEt	StBu	47	[54]

[a] Method A: DTBP, benzene, 140 °C; olefin/DTBP = 2/1.
 Method B: DTBPO, benzene 60 °C; olefin/DTBPO = 2/1.
 Method C: DTBP, U.V.-irradiation, benzene, 20 °C; olefin/DTBP = 2/1.
 Method D: H-abstraction by DTBPO; substrate/olefin/DTBPO = 20/2/1; 6 h, 60 °C.
[b] Alongside 32% ((Et) (StBu) (CN) C)$_2$.
[c] Alongside 48% ((Et) (StBu) (CN) C)$_2$.
[d] Alongside 41% ((Et) (StBu) (CN) C)$_2$.
[e] Alongside >5% ((Et) (StBu) (CN) C)$_2$

Table 4. Addition of MePhN—CH$_2^{\cdot}$ to captodative olefins CH$_2$=C (c) d[a]

c	d	5 Yield (%)	6 Yield (%)	7 Yield (%)	Ref.
CN	StBu	0	50	0	[54]
CN	SEt	0	46	0	[54]
CN	SMe	0	47	0	[54]
CN	NMe$_2$	0	0	35	[54]
COOMe	StBu	33	0	0	[54]
COOMe	OMe	12	18	0	[54]

[a] Radical generated from N,N-dimethylaniline and DTBP

The adduct-dimers **3** formed in the above reactions are often in equilibrium with their monomeric radicals **8**. The presence of the radicals **8** can be observed by ESR spectroscopy at temperatures as low as room temperature [58–61].

All the above experiments were run using equimolar amounts of radicals and illustrate the extreme stabilization found in captodative radicals. This led Viehe and coworkers to conclude that "the stabilized captodatively substituted radicals … do not undergo typical reactions such as polymerization or hydrogen abstraction but rather they trap another radical R˙ or dimerize" [2].

However, a careful HPLC analysis of the mixture arising from the photo-stimulated reaction between α-methoxydeoxybenzoin and methyl α-methoxy-acrylate at room temperature demonstrated that a considerable number of products are formed during this reaction [62]. Besides the expected products, some were found to result from the addition of a captodative (methoxy)-(methoxycarbonyl)alkyl radical to the captodative olefin itself, i.e. a "propagation" step. The HPLC analysis was not quantitative, but the presence of these products even at high olefin/initiator ratio (2/1) implies that a propagation step cannot be ruled out for all the above experiments only on the basis of the isolation of low-molecular weight products in good yields. As we will show the polymerizability of captodative olefins is confirmed by an analysis of the literature.

4 Radical-Initiated Homopolymerization of Captodative Olefins

Because many electron-donating and electron-withdrawing groups are known and their combinations can generate a very large number of captodative olefins, we restricted our literature search to radical-initiated polymerizations of substituted acrylonitriles $H_2C = C(d)CN$, acrylates $H_2C = C(d)COOR$, and acrylamides $H_2C = C(d)CONH_2$ with d substituents containing either O, S or N. The relevant literature data are summarized in Tables 5–7.

Table 5. Initiated radical homopolymerization of captodative acrylonitriles $CH_2 = C(d)CN$

d	Experimental conditions	Conversion (%)	Ref.
OMe	AIBN(0.2%), benzene, 60 °C, 150 h	8.9	[63]
OEt	$K_2S_2O_8$, H_2O, 4 days	0	[64]
	$(Ph_2COO)_2$, UV light, 2 weeks	0	[64]
	AIBN(1%), benzene, 65 °C, 24 h	13 Low M.W. (NMR)	[65]
OAc	AIBN, benzene	kinetic study	[66]
SMe	AIBN(0.2%), neat, 70 °C, 2 h	92	[67]
	AIBN, 75 °C, several days[a]	N.A. (Dark, glassy polymer)	[68]
Morpholino	$Na_2S_2O_3$(20%), benzene, 15 min	0	[65]

From these tables, it is directly evident that *captodative olefins are capable of undergoing a propagation step leading to oligo- or polymerization.* Some of the olefins even lead to very high molecular weight polymers.

The bulkiness of the alkyl moiety on the heteroatom of the d substituent plays a very important role. α-Methylthioacrylonitrile polymerizes easily, but α-tert-butylthioacrylonitrile does not [67, 68]. The same comparison holds for EtS- and tBuS-substituted acrylates [70].

Because of the general lack of quantitative thermodynamic (ceiling temperature T_c) and kinetic (k_p, $k_p/k_t^{0.5}$) data for the polymerization of the captodative olefins, it is impossible to draw firm conclusions about the importance of electronic factors on their polymerizability. If we compare them with other 1,1-disubstituted olefins by replacing the heteroatom O, S, or N by a CH_2 and check the polymerizability of the resulting olefins, we find that the latter are in fact also difficult to polymerize as shown in Table 8 [77]. Only methyl acrylates and methacrylates give high polymers easily. The polymerizability decreases rapidly with the steric hindrance of the substituent.

A conventional explanation for the difficult radical polymerization of 1,1-disubstituted olefins is the known lowering action of bulky substituents on T_c. The classical example of this effect is α-methylstyrene [78]. Reported T_c values for substituted acrylates are collected in Table 9. Unfortunately, no T_c values have ever been reported for any captodative olefins.

Furthermore the isolation of low-molecular weight products by Viehe and coworkers, especially when bulky α-*tert*-butylthio substituents are placed on the olefin, is also perfectly compatible with classical theories. The steric hindrance of a tBuS-group was calculated to be greater than the steric hindrance of a neopentyl $(CH_2C(CH_3)_3)$, a cyclohexyl, or a t-butyl group [84].

Table 6. Initiated radical homopolymerization of captodative acrylates $CH_2=C(d)COOR$

d	R	Experimental conditions	Conversion (%)		Ref.
OMe	Me	AIBN(0.14%), benzene, 60 °C, 150 h	25.7	$\eta_{sp}/c = 0.197$ (c = 1.05 g/dL)	[63]
		$(PhCOO)_2$	N.A.		[69]
		AIBN	N.A.		[69]
		α-methoxydeoxybenzoin, hv, RT	N.A.	oligomers (HPLC)	[62]
OMe	tBu	AIBN (5 mM), bulk, 60 °C, 8.5 h	7.4	$M_n = 5.6 \times 10^4$ (SEC[a])	[70]
OAc	Me	AIBN (5 mM), bulk, 60 °C, 10 h	6.2	$M_n = 5.3 \times 10^4$ (SEC[a])	[70]
SEt	Me	AIBN (5 mM), bulk, 60 °C, 0.8 h	5.2	$[\eta] = 1.56$ dL/g	[70]
StBu	Me	AIBN (10 mM), bulk, 60 °C, 11 h	25.2	$M_n = 3.6 \times 10^4$ (SEC[a])	[70]
N(Me)COPr	Me	AIBN (10 mM), bulk, 60 °C, 10 h	0		[70]
NHCOPr	Me	Free radical initiators	0		[70]
NHCOR'	Me	VAZO-67, hexane/MeOH(95/5), 60 °C	90	$[\eta] = 0.64$ dL/g	[71, 72]
R' = Me		AIBN or VAZO-67, 60 °C			
C_5H_{11}		hexanes	51	$[\eta] = 0.56$ dL/g	[72]
C_6H_{13}		hexanes/MeOH(80/20)	46	$[\eta] = 0.41$ dL/g	[72]
C_7H_{15}			N.A.	$[\eta] = 1.0$ dL/g	[72]
C_8H_{17}		hexanes	46	$[\eta] = 1.48$ dL/g	[72]
C_9H_{19}			74	$[\eta] = 2.07$ dL/g	[72]
$C_{11}H_{23}$			66	$[\eta] = 2.28$ dL/g	[72]
			variable	$M_w = 0.10{-}15.2 \times 10^5$	[73]
$C_{13}H_{27}$			N.A.	$[\eta] = 2.30$ dL/g	[72]
$C_{15}H_{31}$			74	$[\eta] = 1.30$ dL/g	[72]
			57	$[\eta] = 0.91$ dL/g	[72]
$C_{17}H_{35}$			73	$[\eta] = 1.59$ dL/g; $M_w = 1.74 \times 10^6$ (LALLS)	[72, 74]
$C_7F_{15}(CH_2)_8CH=CH_2$			48	$[\eta] = 0.54$ dL/g	[72]
NHCONHR'	Me		N.A.	$[\eta] = 2.70$ dL/g	[72]
R' = Me		$K_2S_2O_8$, H_2O, 35 °C, overnight	60	$[\eta] = 0.40$ dL/g	[75]
Et			N.A.	$[\eta] = 0.22$ dL/g	[75]
Pr		VAZO-67, benzene, 60 °C, overnight	40–60	$[\eta] = 0.20$ dL/g	[75]
Bu			40–60	$[\eta] = 0.54$ dL/g	[75]

[a] By comparison with polystyrene standards

Table 7. Initiated radical homopolymerization of captodative acrylamides $CH_2=C(d)CONH_2$

d	Experimental conditions	Conversion (%)	Ref.
OMe	Cumyl peroxide (0.5–5%), bulk, 100 °C, 20 h	54–87 [η] = 0.50–0.83[a]	[76]
	AIBN (2%w), bulk, 115 °C, 19 h	4	[76]
	$KBrO_3$ (0.06%), Na_2SO_3 (0.02%), H_2O, H_2SO_4, 90 °C, 20 h	6.5	[76]
	$K_2S_2O_8$ (0.18%), iPrOH, H_2O, 75 °C, 3 h	8	[76]
SMe	AIBN (0.17%), bulk, 70 °C, 2 h	82	[67]

[a] c = 0.5%, 1 N sodium citrate, 30 °C

Table 8. Radical polymerizability of methyl α-alkylsubstituted acrylates $H_2C=C(R)COOMe$ [77]

R	Polymerizability[a]
H	#
CH_3	#
CH_2CH_3	+
$CH_2CH_2CH_3$	+
$CH(CH_3)_2$	−
$CH_2CH_2CH_2CH_3$	○
$CH(CH_3)(CH_2CH_3)$	−
$CH_2CH(CH_3)_2$	−
C_6H_5	+
$CH_2C_6H_5$	+
c-C_6H_{11}	−

[a] #: Polymerizes to high polymer.
   ○: Some indications for oligomer formation.
   +: Polymerizes slowly to low polymers.
   −: Does not polymerize

Table 9. Ceiling temperatures for substituted acrylates

Structure	T_c (°C) [olefin] = 1 M	T_c (°C) [olefin] = 5 M	T_c (°C) [olefin] = 8.35 M	Ref.
$H_2C=C(CH_3)COOMe$	160	218[a]	241[a]	[79]
$H_2C=C(CH_2CH_3)COOMe$			73–78	[80]
$H_2C=C(CH_2Ph)COOMe$		67		[81]
$H_2C=C(Ph)COOMe$	−40[b]			[82]
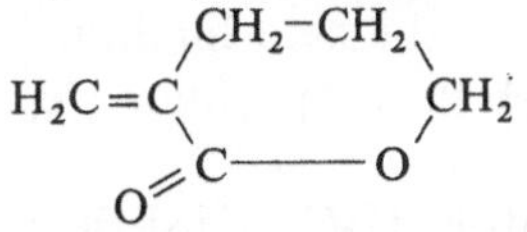 $H_2C=C$ (ring: CH_2–CH_2 / CH_2, $C=O$, O)	83	135[a]	154[a]	[83]

[a] Calculated from the reported ΔHp and ΔSp.
[b] Anionic polymerization

Experimental conditions are also a very important factor to consider: reaction of α-methylthioacrylonitrile with AIBN was found to give either 69% of isolated low-molecular weight products [2] or 92% of polymer [67], depending on the conditions. The nature and concentration of the initiator also influence the polymerization. The rate of addition of the primary radical (initiator radical) to the captodative olefin is dependent on the nature of the initiator as observed by ESR [85]. Moreover termination by recombination of the persistent propagating radical with a primary radical becomes very important, as expected from the small molecule chemistry. Therefore the initiator influences both the initiation and termination.

It is interesting to note that Mathias and Hermes have recently proposed a theory opposing Viehe's ideas on the polymerizability of captodative olefins. According to these authors, the polymerizability of a captodative olefin is not reduced because of the stabilization of the captodative radical; on the contrary, the stabilization of the growing radical *favors* the polymerization [72, 75]. This theory was proposed in order to rationalize the very high molecular weights observed during the polymerization of methyl α-acylamidoacrylates $H_2C = C(NHCOR)COOMe$ [72], in particular the spontaneous polymerization of the derivative with $R = n\text{-}C_9H_{19}$ which led to a polymer with an estimated molecular weight of 15 million [73]. In contrast, N-alkyl equivalents $H_2C = C(NMeCOR)COOMe$ do not polymerize at all [71]. A contribution factor to this peculiar behavior could be the strong tendency of amides to form large aggregates in solution due to strong intermolecular hydrogen bonds, leading to high local concentration and favorable alignment of the olefin [86, 87]. Aggregate formation can strongly increase the k_p values, as illustrated by the extraordinarily high k_p value for acrylamide (ca. $10^4 \, M^{-1} \, s^{-1}$ at 25 °C in H_2O) [88]. In addition this long chain alkyl α-acylamidoacrylate is known to form monolayers, indicating its tendency to align [74].

Compared with other 1,1-disubstituted olefins, the captodative olefins do not seem to present abnormal behavior that could be related to the captodative stabilization of the transient growing radical. Both conversions and molecular weights can be rationalized by classical theories linking the polymerizability of these olefins to steric and dipole-dipole repulsions.

5 Radical-Initiated Copolymerization of Captodative Olefins

Typical captodative olefins have been copolymerized with various olefins (Table 10). Some olefins which do not homopolymerize, as $H_2C = C[N(Me)COnPr]$ COOMe for example did copolymerize [71]. This is not a surprising result; such an effect is very often described in the literature. The reason is that the steric hindrance affects the propagation steps of a copolymerization less than the propagation step of the corresponding homopolymerization.

The claimed copolymerizability of α-ethylthioacrylonitrile $H_2C = C(SEt)CN$ with acrylonitrile at 70 °C in the patent literature is however doubtful [67]. α-Ethylthioacrylonitrile acts as a chain transfer agent during the polymerization

Table 10. Copolymerizations of captodative olefins, $CH_2 = C(c)d$

c	d	Copolymerizable with
CN	OMe	Styrene [63], ethyl α-chloroacrylate [89], acrylonitrile [63]
CN	OEt	Methyl methacrylate [65], butadiene [90]
CN	OAc	Styrene [91, 92], acrylonitrile [92]
CN	SEt	Acrylonitrile [67]
COOMe	OMe	Styrene [63, 70], ethyl α-chloroacrylate [89], acrylonitrile [63]
COOMe	OAc	Styrene [70]
COOMe	SEt	Styrene [70]
COOMe	N(Me)COnPr	Methyl acrylate [71]
COOtBu	OMe	Styrene [70]
CONH₂	SMe	Styrene [67]

of acrylonitrile, styrene and methyl methacrylate, and no copolymerization is observed at 60 °C [93].

Similarly, acrylonitrile, methyl acrylate and acrylamide α-substituted with a benzyloxy group act as chain transfer agents during the polymerization of MMA, St, MA, and VA, which is due to the following fragmentation reaction [94]:

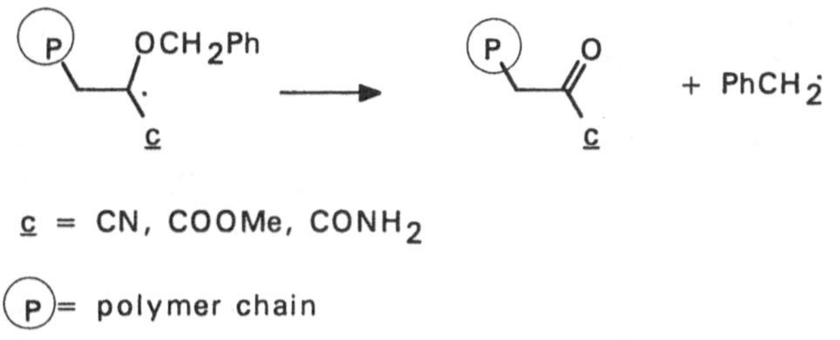

$\underline{c}$ = CN, COOMe, CONH₂

(P)= polymer chain

Some quantitative data (r_1, r_2, Q, e values) for the copolymerization of captodative olefins are available, mainly for the copolymerization with styrene. They are compiled in Tables 11–14. In Tables 11 and 12, the data for the copolymerization of captodative olefins with styrene are given for comparison together with the values for other 1,1-disubstituted olefins. However, these data have to be considered with caution because of the possibility that some propagation steps could be reversible and because of possible penultimate effects [95].

It can be seen from Tables 11 and 12 that the reactivity of the α-substituted acrylonitriles and acrylates toward the polystyryl radical ($1/r_2$) increases slightly with the bulkiness of the substituent if the electronic factors are kept constant (COOMe < COOEt or COOtBu), and also increases regularly with the electrophilicity of the olefin in each series. A noticeable exception to the last rule is methyl α-ethylthioacrylate $H_2C=C(SEt)COOMe$, whose $1/r_2$ value of >100 [70] is higher than the values for the very electrophilic methyl α-cyanoacrylate $H_2C=C(CN)COOMe$ ($1/r_2 = 100$) [101] and diethyl methylenemalonate $H_2C=C(COOEt)_2$ ($1/r_2 = 33$) [96, 98]. As mentioned above its equivalent in the

Table 11. Copolymerization parameters of α-substituted acrylonitriles $CH_2 = C(d)CN$ (M_1) in their radical copolymerization with styrene (M_2) at 60 °C

d	r_1	r_2	Q_1	e_1	$1/r_2$	Ref.
OMe	0.35	0.53	0.72	0.40	1.9	[63]
F	0.03	0.44				[96]
H	0.04[c]	0.40[c]	0.60[c]	1.20[c]	2.5[c]	[97]
Me	0.16[c]	0.30[c]	1.12[c]	0.81[c]	3.3[c]	[97]
OAc	0.20	0.16	1.17	1.06	6.3	[91]
	0.20	0.19			5.3	[92]
Cl	0.13	0.06	2.83	1.40	16.7	[98]
	0.075	0.055			18.2	[99]
Ph	0.7[b]	0.02[b]	9.60[b]	1.26[b]	50[b]	[98, 100]
COOMe	0.03	0.01	12.6	2.1	100	[101]
CN	0.001[a]	0.005[a]	12.6[a]	2.7[a]	200[a]	[102]

[a] At 45 °C.
[b] At 80 °C.
[c] Average value

Table 12. Copolymerization parameters of α-substituted acrylates $H_2C = C(d)COOR$ (M_1) in their radical copolymerization with styrene (M_2) at 60 °C

d	R	r_1	r_2	Q_1	e_1	$1/r_2$	Ref.
SEt	Me	3.8	<0.01			>100	[70]
OMe	Me	0.53	1.15			0.9	[70]
		0.51	1.10	0.47	0.04	0.9	[63]
OMe	tBu	0.69	0.77			1.3	[70]
Me	Me	0.50	0.50	0.74	0.40	2	[97]
Ph	Me	1.0	0.055	5	0.8	18	[103]
		0.45	0.06			17	[104]
		0.4[a]	0.03[a]			33[a]	[105]
Ph	Et	0.19[a]	0.04[a]			25	[105]
OAc	Et	0.20	0.57	0.54	0.67	1.8	[106]
OAc	Me	0.34	0.45			2.2	[70]
H	Me	0.18	0.75	0.42	0.60	1.3	[96]
CH_2F	Et	0.091	0.341			2.9	[107]
Cl	Et	0.30	0.08	2.65	1.13	12.5	[94, 98]
		0.33	0.07			14	[108]
Br	Et	0.50	0.06	3.70	1.07	17	[98]
COOEt	Et	0.08	0.03	4.78	1.66	33	[98]
CN	Me	0.03	0.01	12.6	2.1	100	[101]

[a] At 65 °C.
[b] At 70 °C

acrylonitrile series, α-ethylthioacrylonitrile, does not copolymerize with styrene and acts as a chain transfer agent.

Correlations similar to Hammett relationships have been described in the literature for some 1,1-disubstituted olefins including captodative ones [96, 98]. A correlation between $1/r_2$ and relative rate constants for the addition of the cyclohexyl radical has also been presented [109].

Tables 13 and 14 summarize the quantitative data concerning the copolymerizations of captodative olefins with comonomers other than styrene. From these data, it can be concluded that captodative olefins copolymerize easily with various comonomers.

Table 13. Copolymerization parameters of captodative acrylonitriles $H_2C = C(d)CN$ (M_1) in their radical copolymerization with monomers (M_2) other than styrene

d	Comonomer	r_1	r_2	Conditions	Ref.
OMe	Ethyl α-chloroacrylate	0.30	0.90	30 °C	[89]
	Acrylonitrile	1.93	0.37	60 °C	[63]
OEt	Methyl methacrylate	5.29	0.03	65 °C	[65]
OAc	Acrylonitrile	7.4	0.09	60 °C	[92]

Table 14. Copolymerization parameters of captodative acrylates $H_2C = C(d)COOR$ (M_1) in their radical copolymerization with monomers (M_2) other than styrene

d	R	Comonomer	r_1	r_2	Conditions	Ref.
OMe	Me	Ethyl α-chloroacrylate	0.11	0.58	30 °C	[89]
		Acrylonitrile	0.30	0.15	60 °C	[63]
OAc	Et	Ethyl α-chloroacrylate	0.30	1.71	30 °C	[89]
		Ethyl acrylate	1.0	1.0	60 °C	[106]
		Methyl methacrylate	1.65	0.65	60 °C	[106]
		Vinyl acetate	0.08	5.4	60 °C	[106]

6 Spontaneous Polymerization of Captodative Olefins

Significant spontaneous thermal polymerizability is one of the most remarkable characteristics of captodative olefins. As mentioned in Sect. 4, methyl α-acylamido-acrylate with a C_9 alkyl substituent spontaneously homopolymizes to high molecular weight, even at relatively low temperature (0–20 °C) [73].

Table 15 compares the rates of the spontaneous and AIBN-initiated homopolymerizations of typical captodative acrylates at 60 °C [70]. All monomers, except methyl α-*tert*-butylthioacrylate, give polymers with molecular weights above 10 000 in significant yield in the absence of a radical initiator, while methyl methacrylate gives no polymer under the same conditions. In particular, methyl α-acetoxy-acrylate displays high polymerizability. Even at higher temperature, namely at 180 °C in bulk, α-acetoxy- and α-methoxyacrylates easily polymerize spontaneously indicating low steric hindrance and a moderately high ceiling temperature. Effective generation of initiating species by the captodative effect and an appropriate $k_p/k_t^{0.5}$ value have to be the cause of this thermal polymerizability. Spontaneous polymerizations of α-acetoxy, α-methoxy, and α-ethylthio acrylates are inhibited by the addition of a stable radical, 2,2,6,6-tetramethyl-1-piperidinoxyl, suggesting a radical mechanism of polymerization although more detailed kinetic studies would be desirable. Methyl α-*tert*-butylthioacrylate only gives dimers,

Table 15. Spontaneous polymerizations of captodative substituted acrylates
$CH_2 = C(d)COOR$ in bulk at 60 °C

d	R	[AIBN] (mM)	Time (h)	Yield (%)	M_n ($\times 10^{-4}$)
Me	Me	0	10.0	0	—
Me	Me	5	0.8	5.5	—
OAc	Me	0	3.0	9.0	—
OAc	Me	5	0.8	5.2	—[a]
OMe	Me	0	8.5	5.5	1.7
OMe	Me	5	8.5	7.4	5.6
OMe	tBu	0	10.0	4.1	4.4
OMe	tBu	5	10.0	6.2	5.3
SMe	Me	0	3.0	7.4	—
SMe	Me	10	11.0	25.2	3.6
StBu	Me	0	10.0	0	—
StBu	Me	10	10.0	0	—

[a] Intrinsic viscosity $[\eta] = 1.56$ dL/g in CH_2Cl_2 at 6 °C

trimers, and tetramers and no polymer. This is due to steric hindrance, and is also the case for α-phenylthioacrylates [85] and α-trimethylsilylacrylic acid and its methyl ester [110].

Another noticeable characteristic of captodative olefins is the influence of the reaction medium. The stabilizing effect of solvent on the persistency of a captodatively radical has been reported experimentally for the bond homolysis of bis(3,5,5-trimethyl-2-oxomorpholin-3-yl) [111], but was not found for the 2,3-diphenyl-2,3-dimethoxysuccinonitrile homolysis [112]. Theoretically the solvent-assisted stabilization has been predicted for the captodative substituted nitriles in solvent with large dielectric constants [113–114]. Table 16 illustrates the solvent effect on the spontaneous thermal polymerizations [115]. The polymer yields are

Table 16. Solvent effect on spontaneous polymerizations of methyl α-acetoxyacrylate and methyl α-methoxyacrylate at 60 °C

Solvent	MMA[a]		MMOA[a]	
	Time (h)	Yield (%)	Time (h)	Yield (%)
none	3.0	9.0	15	9.8
Acetic acid[b]	3.0	4.8	15	4.1
Benzene[c]	5.0	5.2	15	3.6
Acetonitrile[d]	3.0	2.8	15	2.1
N-Methyl formamide[e]	5.0	0.2	15	0

[a] MAA: Methyl α-acetoxyacrylate, MMOA: methyl α-methoxyacrylate.
[b] $E_T = 51.9$, $\varepsilon = 6.19$, $\eta = 0.700$ Cp.
[c] $E_T = 34.5$, $\varepsilon = 2.27$, $\eta = 0.390$ Cp.
[d] $E_T = 46.0$, $\varepsilon = 37.5$, $\eta = 0.263$ Cp.
[e] $E_T = 54.1$, $\varepsilon = 182.4$, $\eta = 0.996$ Cp

Table 17. Copolymerization parameters for the copolymerization of captodative substituted acrylates $CH_2 = C(d)COOR$ (M_1) and styrene (M_2) at 60 °C

d	R	[AIBN] (mM)	r_1	r_2
OAc	Me	5	0.34	0.45
OAc	Me	0	0.27	0.40
OMe	Me	5	0.53	1.15
OMe	Me	0	0.56	1.18
OMe	tBu	5	0.69	0.77
OMe	tBu	0	0.68	0.79
SEt	Me	5	3.8	<0.01
SEt	Me	0	5.8	<0.01

substantially reduced in solution compared to the bulk reactions, which may be explained by the effect of dilution on the initiation step, i.e. the initiation might proceed according to second or higher order kinetics. Moreover the yields decrease with increasing solvent polarity, due to increased stability of the captodative radical and a decreased tendency to propagate.

Spontaneous thermal copolymerizations of captodative acrylates with styrene lead to a copolymer with higher molecular weight than the homopolymer. Copolymerization parameters are summarized in Table 17 [70]. Both parameters r_1 and r_2 in the spontaneous copolymerizations are in agreement with those in the AIBN-initiated copolymerizations within experimental error, supporting a radical mechanism for the spontaneous copolymerizations.

These copolymerization parameters are only slightly influenced by the solvent used (Table 18) [116], suggesting a small solvent effect on the propagation reaction. The reactivity of methyl α-methoxyacrylate towards a polystyryl radical ($1/r_2$) however tends to increase with increasing E_T value or dielectric constant of the solvent. Here again it appears that increased solvent polarity leads to an increased persistency of the captodative radical.

Table 18. Copolymerization parameters for the copolymerization of methyl α-methoxyacrylate (M_1) and styrene (M_2) at 60 °C

Solvent	r_1	r_2	$1/r_2$	Q	e
Benzene	0.53	1.15	0.87	0.50	−0.10
1,4-Dioxane	0.63	1.30	0.77	0.54	−0.35
Acetonitrile	0.59	1.08	0.93	0.54	−0.13
Acetic acid	0.55	0.98	1.02	0.54	−0.01
Ethanol	0.61	0.94	1.06	0.59	−0.05
N-Methyl formamide	0.71	0.92	1.09	0.64	−0.15

7 Lewis Acid-Assisted Radical Polymerization of Captodative Olefins

7.1 Polymerization

Lewis acids have long been known to influence free radical polymerizations [117]. They have been particularly important in copolymerizations of hydrocarbon olefins with electron-poor monomers such as acrylates or acrylonitriles. In this way strictly alternating copolymers can be synthesized from monomer pairs which in the absence of Lewis acids would give more random copolymers. The Lewis acid complexes with the electron pair of the acceptor group of the acrylate or acrylonitrile to form the more electrophilic complexed monomer, which then copolymerizes in alternating fashion with the electron-rich hydrocarbon olefin.

The homopolymerization of methyl α-methoxyacrylate is influenced by Lewis acids [116]. Weaker Lewis acids such as $ZnCl_2$ accelerate, but stronger Lewis acids ($SnCl_4$, $EtAlCl_2$, and $BF_3.OEt_2$) inhibit both AIBN-initiated and spontaneous homopolymerizations at 25 and 60 °C.

Copolymerization of this acrylate (M_1) with styrene (M_2) by AIBN at 60 °C in molar ratio of $M_1/ZnCl_2 = 9$ gives a copolymer, possessing about 1:1 molar ratio of both monomer units in a wide range of feed composition, suggesting the production of the copolymer containing an alternating sequence (Fig. 1). The copolymerization shows the highest rate at about 1:1 molar feed composition, as is often seen for complexed free radical copolymerizations. Apparent monomer reactivity ratios obtained from $ZnCl_2$-assisted copolymerization are $r_1 = 0.32$ and $r_2 = 0.05$ ($Q_1 = 3.9$ and $e_1 = 1.23$) in contrast to those obtained from the copolymerization without $ZnCl_2$, $r_1 = 0.53$ and $r_2 = 1.15$. ($Q_1 = 0.47$ and $e_1 = 0.04$). Clearly the addition of $ZnCl_2$ results in increased resonance and decreased electron-density of the vinyl group of the acrylate, causing the generation of an alternating sequence in the copolymer.

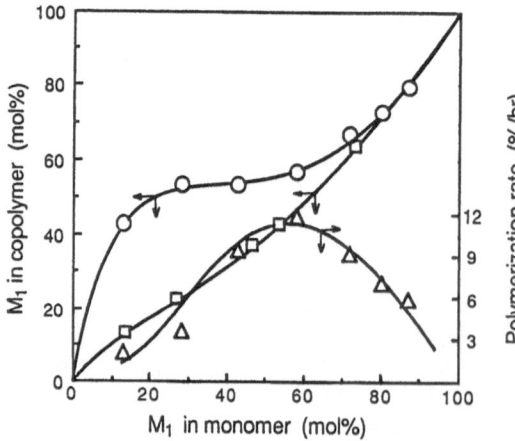

Fig. 1. Copolymer composition curves for the copolymerization of methyl α-methoxyacrylate and styrene by AIBN in the presence ($\circ$) and absence ($\square$) of $ZnCl_2$ at 60 °C and the polymerization rate in the presence of $ZnCl_2$ ($\triangle$).

7.2 Structure of Lewis Acid-Coordinated Captodative Radicals

The coordination of a functional group of the monomer to the Lewis acid is the first step in all acid-assisted polymerizations, but the resulting complexed radical also should play an important role in the peculiarity of these polymerizations. To date, the mode of coordination or the concrete role of the Lewis acid has not been defined in detail because of the instability of the complexed radical intermediates.

Captodatively substituted radicals are well suited for such studies because of the persistency of such radicals. The following complexed captodative radicals have been detected by ESR: $Me_2(CN)CCH_2C^.(CN)SEt.SnCl_4$ [58], $Me_3CCH_2C^.(CN)SEt.SnCl_4$ [59], $Me_2C^.CN.AlMe_3$ [118], and $Me_2C^.COOMe$. $AlMe_3$ [118]. The metal-coordinated radicals possess a negligible splitting by the tin nucleus or a very small splitting by the aluminium nucleus of 0.49 and 1.70 G, in addition to the increased coupling constants of the cyano nitrogen or methoxy hydrogens.

$$Me_3CCH_2-\underset{\underset{COOMe}{|}}{\overset{\overset{OMe}{|}}{C}}_{1/2} \quad \rightleftharpoons \quad 2 \ Me_3CCH_2-\underset{\underset{COOMe}{|}}{\overset{\overset{OMe}{|}}{C}}^.$$

9   **10**

A clear variation of radical structure with complexation can be observed for radical **10** [60, 61]. Figure 2 shows the ESR spectrum of the radical generated by dissociation of dimer **9** in the presence or absence of $SnCl_4$. The uncomplexed radical **10** exhibits splitting by the two equivalent β-hydrogens (9.31 G), while the Sn-complexed radical **10** gives nonequivalent $a_{β-H}$ values (7.46 and 10.30 G) and $a_{Sn} = 8.16$ G. Therefore, we can conclude that the complexation results in the restriction of the rotation of the $C_α$-$C_β$ bond because of steric hindrance and significant delocalization of spin, even onto the tin atom at the δ-position. The schematic structure of the complexed radical **10** is shown considering the ESR and stoichiometric results. Moreover, reduction of spin density on a reactive central carbon by complexation with $SnCl_4$ was semiquantitatively demonstrated by the decrease in the splitting constant of the ^{13}C satellite at the α-position of $Me_3CCH_2C^.(CN)SEt$ from 29.12 to 24.17 G, suggesting about 20% reduction of the spin density [59].

These spectral data support the qualitative experimental results showing the increase of the e-value for the Lewis acid-assisted copolymerization of methyl α-methoxyacrylate [116].

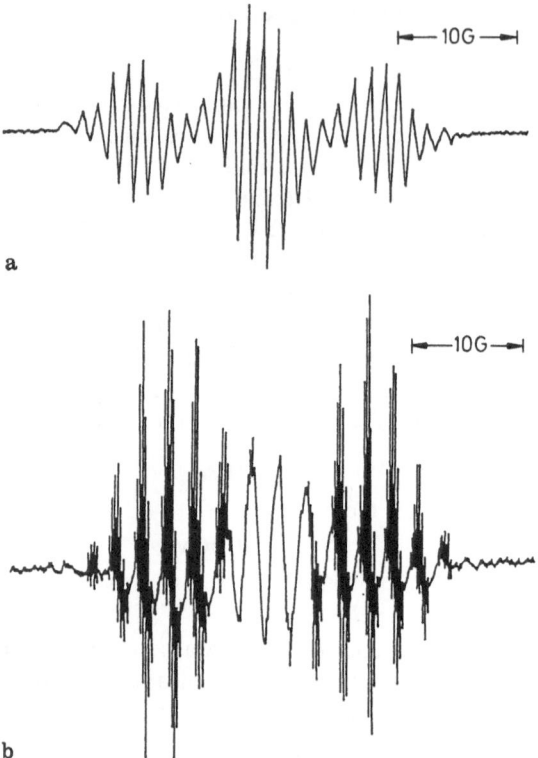

Fig. 2a, b. ESR spectra of **10** in the absence (**a**) and presence (**b**) of SnCl$_4$ at 23 °C. Microwave power 3(**a**) and 0.2 (**b**) mW, and modulation amp. 0.5 G (**a**) and 0.1 G (**b**)

In an actual polymerization system, steric compression between adjacent complexed monomer units in a polymer chain also plays a role in the termination as well as the propagation of vinyl monomers [119].

8 The Bond-Forming Initiation Theory for Spontaneous Polymerizations

The Bond-Forming Initiation Theory was proposed to explain spontaneous "charge-transfer" polymerizations of vinyl monomers [9–10]. This theory will be applied to the spontaneous polymerizations of captodative olefins.

8.1 Background on Thermal (Spontaneous) Polymerizations of Vinyl Monomers

Spontaneous homopolymerization of a single monomer must occur by a free radical mechanism. The most studied example is styrene [120–121]. Flory originally proposed a 1,4-diphenylbutanediyl as the initiating species [122]. More recent work has supported a later proposal by Mayo involving molecule-assisted homolysis of an initially formed cycloadduct [123]. Nevertheless the *cis*- and *trans*-1,2-diphenylcyclobutanes which are also formed in these polymerizations speak for the presence of the Flory diradical.

Spontaneous copolymerizations are encountered much more frequently, particularly when monomers of opposite polarity are mixed [9–10]. Early workers noticed that, upon mixing of certain electron-rich and electron-poor olefins, spontaneous polymerizations occurred without added initiator [99, 124–128]. Mixing electron-rich olefins with electron-poor olefins almost always results in brightly colored solutions. The colors are due to the CT excitation (hv_{CT}) of the electron-donor-acceptor (EDA) complex [129]. Theories for these spontaneous polymerizations mostly center around the charge-transfer complexes (CT or EDA complexes) [128].

The outcome of "charge-transfer" polymerizations has been systematized by Iwatsuki and Yamashita in their penetrating early review [130]. They arrived at a correlation of polymerization behavior with the value of the EDA complex equilibrium constant, K_{eq}. With weak donor and acceptor olefins, no spontaneous polymerization takes place, while the addition of a radical initiator results in a random or an alternating copolymer depending on the value of K_{eq}. As the donor and acceptor strength of the olefins increases, spontaneous initiation rates for radical copolymerization increase and with even stronger donor and acceptor olefins, ionic homopolymerization takes place (cationic and/or anionic).

8.2 Proposed Theory: The Bond-Forming Initiation Theory

From organic chemistry it is known that cycloaddition reactions leading to cyclobutanes are required to be stepwise reactions, according to the Woodward-Hoffmann rules [131]. A bond is formed between the two olefins, leading to a tetramethylene intermediate (T). In a subsequent step, the second bond is formed, yielding the cycloadduct. Depending on the reactants, either zwitterionic or diradical tetramethylenes can be proposed as intermediates [132, 133].

A unifying hypothesis for the observed organic chemistry was advanced by Huisgen [132], who suggested that all tetramethylenes lie on a continuous scale between zwitterionic and diradical structures and may be regarded as resonance hybrids of the two extreme forms. The predominant nature of the tetramethylene intermediate is determined by the terminal substituents, and the termini can interact with each other by through-bond interaction [132, 134].

The Bond-Forming Initiation theory (Scheme 1), originally proposed in 1983, extends the Huisgen hypothesis and proposes that these same tetramethylenes are the true initiators for the observed spontaneous polymerizations, and that this concept is valid for both ionic and radical polymerizations [9, 10]. The tetramethylenes offer a lower energy pathway for initiation than ion radicals.

The nature of the polymer is *diagnostic* of the nature of the initiating intermediate. The homopolymer of the donor olefin in the presence of an acceptor monomer can only be obtained by a cationic propagation mechanism, and not by a free-radical mechanism. Correspondingly, the homopolymer of the acceptor olefin

Cycloaddition

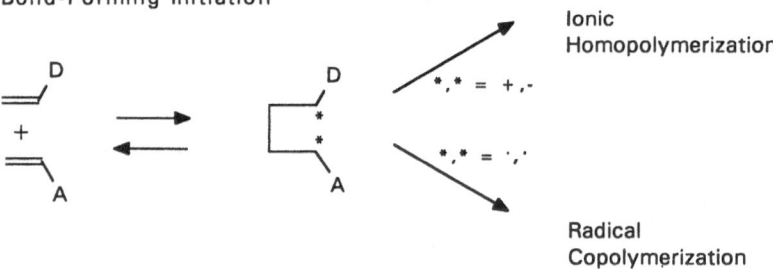

, = +,- or ','

Bond-Forming Initiation

Ionic
Homopolymerization

, = +,-

Radical
Copolymerization

Scheme 1

in the presence of a donor monomer can only be obtained by an anionic mechanism. In contrast, the free-radical propagation leads to copolymerization with a strong tendency to alternate.

A thorough study of the spontaneous copolymerization of *p*-methoxystyrene and dimethyl cyanofumarate was carried out [135] (Scheme 2). In this system the

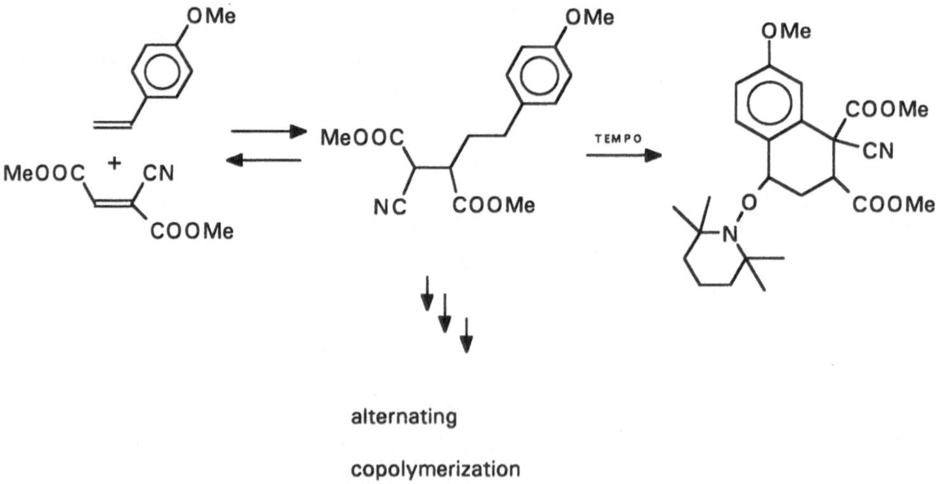

alternating

copolymerization

Scheme 2

kinetics were consistent with the postulate that the initiating species is a diradical tetramethylene. Moreover the diradical tetramethylene was trapped with TEMPO (2,2,6,6-tetramethylpiperidine-*N*-oxyl). The trapping in conjunction with the kinetics provide powerful evidence for bond-forming initiation in this radical copolymerization.

As an illustration of initiation of a cationic polymerization by a zwitterionic tetramethylene, the polymerization of *N*-vinylcarbazole (NVCz) in the presence of dimethyl 2,2-dicyanoethylene-1,1-dicarboxylate was studied in great detail [136] (Scheme 3). The cationic homopolymerization of NVCz could be initiated by adding either the electrophilic olefin or the cyclobutane adduct. The proposed mechanism involves bond formation to the zwitterionic tetramethylene, which closes reversibly to the cyclobutane adduct, and can be trapped with methanol.

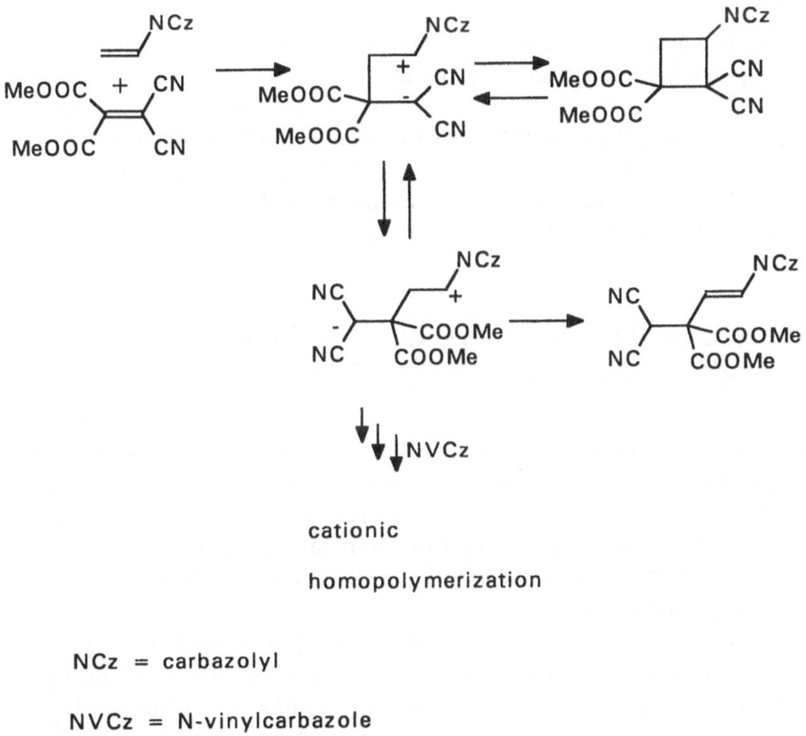

cationic

homopolymerization

NCz = carbazolyl

NVCz = N-vinylcarbazole

Scheme 3

8.3 Generation of Tetra- and Trimethylene Diradicals from Cyclic Compounds

Polymerizing a third monomer using the tetramethylenes from donor-acceptor olefin pairs as initiators is generally not practical. For example, initiation of styrene or acrylonitrile polymerizations using combinations of strong donor and strong acceptor monomers to generate the tetramethylenes results in terpolymerization. The donor and acceptor monomers are swept out, leaving no initiating species.

One way around this difficulty is to generate the tetramethylenes from the cyclobutane adducts. The cyclobutane adduct of NVCz and tetracyanoethylene placed in a solution of excess N-vinylcarbazole causes cationic homopolymerization of the latter [136]. However a cyclobutane whose substitution pattern will lead on cleavage to a tetramethylene diradical at reasonable temperatures has not yet been found. A possible explanation is that a tetramethylene diradical has one less bond than a tetramethylene zwitterion, and so is less stable [137]. Another explanation may be that tetramethylene zwitterions prefer to exist in the *cis* form for coulombic reasons, but tetramethylene diradicals appear to prefer a *trans*, extended conformation and are difficult to generate from cyclic precursors.

Although cyclobutanes with varying substitution patterns are known, cyclopropanes present a much wider variety and much greater ease of synthesis. Ethyl 2-(*p*-methoxyphenyl)-1-cyanocyclopropanecarboxylate has been shown to thermally initiate the diradical polymerization of acrylonitrile [138]. In the presence of zinc chloride as activator, it also initiates the diradical polymerization of styrene [139]. On the other hand, this same initiator also initiates the thermal cationic polymerization of N-vinylcarbazole [140]. This direction of tetra- and trimethylene chemistry is currently under active investigation.

9 Application of the Bond-Forming Initiation Theory to the Spontaneous Polymerization of Captodative Olefins

The Bond-Forming Initiation theory is based on the fact that cycloaddition and initiation of copolymerization compete in the reactions of donor and acceptor monomers. Experimental conditions played a great role in that dilute solutions favored cycloadditions, while high concentration favored polymerization. The main concept is that the intermediates in the cycloaddition reactions are also the initiators of the polymerization.

The cycloaddition reactions of captodative olefins all are considered to proceed through the intermediacy of a 1,4-diradical, due to the captodative stabilization of the terminal radicals. In cross-cycloadditions captodative olefins easily give cyclobutanes when heated with fluoroolefins [141]. They also react with allenes to give methylenecyclobutanes [142], and with methylenecyclopropane to give spiro[2.3]hexanes [143].

Some captodative olefins homodimerize near room temperature [141, 144–146] and an intramolecular example is also known (Scheme 4) [147]. These [2+2]-cycloadditions are often reversible. The unfavorable entropic contribution and the strain in the cyclobutane derivative can operate in the same direction so that the cycloaddition becomes thermodynamically impossible. Even when substituted with relatively small groups, a cyclobutane can cleave back into the starting

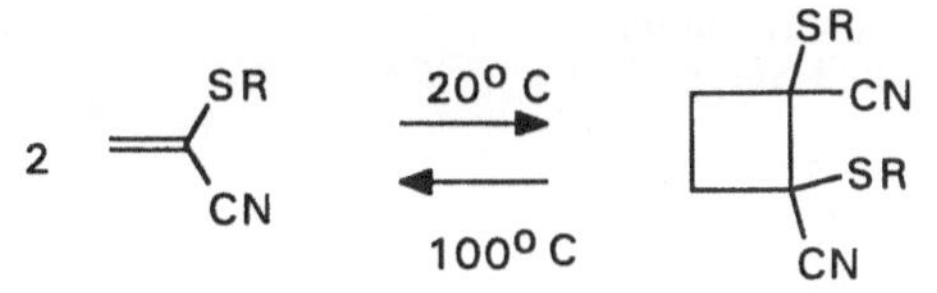

[145]

[141]

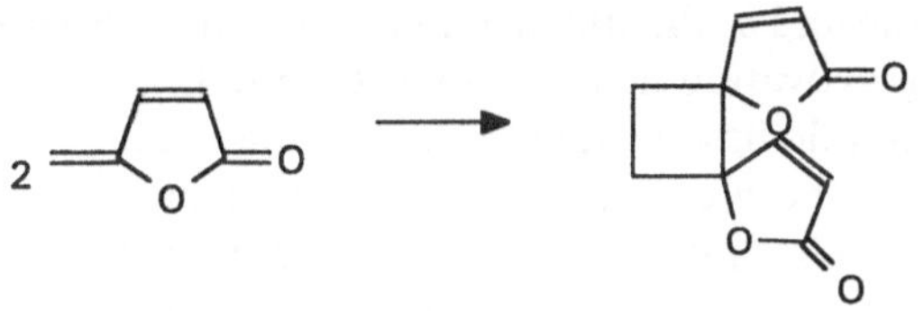

[145]

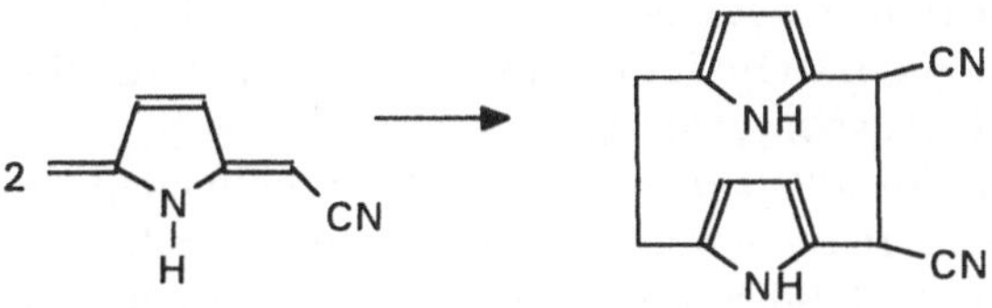

[146]

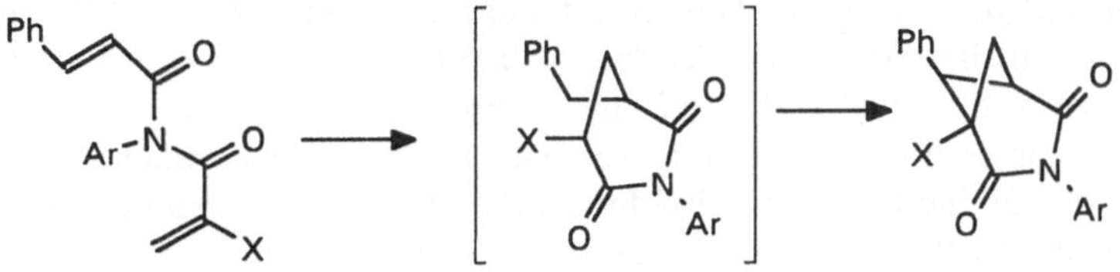

[147]

X = Me   140°C , 4h   75% yield

X = SMe   140°C , 5 min   91% yield

140°C , 3 days   90% yield

Scheme 4

olefins. An example is 1,2-diamino-1,2-dicyanocyclobutane which reverts back to olefin at 100 °C [148]:

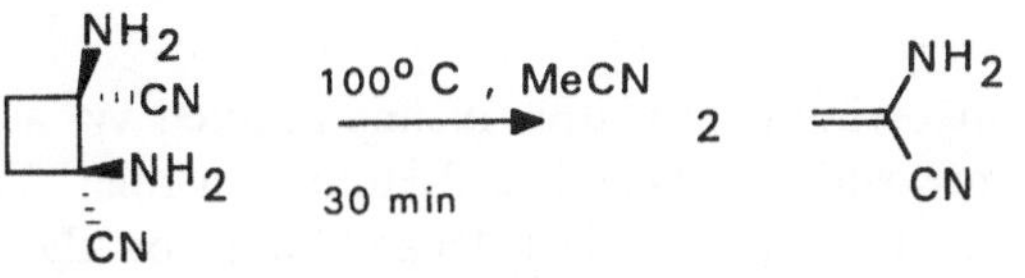

Table 19. Activation energy and enthalpy of reaction for the [2+2]-cyclodimerization of α-substituted acrylonitriles $H_2C=C(d)CN$ (in kcal mol^{-1})

d	E_a	ΔH	Ref.
H	26.1	–	[149]
SMe	12	–14	[150, 151]
StBu	15.4	–7.6	[65]

Table 19 summarizes the kinetic and thermodynamic data concerning the [2+2]-cycloaddition of acrylonitrile and α-alkylthioacrylonitriles. The presence of the alkylthioether group lowers the activation energy by more than 10 kcal mol^{-1}.

The mechanism of the [2+2]-cyclodimerization of α-alkylthio- and α-aryl-thioacrylonitriles was investigated in great detail by Gundermann and his collaborators [150]. In the case of α-(p-methoxyphenylthio)-acrylonitrile, the electron paramagnetic resonance spectrum corresponding to the singlet state of the 1,4-diradical was detected after heating the corresponding cyclobutane at 150–160 °C. Even if this experiment does not prove the intermediacy of the 1,4-diradical during the cyclodimerization, it demonstrates that its formation is easy and strongly supports the above mechanism.

As mentioned above, α-alkylthioacrylates form clear resins at room temperature but on heating at 150 °C, they dimerize to acyclic α,β-dihydromuconate derivatives [150]. 1,2-Bis(alkylthio)cyclobutane-1,2-dicarboxylic esters obtained by an indirect way (hydrolysis and subsequent esterification of the dinitriles) on heating give the same α,β-dihydromuconates [150]. This result suggests that the captodative diradical is the intermediate in both reactions. A disproportionation through an intramolecular hydrogen transfer then occurs in preference to a ring closure, probably for steric or conformational reasons. This captodative tetramethylene diradical formed by heating methyl α-ethylthioacrylate has indeed been observed in the ESR spectrum, as shown in Fig. 3 [70].

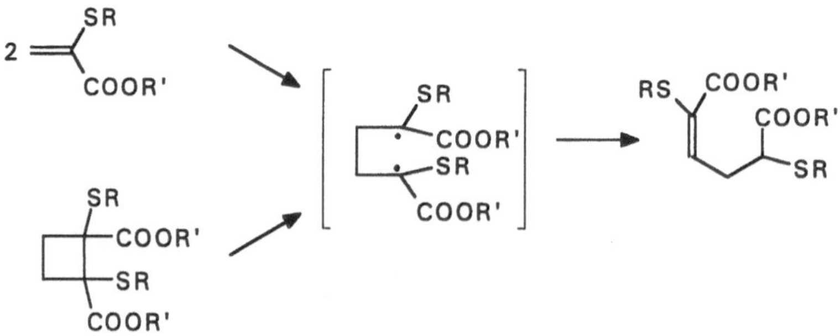

Thus cycloadditions involving captodative olefins proceed via a 1,4-diradical intermediate in accordance with the Woodward-Hoffmann rules. The ease with which an olefin can generate a 1,4-diradical depends both on the ability of its

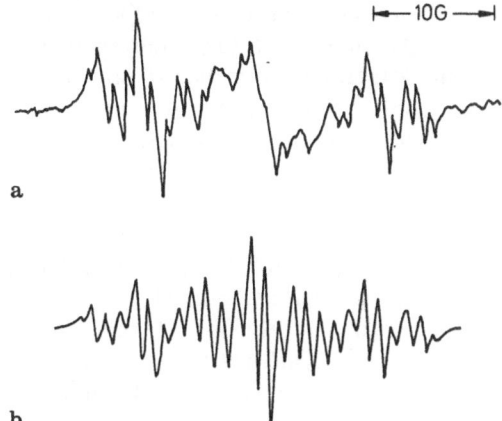

|←—10G—→|

a

b

Fig. 3a, b. Experimental ESR spectrum of the radical produced by the heating of methyl α-ethylthioacrylate at 85 °C (**a**), and simulated ESR spectrum of the radical $-CH_2C^{·}(SEt)COOMe$ (**b**)

substituents to stabilize a radical and on the stabilizing or destabilizing effect the same substituents exert on the starting olefin. Fluoroolefins are very prone to diradical [2 + 2]-cycloaddition because of the second reason. Some captodative olefins easily dimerize via 1,4-diradicals because of the first reason.

We now propose that the diradical intermediates in the cycloaddition reactions of captodative olefins are also the initiators in the observed spontaneous polymerizations of these olefins. Tail-to-tail combination of captodative olefins are expected to provide a low but constant concentration of diradicals, which are capable of initiation. Whether or not polymerization ensues must depend on the experimental conditions, propagation equilibria and rates.

10 Conclusions and Perspectives

Polymerization of captodative olefins is seen to be a potentially wide area of interest in polymer chemistry. Contrary to impressions in the literature, such olefins can homo- and copolymerize to high molecular weight polymers. Results to the contrary can be explained in terms of excessive steric hindrance by large electron-donating substituents, which should lead to slow propagation rates and low ceiling temperatures.

Much remains to be done in this new field. Absolute propagation rate constants need to be determined for those captodative olefins which yield high polymers. No data on ceiling temperatures of captodative olefins have yet been reported. Such data for olefins with substituents with systematically varying steric requirements would be of great interest.

In the area of free radical copolymerization, r_1, r_2 values for systematically varied pairs of comonomers should be determined. These data will permit more definitive structure-reactivity relationships to be deduced. Varying the comonomer from strongly electron-donating to strongly electron-accepting structures would be of particular value.

The Bond-Forming Initiation Theory gives a good interpretation of the observed spontaneous polymerizations of captodative monomers. The tetramethylene diradicals already implicated as initiators in the thermal (spontaneous) polymerizations of vinyl monomers can be particularly stabilized by captodative substituents. For comparison, and to initiate the polymerization of third monomers, captodative cyclobutanes and cyclopropanes are particularly appropriate precursors for generating tetra- and trimethylene diradicals. In particular the extensive work of Viehe [3, 45, 46] showed that thermolysis of captodative substituted cyclopropanes leads to trimethylene captodative diradicals at reasonable temperatures. Their initiating abilities for polymerization have not yet been determined.

To broaden our understanding of the chemical behavior of these novel monomers, it would be appropriate to try anionic polymerizations of captodative monomers. Inasmuch as sulfur is able to stabilize adjacent carbanions, α-alkylthioacrylates and -acrylonitriles should respond well to anionic initiators. Cationic polymerization of certain captodative monomers may also be of interest as alkylthio- and cyano-substituents can stabilize a cationic propagating center.

Acknowledgements: The authors acknowledge partial financial support from NATO Scientific Affairs Division, and the National Science Foundation, Division of Materials Research.

11 References

1. Viehe HG, Janousek Z, Merényi R, Stella L (1985) Acc Chem Res 18: 148
2. Viehe HG, Merényi R, Stella L, Janousek Z (1979) Angew Chem, Int Ed Engl 18: 917
3. Sustmann R, Korth HG (1990) Adv Phys Org Chem 26: 131
4. Viehe HG, Merényi R, Janousek Z (1988) Pure Appl Chem 60: 1635
5. Viehe HG, Janousek Z, Merényi R, Stella L (1982) Kagaku-Dojin 46: 760
6. Carey FA, Sundberg RJ (1990) Advanced organic chemistry, part A, 3rd ed, Plenum, New York, p 681
7. Weixing L (1988) Huaxue Tongbao 6: 19
8. Somsak L (1989) Magy Kem Lapja 154: 22
9. Hall HK Jr, Padias AB (1990) Acc Chem Res 23: 3
10. Hall HK Jr, (1983) Angew Chem Int Ed Engl 22: 440
11. Crans D, Clark T, Schleyer PVR (1980) Tetrahedron Lett. 21: 3681
12. Fossey J (1987) Proceedings of the Fifth European Symposium on Organic Chemistry (Jerusalem), p 179
13. Leroy G, Peeters D (1981) J Mol Struct (Theochem) 85: 133
14. Clark T, Unpublished results described in ref 3
15. Pius K, Chandrasekhar J (1990) J Chem Soc Chem Commun 41
16. Louw R, Bunk JJ (1983) Recl Trav Chim Pays-Bas 102: 119
17. Pasto DJ (1988) J Am Chem Soc 110: 8164
18. Zamkamei M, Kaiser JH, Birkhofer H, Beckhaus HD, Ruechardt C (1983) Chem Ber 116: 3216
19. Bordwell FG, Lynch TY (1989) J Am Chem Soc 111: 7558
20. Bordwell FG, Bausch MJ, Cheng JP, Lynch TY, Mueller ME (1990) J Org Chem 55: 58
21. Ruechardt C, Beckhaus HD (1986) Top Curr Chem 130: 1
22. McMillen DF, Golden DM (1982) Ann Rev Phys Chem 33: 493
23. Delbecq F (1983) Chem Phys Lett 99: 21; (1983) J Mol Struct (Theochem) 93: 353
24. Leroy G (1988) J Mol Struct (Theochem) 168: 77; (1983) Int J Quant Chem 23: 271

25. Leroy G, Peeters D, Wilante C (1982) J Mol Struct 88: 217
26. Szwarc M (1948) J Chem Phys 16: 128
27. Benson SW (1965) J Chem Ed 42: 502
28. Nicholas AMdeP, Arnold DR (1984) Can J Chem 62: 1850
29. Rodgers AS, Wu MCR, Kuitu L (1972) J Phys Chem 76: 918
30. Sylvander L, Stella L, Korth HG, Sustmann R (1985) Tetrahedron Lett 26: 749
31. Korth HG, Lommes P, Sustmann R, Sylvander L, Stella L (1985) Nouv J Chimie 11: 365
32. Arnold DR, Nicholas AMdeP, Snow MS (1985) Can J Chem 63: 1150
33. Lehnig M, Stewen U (1989) Tetrahedron Lett 30: 63
34. Rhodes CJ, Roduner E (1988) Tetrahedon Lett 29: 1437
35. Aurich HG, Deuschle E (1981) Liebigs Ann Chem 719
36. Nootens C, Merényi R, Janousek Z, Viehe HG (1988) Bull Soc Chim Belg 97: 1045
37. Sakurai H, Kyushin S, Nakadaira Y, Kira M (1988) J Phys Org Chem 1: 197
38. Stella L, Pochat F, Merényi R (1981) Nouv J Chimie 5: 55
39. McInnes I, Walton JC (1987) J Chem Soc, Perkin Trans II 1077 and 1789
40. Leigh WJ, Arnold DR (1981) Can J Chem 59: 609
41. Arnold DR, Yoshida M (1981) J Chem Soc Chem Commun 1203
42. Creary X, Sky AF, Mehrsheikh-Mohammadi ME (1988) Tetrahedron Lett 29: 6839
43. Creary X, Mehrsheikh-Mohammadi ME (1986) J Org Chem 51: 1110; 51: 2664
44. Merényi R, Daffe V, Klein J, Masamba W, Viehe HG (1982) Bull Soc Chim Belg 91: 456
45. Merényi R, De Mesmaeker A, Viehe HG (1983) Tetrahedron Lett 24: 2765
46. De Mesmaeker A, Vertommen L, Merényi R, Viehe HG (1982) Tetrahedron Lett 23: 69
47. Korth HG, Lommes P, Sustmann R (1984) J Am Chem Soc 106: 663
48. Janousek Z, Bourgeois JL, Merényi R, Viehe HG, Luedtke AE, Gardner S, Timberlake JW (1988) Tetrahedron Lett 29: 3379
49. Hertenstein U, Huenig S, Reichelt H, Schaller R (1986) Chem Ber 119: 699
50. Raucq P, Viehe HG (unpublished results)
51. Mignani S, Merényi R, Janousek Z, Viehe HG (1984) Bull Soc Chim Belg 93: 991
52. Mignani S, Janousek Z, Merényi R, Viehe HG (1985) Bull Soc Chim Fr 1267
53. Beaujean M, Mignani S, Merényi R, Janousek Z, Viehe HG, Kirch M, Lehn JM (1984) Tetrahedron 40: 4395
54. Mignani S, Merényi R, Janousek Z, Viehe HG (1985) Tetrahedron 41: 769
55. Mignani S, Janousek Z, Merényi R, Viehe HG, Riga J, Verbist J (1984) Tetrahedron Lett 25: 1571
56. Mignani S, Beaujean M, Janousek Z, Merényi R, Viehe HG (1981) Tetrahedron Supp 1: 111
57. Neumann WP, Stapel R (1986) Chem Ber 119: 3432
58. Tanaka H, Ota T (1985) J Polym Sci, Polym Lett 23: 93
59. Tanaka H, Yasuda Y, Ota T (1986) J Chem Soc, Chem Commun 109
60. Tanaka H, Sakai I, Ota T (1986) J Am Chem Soc 108: 2208
61. Tanaka H, Sakai I, Sasai K, Sato T, Ota T (1988) J Polym Sci, Polym Lett 26: 11
62. Hageman HJ, Oosterhoff P, Overeem T, Polman RJ, van der Werf S (1985) Makromol Chem 186: 2483
63. Otsu T, Yamada B, Yoneno H (1969) Bull Chem Soc Jpn 42: 3207
64. Price CC, Coyner EC, DeTar D (1941) J Am Chem Soc 63: 2796
65. Leest Y (1978) Licentiaatsproefschrift, Katholieke Universiteit te Leuven, Laboratorium voor Makromolekulaire en Organische Chemie (Prof. G. Smets)
66. Ohta T, Kobayashi M, Okuda A (1968) Kogyo Kagaku Zasshi 71: 899; Chem Abstr (1968) 69: 77809c
67. Naarmann H, Wittmer P, Stella L, Janousek Z, Merényi R, Viehe HG Ger Offen 2,815,250; Eur Pat 0,004,607 B1; Chem Abstr (1980) 92, 42607h
68. Gundermann KD (1963) Angew Chem 75: 1194
69. Ivanov SS (1961) Vysokomol Soedin 3: 368, Chem Abstr (1962) 56, 1590i
70. Tanaka H, Kameshina T, Sasai K, Sato T, Ota T (1991) Makrom Chem 192: 427

71. Hermes RE, Mathias LJ (1987) Polym Bull 17: 189
72. Mathias LJ, Hermes RE (1988) Macromolecules 21: 11
73. Mathias LJ, Hermes RE (1986) Macromolecules 19: 1536
74. Hermes RE, Mathias LJ, Virden JW Jr (1987) Macromolecules 20: 901
75. Mathias LJ, Kurz DW, Viswanathan T (1988) J Polym Sci, Polym Lett 26: 233
76. Platt JL Jr (1981) U.S.Pat 4,282,343; Chem Abstr (1981) 95: 187 892 u
77. Chikanishi K, Tsuruta T (1965) Makromol Chem 81: 198
78. Ivin KJ, Busfield WK (1989) Encyclopedia of polymer science and technology, 2nd edn, vol 12, Wiley, New York, p 555
79. Schulz GV, Wittmer P (1964) Z Phys Chem N Folge 41: 267
80. Penelle J, Rufflard G (unpublished results)
81. Madruga EL, San Roman J, Lavia MA, Monreal MCF (1984) Macromolecules 17: 989
82. Hopff H, Luessi H, Borla L (1965) Makromol Chem 81: 268
83. Ueda M, Takahashi M, Imai Y, Pittman CU Jr (1983) Macromolecules 16: 1300
84. Charton M (1983) Top Curr Chem 114: 107
85. Tanaka H (unpublished results)
86. Chapiro A (1972) Pure Appl Chem 30: 77; (1973) Eur Polym J 9: 417
87. Chapiro A, Perec-Spritzer L (1975) Eur Polym J 11: 59
88. Pascal P, Napper DH, Gilbert RG, Piton MC, Winnik MA (1990) Macromolecules 23: 5161
89. Yamada B, Hayashi T, Otsu T (1983) J Macromol Sci-Chem A 19: 1023
90. Clifford AM (1945) U.S.Patent 2,385,258; Chem Abstr (1946) 40: 767(4)
91. Oota T, Kobayashi M, Ogawa H (1968) Kogyo Kagaku Zasshi 71: 1542; quoted in Yamada B, Otsu T (1969) J Macromol Sci-Chem A 3:1551
92. Nowak RM, Brissette PL (unpublished results) quoted in Young LJ (1961) J Polym Sci 54: 411
93. Tanaka H, Miyake H, Ota T (1984) J Macromol Sci-Chem A 21: 1523
94. Meijs GF, Rizzardo E (1990) Makromol Chem 191: 1545
95. Tanaka H, Sasai K, Sato T, Ota T (1988) Macromolecules 21: 3534
96. Ueda M, Yazana M (1985) Nippon Kagaku Kaishi 1862; quoted in [97 b]
97. a) Polymer Handbook, 2nd edn (1975) Brandrup J, Immergut EH (eds) Wiley-Interscience, New York
    b) Polymer Handbook, 3rd edn (1989) Branrup J, Immergut EH (eds) Wiley-Interscience, New York
98. Yamada B, Otsu T (1969) J Macromol Sci-Chem A 3: 1551; (1969) J Polym Sci A 1 7: 2439
99. Grassie N, Grant EM (1966) Eur Polym J 2: 255
100. Kreisel M, Garbatski U, Kohn DH (1964) J Polymer Sci A 2: 105
101. Kinsinger JB, Panchak JR, Kelso RL, Bartlett JS, Graham RK (1965) J Appl Polym Sci 9: 429
102. Gilbert H, Miller FF, Averill SJ, Carlson EJ, Folt VL, Heller HJ, Stewart FD, Schmidt RF, Trumbull HL (1956) J Am Chem Soc 78: 1669
103. Luessi H (1967) Makromol Chem 103: 62
104. Otsu T, Yamada B, Nozaki T (1967) Kogyo Kagaku Zasshi 70: 1941; quoted in [97 b]
105. Chikanishi K, Tsuruta T (1965) Makromol Chem 81: 211
106. Unruh CC, Laakso JM (1958) J Polym Sci 33: 87
107. Powell JA, Graham RK (1965) J Polym Sci A 3: 3451
108. Yamada B, Otsu T (1968) J Macromol Sci-Chem A 3: 1325
109. Giese B, Meixner J (1980) Polym Bull 2: 805
110. Ottolenghi A, Fridkin M, Zilkha A (1963) Can J Chem 41: 2977
111. Olson JB, Koch TH (1986) J Am Chem Soc 108: 756
112. Beckhaus MD, Ruechardt C (1987) Angew Chem Int Ed Engl 26: 770
113. Katritzky AR, Zerner MC, Karelson MM, (1986) J Am Chem Soc 108: 7213
114. Karelson M, Katritzky AR, Zerner MC (1991) J Org Chem 56: 134
115. Tanaka H, Kikukawa Y, Kameshima T, Sato T, Ota T (1989) Polymer Preprints, Japan 38: 1389

116. Tanaka H, Kikukawa Y, Kameshima T, Sato T, Ota T (1991) Makromol Chem, Rapid Commun 12: 535
117. Bamford CH, Jenkins AD, Johnston R (1957) Proc Royal Soc A241: 364
118. Brumby J (1982) J Chem Soc, Chem Commun, 677
119. Tanaka H, Kato H, Sakai I, Sato T, Ota T (1987) Makromol Chem, Rapid Commun 8: 223
120. Pryor WA, Lasswell L (1975) Adv Free Radical Chem 5: 27 London p 27
121. Pryor WA (1978) ACS Symp Ser 69: 33
122. Flory PJ (1937) J Am Chem Soc 59: 241
123. Mayo FR (1953) J Am Chem Soc 75: 6133; (1968) ibid 90: 1289
124. Stille JK, Chung DC (1975) Macromolecules 8: 114
125. Shirota Y, Mikawa H (1977) J Macromol Sci, Rev Macromol Chem C16: 129
126. Shirota Y (1985) Encyclopedia of polymer science and technology, 2nd ed, vol 3 Wiley, New York, p 327
127. Hill DTJ, O'Donnell Jr, O'Sullivan PW (1982) Prog Polym Sci 8: 215
128. Cowie JMG, Ed (1985) Alternating Copolymers, Plenum Press, New York
129. Mulliken RS (1952) J Phys Chem 56: 801
130. Iwatsuki S, Yamashita Y (1971) Prog Polym Sci, Jpn 2: 1
131. Woodward RB, Hoffmann R (1969) Angew Chem, Int Ed Engl 8: 781
132. Huisgen R (1977) Acc Chem Res 10: 199
133. Bartlett PD (1970) Q Rev, Chem Soc 24: 473
134. Salem L, Rowland C (1972) Angew Chem, Int Ed Engl 11: 92
135. Hall HK Jr, Padias AB, Pandya A, Tanaka H (1987) Macromolecules 20: 247
136. Gotoh T, Padias AB, Hall HK Jr (1986) J Am Chem Soc 108: 4920
137. Jug K (1987) J Am Chem Soc 109: 3534
138. Li T, Willis TJ, Padias AB, Hall, HK Jr (1991) Macromolecules 24: 2485
139. Li T, Hall, HK Jr (1991) Polym Bull 25: 537
140. Li T, Padias AB, Hall, HK Jr (1992) Macromolecules 25: (in press)
141. De Cock, Piettre S, Lahousse F, Janousek Z, Merényi R, Viehe HG (1985) Tetrahedron 41: 4183
142. Coppe-Motte G, Borghese A, Janousek Z, Merényi R, Viehe HG (1986) Substituent Effects in Radical Chemistry Viehe HG, Janousek Z, Merényi R Eds, NATO ASI Series C189 D Reidel, Dordrecht, p 371
143. De Meijere A, Wenck H, Seyed-Mahdavi F, Viehe HG, Gallez V, Erden I (1986) Tetrahedron 42: 1291
144. Gundermann KD, Loesler A (1972) Liebigs Ann Chem 1895
145. Moriarty RM, Romain CR, Karle IL, Karle J (1965) J Am Chem Soc 87: 3251
146. Flitsch W, Kneip HG (1985) Liebigs Ann Chem 758: 155
147. Alder A, Bellus D (1983) J Am Chem Soc 105: 6712
148. Ksander G, Bold G, Lattmann R, Lehmann C, Frueh T, Xiang YB, Inomata K, Buser HP, Schreiber J, Zass E, Eschenmoser A (1987) Helv Chim Acta 70: 1115
149. Shuraeva VN, Katsobashvili VYa (1975) J Org Chem U.S.S.R. 11(5): 914
150. Gundermann KD (1972) Intra-Science Chem Rept 6: 91
151. Gundermann KD, Huchting R (1959) Chem Ber 92: 415

Editor J. P. Kennedy
Received October 2, 1991

Synthesis and Properties of Fluorinated Diols

B. Boutevin and J. J. Robin
Ecole Nationale Supérieure de Chimie de Montpellier, Laboratoire de Chimie
Appliquée, U.R.A. C.N.R.S. D11930/France

This review summarizes different synthetic methods of obtaining either molecular or macromolecular fluorinated telechelic compounds. These compounds may be obtained either by direct synthesis or by syntheses which need many steps. Thus, reactions of polycondensation, polymerization, telomerization, fluorination of oligomers or the use of functional initiators are described in detail. On the other hand, the emphasis is also focussed on the interest of each method of preparation. Specific properties (solubility, thermal stability, surface properties ...) of the obtained fluorinated telechelic products are discussed with regard to modern industrial requirements. We can observe that the compounds obtained by recently developed methods may be listed in two classes: those which exhibit better surface properties and those which have better thermal stability.

1 Introduction 106

2 Results 106

3 Synthesis of Macrodiols 109
   3.1 Condensation of Fluorinated Diols with Fluorinated
       or Non Fluorinated Diacids 110
   3.2 Polymerization of Monoepoxide with a Fluorinated Chain 111
   3.3 Fluorinated Diglycidyl Ethers 112
   3.4 Polymerization of Perfluorinated Epoxides 113
   3.5 Fluorination of Non Fluorinated Polymers 118
   3.6 Polymerization of Oxethanes 119
   3.7 Use of Functional Initiators 120
   3.8 Reaction Between a Dialcoholate with a Dihalogenide 122
   3.9 The Telomerization with Functional Telogens 124
   3.10 Polyethers with Fluorinated Grafts 126
   3.11 Polybutadienes with Fluorinated Grafts 128

4 Conclusion 128

5 References 129

1 Introduction

The fluorinated polymers exhibit particularly interesting properties which lead to very specific applications in spite of their high prices. Their chemical resistance due to very low molecular interactions, gives them thermostable properties and properties of resistance towards concentrated bases and acids but also excellent surface properties. But, the processing of these polymers is still difficult since several of them are not usually soluble and others are not meltable or exhibit very high melting points.

Thus, even if their properties at high temperatures are still very good, this is not so at low temperatures because of their high Tg or their high crystallinity. However, the polyethers for which the $C-O-C$ bands bring free spinning and decrease the hardness of the perhalogenated $-C-C$ sequence are exceptions. These are the Fomblin polyethers from Montefluos and the Krytox ones from Du Pont De Nemours. However, for these products, problems of solubility and reactivity are observed.

In this paper, several methods of obtaining telechelic compounds which contain fluorinated base units or grafts are presented. Actually, numerous laboratories are trying to obtain fluorinated functional oligomers possessing the advantageous properties of the fluorinated compounds without having their drawbacks.

2 Results

Two kinds of diol are required to produce polyurethanes: those called "short" or molecular (they exhibit low molecular weights) and those called "long" or macromolecular, the $\bar{M}_n$ of which are in the range 1,000–3,000.

In fluorine chemistry, both these diols are also essential, but their number and the variety of their structures are relatively low. The chemistry of "short" diols is more important than that of their homologues which exhibit greater molecular weights.

First, the fluorination of diacids leads to perfluorinated acid difluorides, according to the Simons' process [1] and they can be changed into diols by well-known classical methods described in general books such as the one by Lovelace [2].

Another technique consists of oxidizing the double bonds on fluorinated chains with $KMnO_4$ whatever the nature of the chains, linear or cyclic [3–6].

The last method is telomerization, followed by the chemical change of the end-groups. Thus, from telomers of chlorotrifluoroethylene (C.T.F.E.) with Cl_3CBr. Barnhart et al. [7] prepared diacids according to the following scheme:

$$Cl_3CBr \quad + \quad n\,F_2C{=}CFCl \quad \xrightarrow[\text{or U.V.}]{\text{peroxide}} \quad Cl_3C-(CF_2-CFCl)_n-Br$$

$$Cl_3C-(CF_2-CFCl)_n-Br \quad \xrightarrow{\text{oleum}} \quad HO_2C-(CF_2-CFCl)_{n-1}-CF_2CO_2H$$

In the same way, we performed the synthesis of diacids in three streps: first, the telomerization of CTFE with CCl_4 by redox catalysis [8], followed by quantitative change of the $CFCl_2$ end-group into CCl_3 with aluminium trichloride [9] and finally, the oxidation with oleum [10].

$$CCl_4 \quad + \quad F_2C=CFCl \quad \xrightarrow{\text{FeCl}_3/\text{benzoin}} \quad Cl_3C-(CF_2-CFCl)_n-Cl$$

$$Cl_3C-(CF_2-CFCl)_n-Cl \quad \xrightarrow{\text{AlCl}_3} \quad Cl_3C-(CF_2-CFCl)_{n-1}-CF_2-CCl_3$$

$$Cl_3C-(CF_2-CFCl)_{n-1}-CF_2CCl_3$$

$$\xrightarrow{\text{oleum}} \quad HO_2C-(CF_2-CFCl)_{n-1}-CF_2-CO_2H$$

Furthermore, from iodinated compounds, Hauptschein [11] brought interesting chemical changes of the $C_nF_{2n+1}-I$ products.

Besides, the $I-C_nF_{2n}-I$ compounds and their homologues

$$I-C_2H_4-C_nF_{2n}-C_2H_4-I$$

can be prepared from the following scheme:

$$I_2 \quad + \quad C_2F_4 \quad \longrightarrow \quad I-C_2F_4-I \quad \xrightarrow{\text{C}_2\text{F}_4} \quad I-(C_2F_4)_n-I$$

$$I-(C_2F_4)_n-I \quad \xrightarrow{\text{C}_2\text{H}_4} \quad I-C_2H_4-(C_2H_4)_n-C_2H_4-I$$

Also, such a reaction can be used with bistrichloromethyles compounds [12]

$$Cl_3C-(CF_2CFCl)_n-CF_2-CCl_3$$
$$+ \quad H_2C=CH-CH_2-O-COCH_3 \quad \xrightarrow[\text{MeCN}]{\text{CuCl}}$$

$$AcOCH_2-CHCl-CH_2-CCl_2(CF_2CFCl)_nCF_2-CCl_2-CH_2-CHCl-CH_2OAc$$

$$+ \quad AcOCH_2-CHCl-CH_2-CCl_2(CF_2CFCl)_nCF_2-CCl_3$$

In the same way, with ethylene, we obtained the corresponding dichlorides [13]:

$$Cl_3C-R_F-CCl_3 \quad + \quad H_2C=CH_2$$
$$\longrightarrow \quad Cl-C_2H_4-CCl_2-(CF_2-CFCl)_n-CF_2-CCl_2-C_2H_4-Cl$$

$$R_F = C_nF_{2n}$$

Concerning the short diols in which the lateral chains are fluorinated, Smeltz's surveys [14, 15] should be mentioned.

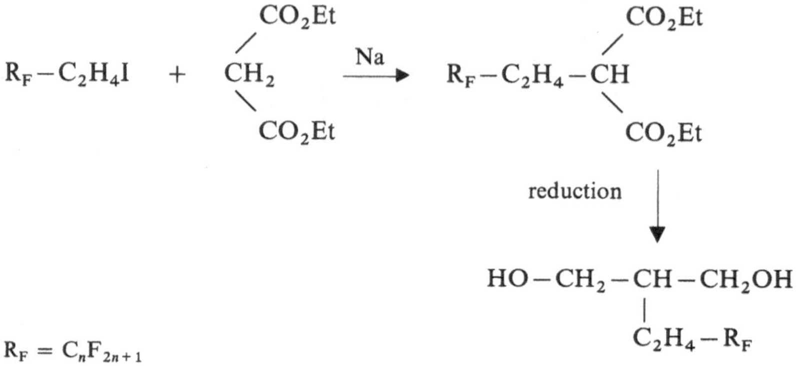

$$R_F - C_2H_4I \quad + \quad \underset{\underset{CO_2Et}{\diagup}}{\overset{\overset{CO_2Et}{\diagup}}{CH_2}} \quad \xrightarrow{\text{Na}} \quad R_F - C_2H_4 - \underset{\underset{CO_2Et}{\diagdown}}{\overset{\overset{CO_2Et}{\diagup}}{CH}}$$

$$\downarrow \text{reduction}$$

$$HO - CH_2 - \underset{C_2H_4 - R_F}{\overset{|}{CH}} - CH_2OH$$

$R_F = C_nF_{2n+1}$

In our laboratory, we prepared several difunctional products from addition of thiols onto olefines [16]:

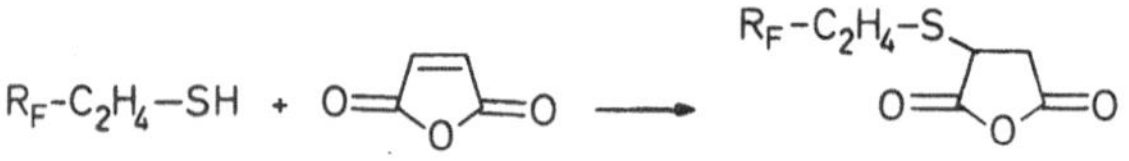

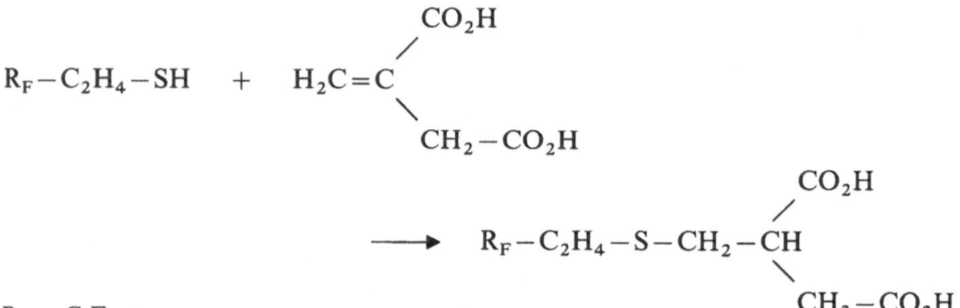

$$R_F\text{-}C_2H_4\text{-}SH \;+\; O=\!\!\overset{\diagup\diagdown}{\underset{O}{\bigcirc}}\!\!=O \;\longrightarrow\; \overset{R_F\text{-}C_2H_4\text{-}S}{O=\!\!\overset{\diagup\diagdown}{\underset{O}{\bigcirc}}\!\!=O}$$

$$R_F - C_2H_4 - SH \quad + \quad H_2C=C\underset{\underset{CH_2 - CO_2H}{\diagdown}}{\overset{\overset{CO_2H}{\diagup}}{}}$$

$$\longrightarrow \quad R_F - C_2H_4 - S - CH_2 - \underset{\underset{CH_2 - CO_2H}{\diagdown}}{\overset{\overset{CO_2H}{\diagup}}{CH}}$$

$R_F = C_nF_{n+1}$

Also of interest are the chemical changes from derivatives of perfluoroalkyl iodides [17]:

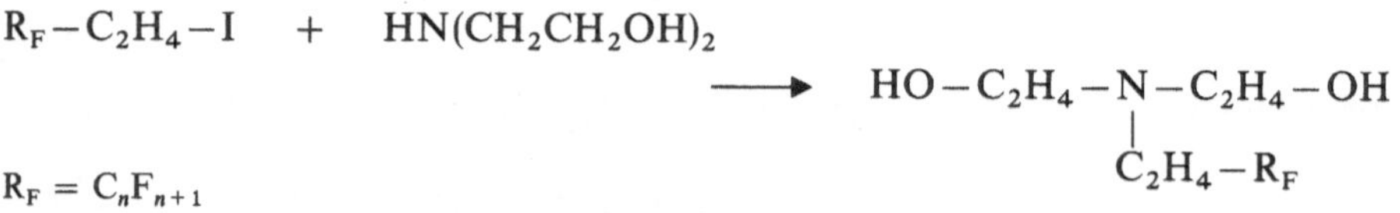

$$R_F - C_2H_4 - I \quad + \quad HN(CH_2CH_2OH)_2$$

$$\longrightarrow \quad HO - C_2H_4 - \underset{C_2H_4 - R_F}{\overset{|}{N}} - C_2H_4 - OH$$

$R_F = C_nF_{n+1}$

Similar diols can be obtained from sulfuryl halogenides [18] and from esters [19]:

$$R_F - (CH_2)_a - SO_2Cl \quad + \quad HN(C_2H_4OH)_2$$

$$\xrightarrow[-HCl]{} \quad R_F - (CH_2)_a - SO_2 - N(C_2H_4OH)_2$$

$a = 2, 4$

The method using diesters [20] is also interesting:

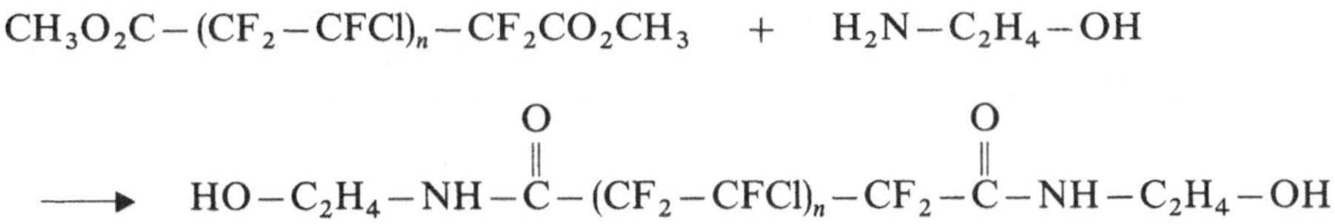

$$CH_3O_2C-(CF_2-CFCl)_n-CF_2CO_2CH_3 \quad + \quad H_2N-C_2H_4-OH$$

$$\longrightarrow \quad HO-C_2H_4-NH-\overset{\overset{\displaystyle O}{\|}}{C}-(CF_2-CFCl)_n-CF_2-\overset{\overset{\displaystyle O}{\|}}{C}-NH-C_2H_4-OH$$

Let us indicate the use of epoxides to obtain the diol [21]:

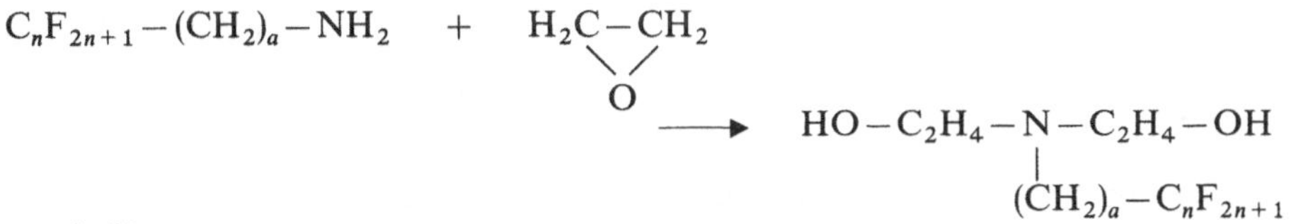

$$C_nF_{2n+1}-(CH_2)_a-NH_2 \quad + \quad \underset{\underset{\displaystyle O}{\diagdown\diagup}}{H_2C-CH_2}$$

$$\longrightarrow \quad HO-C_2H_4-N-C_2H_4-OH$$
$$\underset{(CH_2)_a-C_nF_{2n+1}}{|}$$

$n = 1{-}20$
$a = 2 \text{ or } 4$

We used a method close to the one above in order to prepare primary-secondary diols [22]:

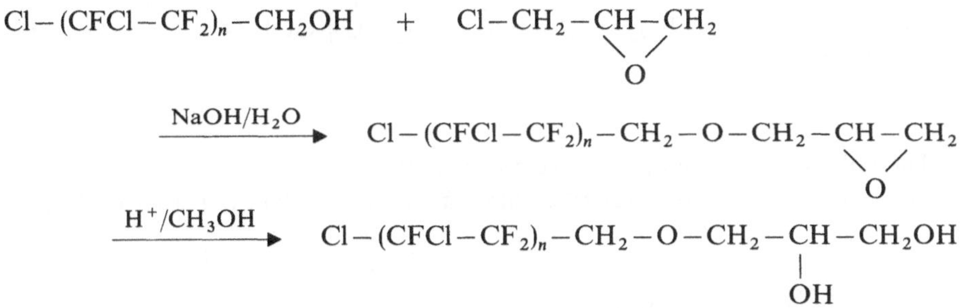

$$Cl-(CFCl-CF_2)_n-CH_2OH \quad + \quad \underset{\underset{\displaystyle O}{\diagdown\diagup}}{Cl-CH_2-CH-CH_2}$$

$$\xrightarrow{NaOH/H_2O} \quad Cl-(CFCl-CF_2)_n-CH_2-O-CH_2-\underset{\underset{\displaystyle O}{\diagdown\diagup}}{CH-CH_2}$$

$$\xrightarrow{H^+/CH_3OH} \quad Cl-(CFCl-CF_2)_n-CH_2-O-CH_2-CH-CH_2OH$$
$$\underset{\underset{\displaystyle OH}{|}}{}$$

After a review on the syntheses of short diols, we can develop the methods to obtain diols called "long" one.

3 Synthesis of Macrodiols

There are about twelve methods to introduce fluorinated grafts or sequences in the diols.

3.1 Condensation of Fluorinated Diols with Fluorinated or Non-Fluorinated Diacids

From 1952, Hauptschein et al. [23] studied the polycondensation of fluorinated acid dichlorides and diols according to the following reaction:

$$n\ Cl-\overset{O}{\overset{\|}{C}}-(CF_2)_3-\overset{O}{\overset{\|}{C}}-Cl \quad + \quad n\ HOCH_2-(CF_2)_x-CH_2OH$$

$$\longrightarrow \quad \left(\overset{O}{\overset{\|}{C}}-(CF_2)_3-\overset{O}{\overset{\|}{C}}-O-CH_2-(CF_2)_x-CH_2O\right)_n$$

$x = 3;\ 4$

More recently, Hollander et al. [24] carried out the synthesis:

$$n+1\ HOCH_2-(CF_2)_3-CH_2OH \quad + \quad n\ EtO-\overset{O}{\overset{\|}{C}}-CH_2-\overset{O}{\overset{\|}{C}}-OEt$$

$$\longrightarrow \quad H\left(OCH_2-(CF_2)_3-CH_2-O\overset{O}{\overset{\|}{C}}-CH_2-\overset{O}{\overset{\|}{C}}\right)_n OCH_2-(CF_2)_3-CH_2OH$$

The polyester has a molecular weight of 1,800.
In our lab., the polytransesterification was used to obtain polyester diols [25]:

$$HO_2C-(CF_2-CFCl)_x-CF_2-CO_2H \quad + \quad HO-C_2H_4-OH$$

$$\xrightarrow{Sb_2O_3/Ac_2Hg} \quad HO-C_2H_4-O\overset{O}{\overset{\|}{C}}-(CF_2-CFCl)_x-CF_2-\overset{O}{\overset{\|}{C}}-O-C_2H_4-OH$$

$$(I)$$

$$(I) \quad \xrightarrow{Ti(Obu)_4/200\ ^{\circ}C}$$

$$HO-CH_2-CH_2-O\left(\overset{O}{\overset{\|}{C}}-(CF_2-CFCl)_x-CF_2-\overset{O}{\overset{\|}{C}}-O-CH_2-CH_2-O\right)_y H$$

3.2 Polymerization of Monoepoxide with a Fluorinated Chain

First, the epoxides are prepared and then they are opened and polymerized as shown in the following examples:

a) 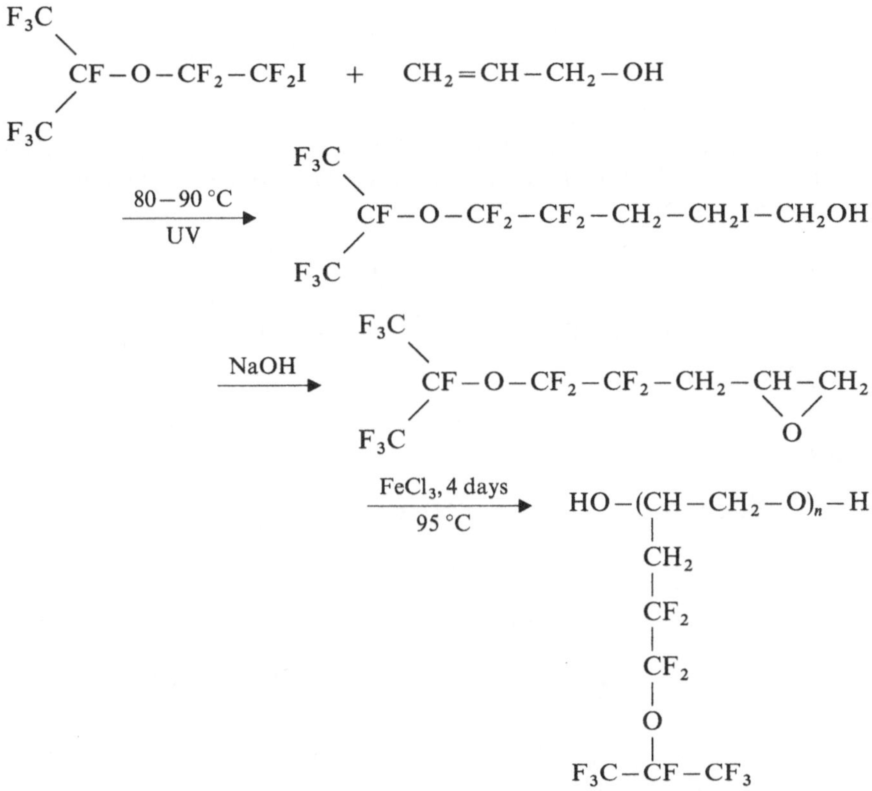 $\begin{array}{c} F_3C \\ \diagdown \\ CF-O-CF_2-CF_2I \\ \diagup \\ F_3C \end{array}$ + $CH_2=CH-CH_2-OH$

$\xrightarrow[\text{UV}]{80-90\,°C}$ $\begin{array}{c} F_3C \\ \diagdown \\ CF-O-CF_2-CF_2-CH_2-CH_2I-CH_2OH \\ \diagup \\ F_3C \end{array}$

$\xrightarrow{\text{NaOH}}$ $\begin{array}{c} F_3C \\ \diagdown \\ CF-O-CF_2-CF_2-CH_2-CH-CH_2 \\ \diagup \qquad\qquad\qquad\quad \diagdown \diagup \\ F_3C \qquad\qquad\qquad\qquad\quad O \end{array}$

$\xrightarrow[95\,°C]{\text{FeCl}_3,\,4\ \text{days}}$ $HO-(CH-CH_2-O)_n-H$
$\qquad\qquad\qquad\qquad\qquad\quad |$
$\qquad\qquad\qquad\qquad\qquad\ \ CH_2$
$\qquad\qquad\qquad\qquad\qquad\quad |$
$\qquad\qquad\qquad\qquad\qquad\ \ CF_2$
$\qquad\qquad\qquad\qquad\qquad\quad |$
$\qquad\qquad\qquad\qquad\qquad\ \ CF_2$
$\qquad\qquad\qquad\qquad\qquad\quad |$
$\qquad\qquad\qquad\qquad\qquad\ \ O$
$\qquad\qquad\qquad\qquad\qquad\quad |$
$\qquad\qquad\qquad\qquad\quad F_3C-CF-CF_3$

The polymerization of these epoxides was performed in Carius tubes [26, 27], eq. (b) by Gosnell and Hollander [24], eq. (c) by Hollander and Trischler [28]:

b) 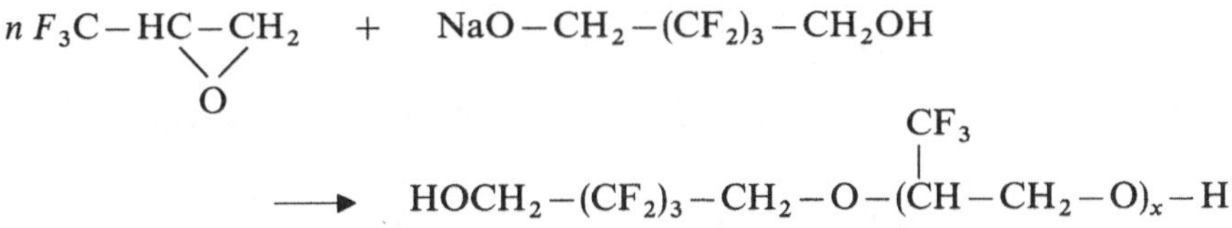 $n\ F_3C-HC-CH_2$ + $NaO-CH_2-(CF_2)_3-CH_2OH$
$\qquad\quad \diagdown \diagup$
$\qquad\qquad O$

$\qquad\qquad\qquad\qquad\qquad\qquad\qquad\qquad\quad CF_3$
$\qquad\qquad\qquad\qquad\qquad\qquad\qquad\qquad\quad |$
$\longrightarrow\ HOCH_2-(CF_2)_3-CH_2-O-(CH-CH_2-O)_x-H$

c) 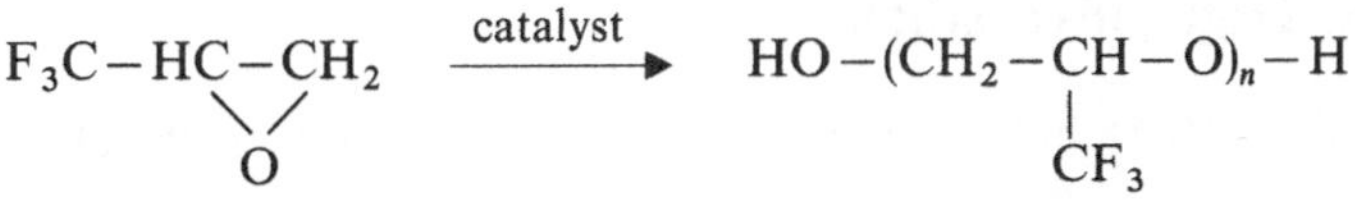 $F_3C-HC-CH_2$ $\xrightarrow{\text{catalyst}}$ $HO-(CH_2-CH-O)_n-H$
$\qquad\ \ \diagdown \diagup$ $\qquad\qquad\qquad\qquad\qquad\qquad\quad |$
$\qquad\qquad O$ $\qquad\qquad\qquad\qquad\qquad\qquad\quad CF_3$

The authors used two kinds of catalysts: either BF_3 (they obtained polymers which exhibit double bond end-groups) or $AlCl_3$ (which led to a polymer which has hydroxylated end-groups).

Knunyants [29] and O'Rear [30] performed the polymerization of epoxides such as:

d) 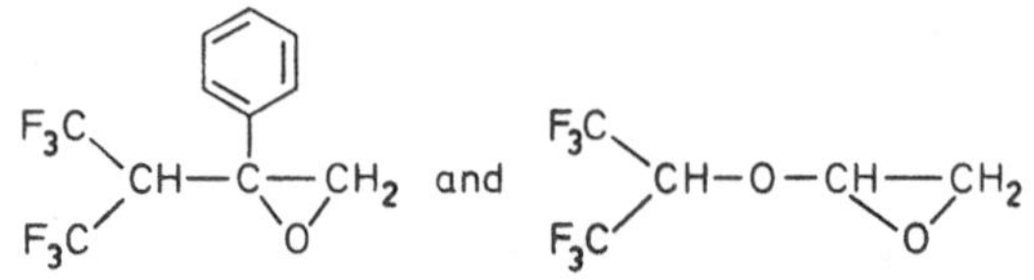 F_3C—CH—C—CH_2 (with phenyl on C, epoxide O) and F_3C—CH—O—CH—CH_2 (with epoxide O)

by using primary and secondary amines such as $C_6H_5-CH_2-NH_2$ and Bu_2NH.

Pittmann et al. [31] used $AlCl_3$ to open epoxides dut the synthesis of these epoxides is original:

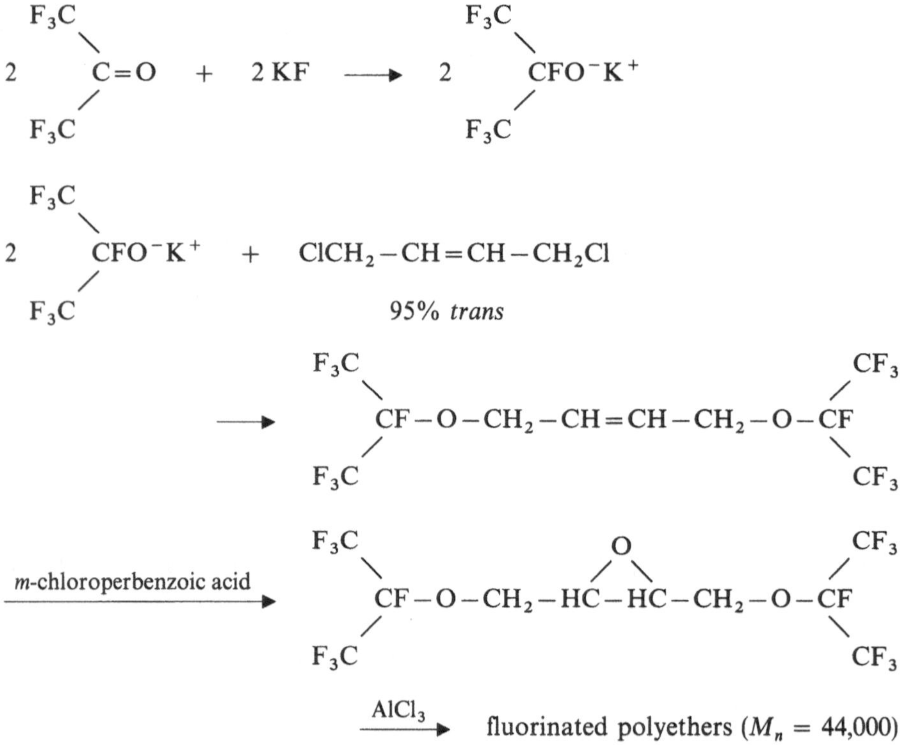

$$2 \quad \begin{array}{c} F_3C \\ \diagdown \\ C=O \\ \diagup \\ F_3C \end{array} + 2\,KF \longrightarrow 2 \quad \begin{array}{c} F_3C \\ \diagdown \\ CFO^-K^+ \\ \diagup \\ F_3C \end{array}$$

$$2 \quad \begin{array}{c} F_3C \\ \diagdown \\ CFO^-K^+ \\ \diagup \\ F_3C \end{array} + ClCH_2-CH=CH-CH_2Cl$$

95% trans

$$\longrightarrow \begin{array}{c} F_3C \\ \diagdown \\ CF-O-CH_2-CH=CH-CH_2-O-CF \\ \diagup \\ F_3C \end{array} \begin{array}{c} CF_3 \\ \diagup \\ \diagdown \\ CF_3 \end{array}$$

$$\xrightarrow{m\text{-chloroperbenzoic acid}} \begin{array}{c} F_3C \\ \diagdown \\ CF-O-CH_2-HC-HC-CH_2-O-CF \\ \diagup \\ F_3C \end{array} \overset{O}{\wedge} \begin{array}{c} CF_3 \\ \diagup \\ \diagdown \\ CF_3 \end{array}$$

$$\xrightarrow{AlCl_3} \text{fluorinated polyethers } (M_n = 44{,}000)$$

3.3 Fluorinated Diglycidyl Ethers

This synthesis consists of the addition of a chlorhydrine in excess onto a diol [32, 33]:

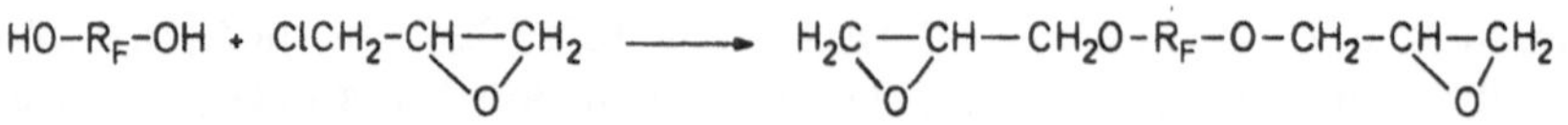

$$HO-R_F-OH + ClCH_2-CH-CH_2 \longrightarrow H_2C-CH-CH_2O-R_F-O-CH_2-CH-CH_2$$

$$R_F = C_nF_{2n}$$

According to the chlorohydrine-amount, diglycidic ethers with varied chain-length were obtained:

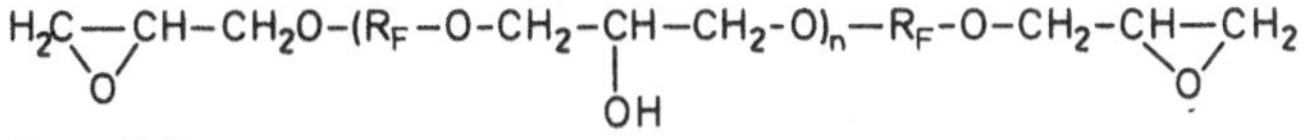

$$H_2C—CH–CH_2O–(R_F–O–CH_2–CH–CH_2-O)_n–R_F–O–CH_2–CH—CH_2$$

$R_F = C_nF_{2n+1}$

Numerous authors have studied such as a reaction according to the nature of the R radical as mentioned below:

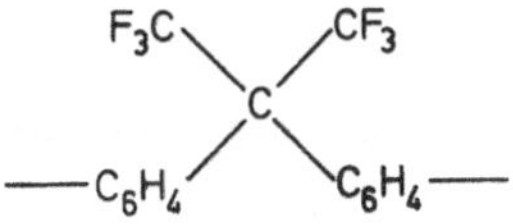

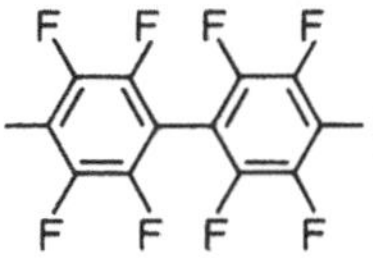

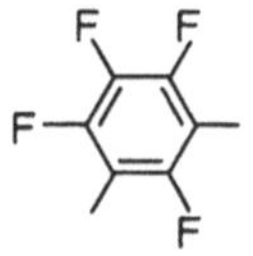

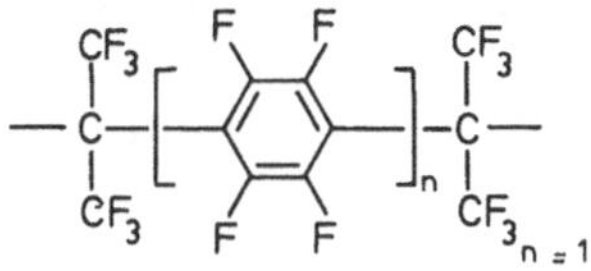

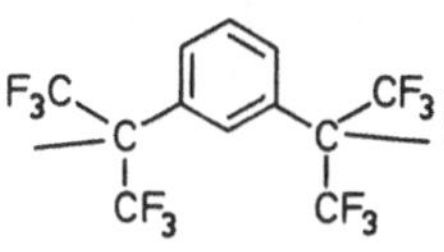

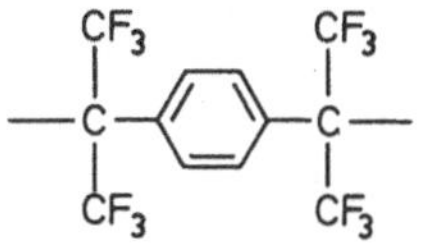

In the same way, O'Rear et al. [32] obtained prepolymers diols from a stoechiometric amount of diol and diglycidic ether and these products led to fluorinated polyurethanes.

3.4 Polymerization of Perfluorinated Epoxides

Two syntheses of polyperfluoroethers have been developed:

   * *anionic polymerization of perfluoroepoxides*

From perfluoroepoxypropylene, Moore [37, 38] obtained polyethers and products manufactured under the trade name Krytox [39, 40] by Du Pont De Nemours.

   * *a direct photooxidation of perfluoroolefines*

This process can be performed on several olefines and leads to product which exhibit different structures and properties. The products are manufactured by the Montefluos Company.
    First, $F_2C=CF–CF_3$ [41] and C_2F_4 [42] can be used. Also, the perfluorobutadienes [43] are interesting reactants.
    From a chemical point of view, several comments can be made on both processes which are totally different:

3.4.1 In an anionic process, a perfluoroalcoholate must be done:

$$\sim CF_2-O^{\ominus} \quad \text{or} \quad \sim CF-O^{\ominus}$$
$$\qquad\qquad\qquad\qquad\qquad |$$
$$\qquad\qquad\qquad\qquad\quad CF_3$$

(for instance, by the reaction between cesium fluoride and acid difluorides).

$$\overset{O}{\overset{||}{F-C}}-(CF_2)_n-\overset{O}{\overset{||}{C}}-F \quad + \quad CsF \quad \longrightarrow \quad Cs^{\oplus} {}^{\ominus}O-(CF_2)_{n+2}-O^{\ominus} {}^{\oplus}Cs$$

The simplest acid difluoride used is FCO−COF but we can use a monofunctional one, R_F-COF and COF_2 (precusors of CF_3O).

The second step concerns the addition of perfluoroethylene oxide (TFEO) or perfluoropropylene oxide (HFPO) according to the following reactions [44, 45, 46]:

$$Cs^{\oplus} {}^{\ominus}O-(CF_2)_n-O^{\ominus} {}^{\oplus}Cs \quad + \quad TFEO$$

$$\longrightarrow \left(\overset{O}{\overset{||}{F-C}}-CF_2-(OC_2F_4)_x-O\right)_2-(CF_2)_n$$

In the same case, HFPO [47, 48] leads to:

$$^{\ominus}O-(CF_2)_n-O^{\ominus} \quad + \quad HFPO$$

$$\longrightarrow \quad {}^{\ominus}O-CF_2-\underset{\underset{CF_3}{|}}{CF}-O-(CF_2)_n-O-\underset{\underset{CF_3}{|}}{CF}-CF_2-O^{\ominus}$$

$$\overset{O}{\overset{||}{F-C}}-\overset{CF_3}{\overset{|}{CF}}-(O-CF_2-\overset{CF_3}{\overset{|}{CF}})_x-O-(CF_2)_n-O-(\overset{CF_3}{\overset{|}{CF}}-CF_2-O)_y-\overset{CF_3}{\overset{|}{CF}}-\overset{O}{\overset{||}{C}}-F$$

HFPO/TFEO blends [49, 50] can be also used.

We can indicate that the fluorinated alcoholates can also initiate the addition of tetrafluoroethylene as Pierce et al. [51] did:

$$\overset{O}{\overset{||}{F-C}}-(CF_2)_x-\overset{O}{\overset{||}{C}}-F \quad + \quad F_2C=CF_2$$

$$\xrightarrow{\text{MF/X}_2} \quad X-C_2F_4-O-(CF_2)_{x+2}-O-C_2F_4X$$

with MF = CsF or KF

Finally, let us mention another method which allow us to prepare fluorinated poly(ethers) by U.V. irradiation. The U.V. radiation leads to the radicals, the duplications of which lead to polyethers with longer chains [52]:

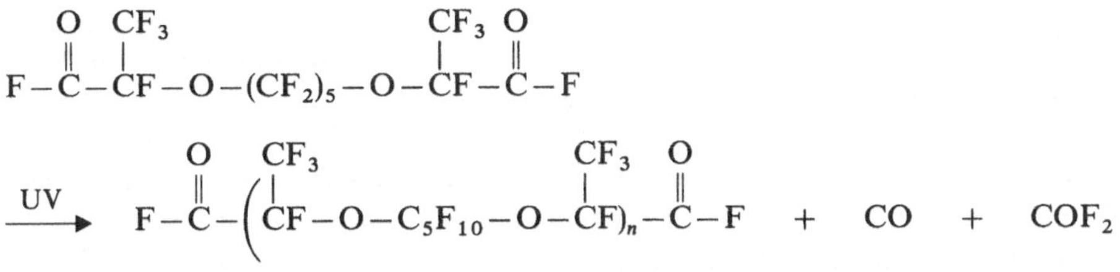

$$F-\overset{\overset{\displaystyle O}{\|}}{C}-\overset{\overset{\displaystyle CF_3}{|}}{CF}-O-(CF_2)_5-O-\overset{\overset{\displaystyle CF_3}{|}}{CF}-\overset{\overset{\displaystyle O}{\|}}{C}-F$$

$$\xrightarrow{\text{UV}} F-\overset{\overset{\displaystyle O}{\|}}{C}-\left(\overset{\overset{\displaystyle CF_3}{|}}{CF}-O-C_5F_{10}-O-\overset{\overset{\displaystyle CF_3}{|}}{CF}\right)_n-\overset{\overset{\displaystyle O}{\|}}{C}-F \quad + \quad CO \quad + \quad COF_2$$

In the same way, Zollinger et al. [53] prepared polyethers by U.V. radiations:

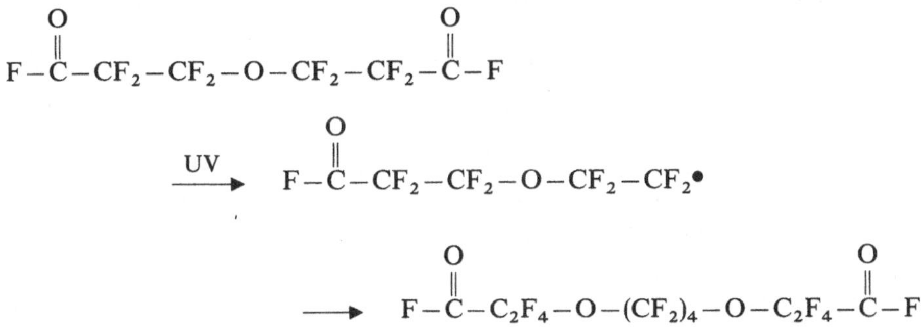

$$F-\overset{\overset{\displaystyle O}{\|}}{C}-CF_2-CF_2-O-CF_2-CF_2-\overset{\overset{\displaystyle O}{\|}}{C}-F$$

$$\xrightarrow{\text{UV}} F-\overset{\overset{\displaystyle O}{\|}}{C}-CF_2-CF_2-O-CF_2-CF_2\bullet$$

$$\longrightarrow F-\overset{\overset{\displaystyle O}{\|}}{C}-C_2F_4-O-(CF_2)_4-O-C_2F_4-\overset{\overset{\displaystyle O}{\|}}{C}-F$$

These acids difluorides were changed into the corresponding diesters and diols.

3.4.2. In the photooxidation process, the simultaneous action of oxygen and U.V. radiations allow us to obtain the following copolymer:

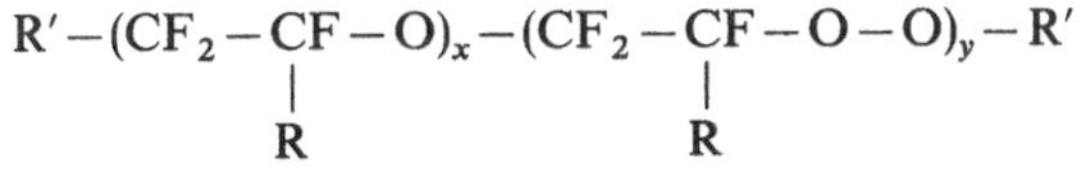

$$R'-(CF_2-\overset{\overset{\displaystyle }{|}}{\underset{\underset{\displaystyle R}{|}}{CF}}-O)_x-(CF_2-\overset{\overset{\displaystyle }{|}}{\underset{\underset{\displaystyle R}{|}}{CF}}-O-O)_y-R'$$

with R = F,CF$_3$ and R' = COF or CF$_2$CFCOF

The authors explained the formation of polyether by the obtaining of poly-peroxides as the first step [41].

The diacids or derivatives were changed into neutral products by several pathways and, above all, by fluorination in order to lead to very stable oil.

Thus, from TFE and HFP, two series of products, Fomblin Z and Y were obtained respectively and produced commercialy by Montefluos [54]:

$$CF_3-O-(C_3F_6O)_p \quad (CF_2O)_q-CF_3 \qquad \text{Fomblin Y}$$

$$CF_3-O-(C_2F_4O)_m \quad (CF_2O)_n-CF_3 \qquad \text{Fomblin Z}$$

$q/p \approx 0.1$ and m/n varies in the range $0.6-1.5$ $10^3 < \bar{M}_n < 4 \times 10^4$

These non functional compounds are used as high performance lubricants which exhibit both very high thermal stability and low viscosity.

Caporiccio [54] gives details of the properties and also the applications of these molecules.

With preference to difunctional compounds, they have a molecular weight of about 2,000. The variety of the functional end-groups is impressive: alcohols, amines, acids, esters and also isocyanates.

The literature describes numerous projects concerning the developing of difunctional products into materials which have interesting physical and mechanical properties. Such surveys were mainly performed in the seventies for applications in space. The research is briefly described here.

The majority of these works concern the synthese of polyurethane from diols described as $HO-R_F-O-R_F-OH$ [45, 55, 57]:

$$HO-R_F-O-R_F-OH \quad + \quad OCN-R-NCO$$

$$\longrightarrow \quad -(O-R_F-O-R_F-O-CO-NH-R-NHCO)-$$

Recently [58], polyesters have been prepared from dimethylterephtalate, ethylene glycol and the diester $CH_3O_2C-R_F-O-R_F-CO_2CH_3$. The copolycondensation by transesterification was studied and the authors investigated the morphology, the thermal and surface properties of these obtained polyesters. We can notice that the PET/copolymer (PET−PFPE) blend thus obtained contains up to 35% in weight of PFPE but 7% only are bond with the prepared PET.

In these two latter cases, the urethanes or esters links do not exhibit chemical or thermal resistances comparable to those of the fluorinated polyethers. So, the researches consist in creating stable bridges between the polyethers chains. Here are listed several exemples:

− Chain-extension thanks to nitrile bonds:
  • From dinitriles [59]:

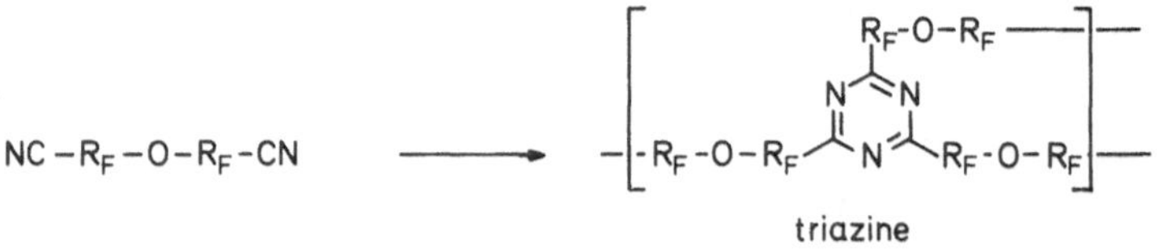

triazine

  • From dialcoholates [60]:

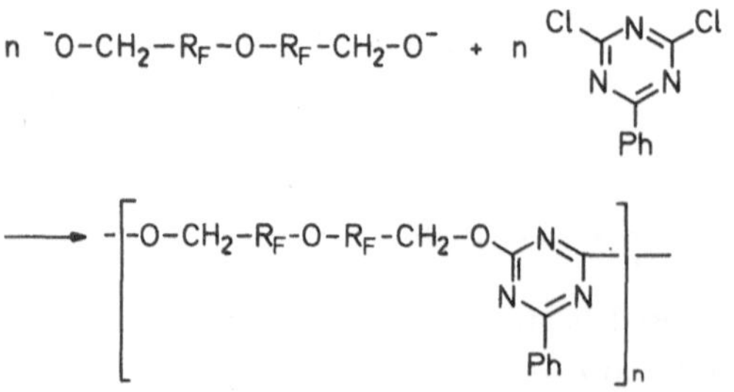

- From a dinitrile/nitrile blend [61]:

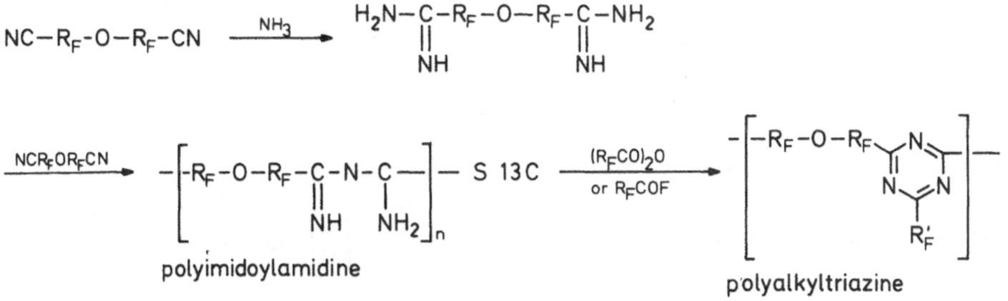

$$NC-R_F-O-R_F-CN \xrightarrow{NH_3} \underset{\overset{\|}{NH}}{H_2N-C-R_F-O-R_F-\underset{\overset{\|}{NH}}{C}-NH_2}$$

$$\xrightarrow{NCR_FOR_FCN} \left[-R_F-O-R_F-\underset{\overset{\|}{NH}}{C}-N-\underset{\overset{\|}{NH_2}}{C}-\right]_n - S\ 13C \xrightarrow[\text{or } R_FCOF]{(R_FCO)_2O} \left[\begin{array}{c}-R_F-O-R_F \\ N \\ N \\ R_F'\end{array}\right]-$$

polyimidoylamidine   polyalkyltriazine

This latter method is interesting since a linear polymer is obtained, contraryly to the direct trimerization of dicyanated compounds.

— Chain-extension thanks to isocyanurate bonds:
  Webster et al. [62] prepared diisocyanated ethers with catalysts such as:

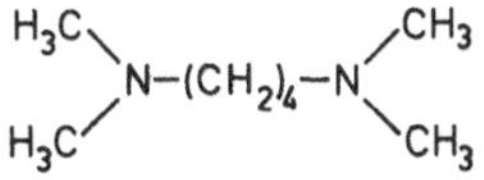

$$\underset{H_3C}{\overset{H_3C}{>}}N-(CH_2)_4-N\underset{CH_3}{\overset{CH_3}{<}}$$

gave polyisocyanurates with the following structure:

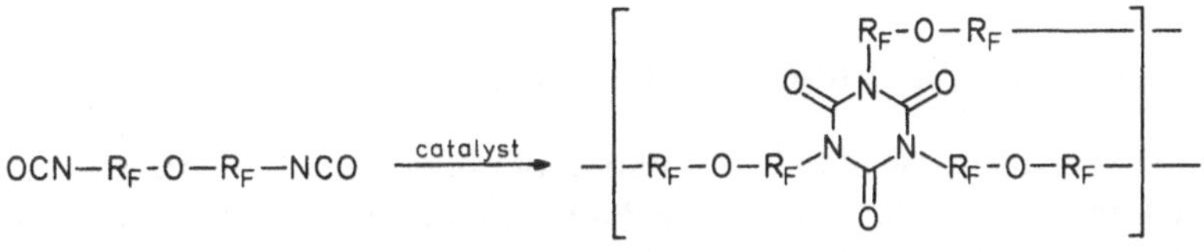

$$OCN-R_F-O-R_F-NCO \xrightarrow{\text{catalyst}} \left[-R_F-O-R_F-N\underset{O}{\overset{O}{\bigtriangleup}}N-R_F-O-R_F-\right]-$$

where: $-R_F-O-R_F-$ is (structure)

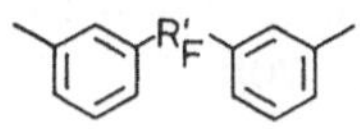

with $R_F' = -(CF_2)_n-;\ -O-(CF_2)_n-O-;$

$$-O-(CF_2-\underset{\overset{|}{CF_3}}{CF}-O)_m-(CF_2)_5-(O-\underset{\overset{|}{CF_3}}{CF}-CF_2)_n-O-$$

— Chain-extension with benzenic bonds:
  Such a reaction was carried out by De Pasquale [63] according to the following scheme:

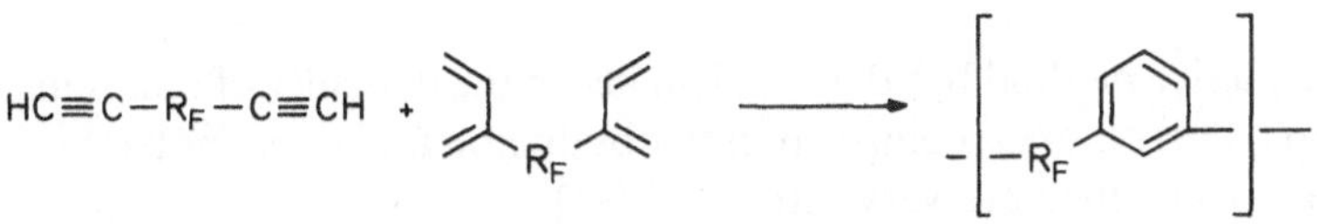

$$HC\equiv C-R_F-C\equiv CH + (diene) \longrightarrow \left[-R_F-\right]-$$

— Chain-extension with oxadiazole, triazoles and benzoxazoles bonds [47]:

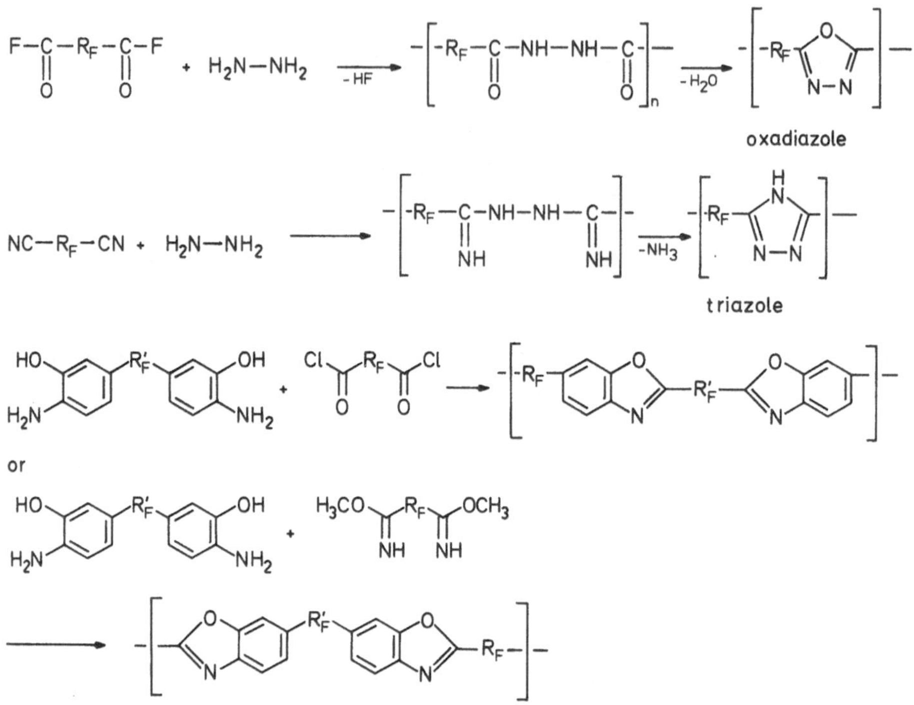

$$F-\underset{\overset{\|}{O}}{C}-R_F-\underset{\overset{\|}{O}}{C}-F \;+\; H_2N-NH_2 \;\xrightarrow{-HF}\; \left[-R_F-\underset{\overset{\|}{O}}{C}-NH-NH-\underset{\overset{\|}{O}}{C}-\right]_n \;\xrightarrow{-H_2O}\; \left[-R_F\overset{O}{\underset{N-N}{\diagup\!\!\diagdown}}\right]-$$

oxadiazole

$$NC-R_F-CN \;+\; H_2N-NH_2 \;\longrightarrow\; \left[-R_F-\underset{\overset{|}{NH}}{C}-NH-NH-\underset{\overset{|}{NH}}{C}-\right]\;\xrightarrow{-NH_3}\; \left[-R_F\overset{\overset{H}{\overset{|}{N}}}{\underset{N-N}{\diagup\!\!\diagdown}}\right]-$$

triazole

(benzoxazole scheme)

or

(second benzoxazole scheme)

3.5 Fluorination of Non Fluorinated Polymers

Very interesting investigations of the fluorination of polymers (H_2/F_2 blend) were carried out by Lagow et al. [60, 71] for more than ten years. These authors worked on polyoxyethylenes [66], on polyoxypropylenes [67] and even on polyesters [68] on which the fluorination of the carbonyl group by SF_4 was performed and finally on several copolymers such as [69, 70]:

$$-\left[\underset{\overset{|}{CF_3}}{\overset{\overset{CF_3}{|}}{C}}-O-CH_2-CH_2-O\right]-\;; \qquad -\left[\underset{\overset{|}{CF_3}}{\overset{\overset{CF_3}{|}}{C}}-O-\underset{}{\overset{\overset{CH_3}{|}}{CH}}-CH_2-O\right]-\;;$$

$$-\left[\underset{\overset{|}{CF_3}}{\overset{\overset{CF_3}{|}}{C}}-O-CH_2-CH_2-CH_2-O\right]-\;;$$

This is a very attractive method but the problem lies in the difficulty of obtaining reactive end-groups. There is great competition to obtain non functional polyethers and researchers at Montefluos are very interested [71].

3.6 Polymerization of Oxethanes

The oxethane $\underset{\underset{\text{CH}_2-\text{O}}{|\quad\quad|}}{\text{CF}_2-\text{CFR}}$ where R = Cl, CF$_3$, F have been known for years and

in 1963, Weinmayr prepared $\overline{CH_2CF_2CH_2O}$ by adding C_2F_4 onto formaldehyde [72]. The Japanese researchers of the Daikin Company also developed these methods. Numerous patents were published and the most pertinent one for the end-groups was applied for 1984 [73]. In this patent, the authors has studied the polymerization of $\underline{CF_2CF_2CH_2O}$ and its copolymerization with $\underset{\underset{O}{\backslash\;/}}{F_2C-CF-CF_3}$

in order to obtain block copolymers according to a catalytic process.

The following examples of the synthesis of monofunctional and difunctional polymers are significant:

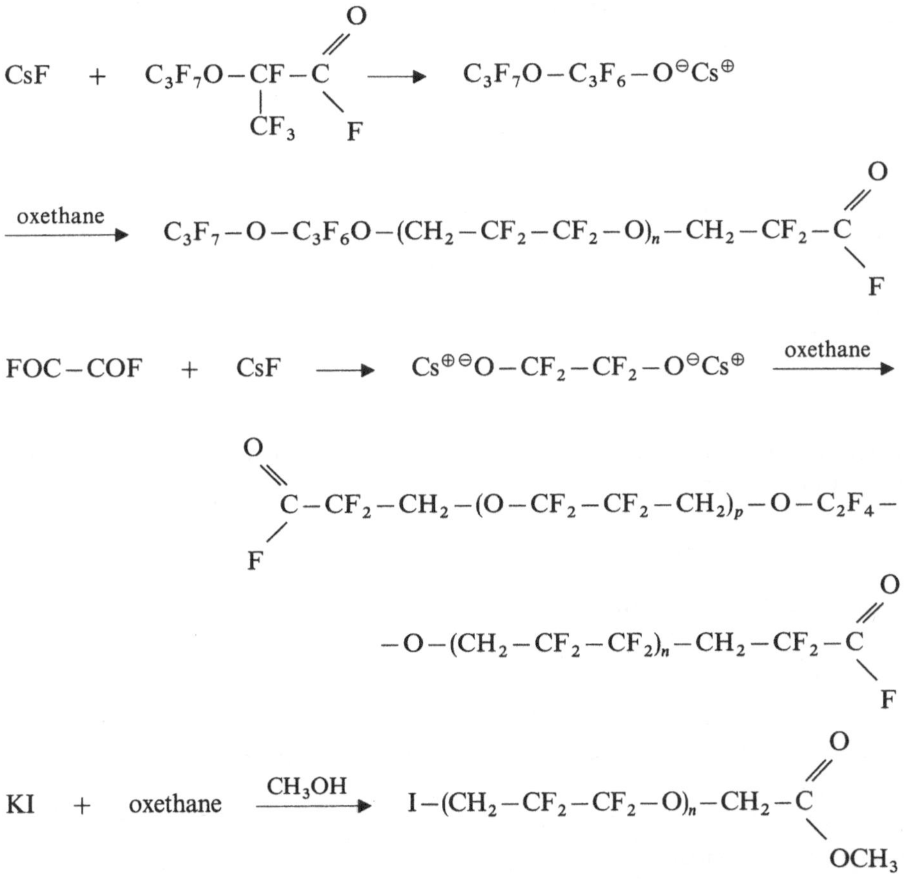

$$CsF \;+\; C_3F_7O-\underset{\underset{CF_3}{|}}{CF}-\overset{\overset{O}{\|}}{C}\underset{F}{\diagdown} \longrightarrow C_3F_7O-C_3F_6-O^{\ominus}Cs^{\oplus}$$

$$\xrightarrow{\text{oxethane}} C_3F_7-O-C_3F_6O-(CH_2-CF_2-CF_2-O)_n-CH_2-CF_2-\overset{\overset{O}{\|}}{C}\underset{F}{\diagdown}$$

$$FOC-COF \;+\; CsF \longrightarrow Cs^{\oplus\ominus}O-CF_2-CF_2-O^{\ominus}Cs^{\oplus} \xrightarrow{\text{oxethane}}$$

$$\underset{F}{\overset{\overset{O}{\diagdown\!\!\diagdown}}{C}}-CF_2-CH_2-(O-CF_2-CF_2-CH_2)_p-O-C_2F_4-$$

$$-O-(CH_2-CF_2-CF_2)_n-CH_2-CF_2-\overset{\overset{O}{\|}}{C}\underset{F}{\diagdown}$$

$$KI \;+\; \text{oxethane} \xrightarrow{\text{CH}_3\text{OH}} I-(CH_2-CF_2-CF_2-O)_n-CH_2-\overset{\overset{O}{\|}}{C}\underset{OCH_3}{\diagdown}$$

Furthermore, these authors performed fluorinations and even chlorinations to increase the thermal stabilities of these products and they mentioned the block copolymers:

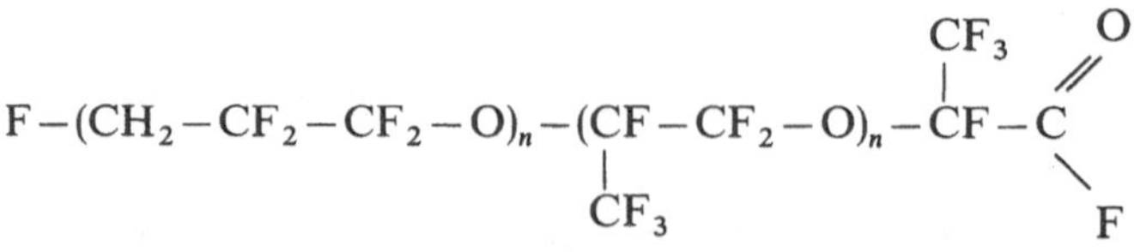

$$F-(CH_2-CF_2-CF_2-O)_n-(CF-CF_2-O)_n-\overset{\overset{\displaystyle CF_3}{|}}{CF}-\overset{\overset{\displaystyle O}{//}}{C}\diagdown F$$

3.7 Use of Functional Initiators

The functional initiators were prepared many years ago and were used for the synthesis of polybutadiene carboxy telechelic by the Thiokol Company [74] according to the following scheme:

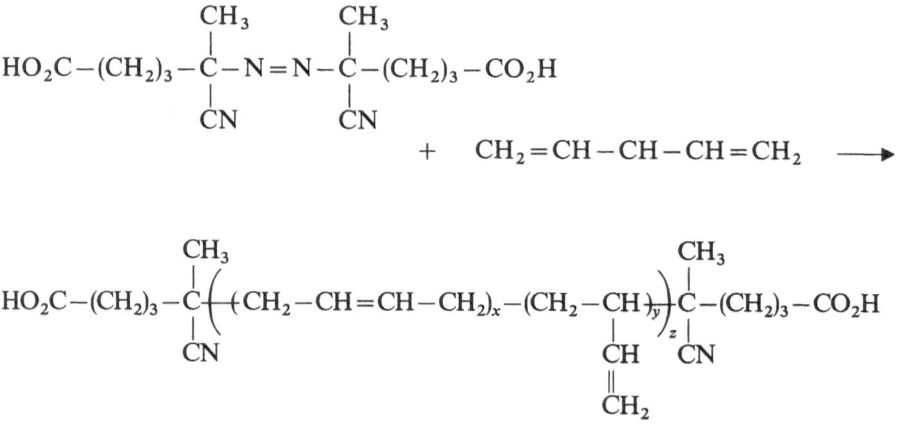

$$HO_2C-(CH_2)_3-\overset{\overset{\displaystyle CH_3}{|}}{\underset{\underset{\displaystyle CN}{|}}{C}}-N=N-\overset{\overset{\displaystyle CH_3}{|}}{\underset{\underset{\displaystyle CN}{|}}{C}}-(CH_2)_3-CO_2H$$

$$+ \quad CH_2=CH-CH-CH=CH_2 \quad \longrightarrow$$

$$HO_2C-(CH_2)_3-\overset{\overset{\displaystyle CH_3}{|}}{\underset{\underset{\displaystyle CN}{|}}{C}}\Big(\!\!\big(CH_2-CH=CH-CH_2\big)_x-\big(CH_2-CH\big)_y\!\!\Big)_z\overset{\overset{\displaystyle CH_3}{|}}{\underset{\underset{\displaystyle CN}{|}}{C}}-(CH_2)_3-CO_2H$$

with side group $\overset{|}{\underset{\parallel}{CH}}$ CH_2

The obtaining of the telechelic structure was performed because the ending-reaction was carried out by recombination of the growing radicals exclusively for such a kind of monomer:

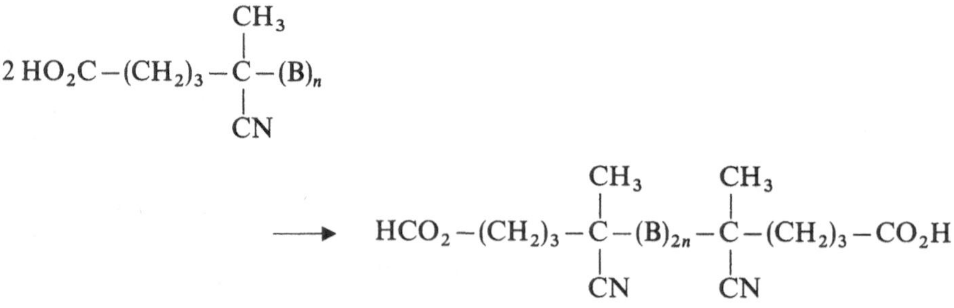

$$2\, HO_2C-(CH_2)_3-\overset{\overset{\displaystyle CH_3}{|}}{\underset{\underset{\displaystyle CN}{|}}{C}}-(B)_n$$

$$\longrightarrow \quad HCO_2-(CH_2)_3-\overset{\overset{\displaystyle CH_3}{|}}{\underset{\underset{\displaystyle CN}{|}}{C}}-(B)_{2n}-\overset{\overset{\displaystyle CH_3}{|}}{\underset{\underset{\displaystyle CN}{|}}{C}}-(CH_2)_3-CO_2H$$

$(B)_n$ = Growing radicals

Also peresters lead to telechelic polymers and we reviewed these surveys recently [75].

The fluorinated initiators have been known for years and, since 1951, Bullit [76] has been preparing products which have the following formula:

$$B-(CF_2)_n-\overset{\overset{\displaystyle O}{\|}}{C}-O-O-\overset{\overset{\displaystyle O}{\|}}{C}-(CF_2)_n-B \qquad \text{with } B = H \text{ and } F$$

In the same way, Miller [77] obtained this kind of initiator:

$$Cl_3C-CO-O_2-OC-CCl_3 \qquad \text{and} \qquad CF_3-CO-O_2-OC-CF_3$$

by the same method = an aqueous mixture of RCOCl with H_2O_2 (or Na_2O_2) at low temperatures (-5 to $-20\,°C$).

More recently, Sokolov et al. [78] prepared, and studied kinetically, initiators of the same kind but they also proposed functional fluorinated initiators:

$$\left(F-O_2S-(CF_2)_2-O-(\underset{\overset{|}{CF_3}}{CF}-CF_2-O)_x-\underset{\overset{|}{CF_3}}{CF}-\overset{\overset{\displaystyle O}{\|}}{C}-O \right)_2$$

with $x = 0$ and 1

These authors noticed that the fluorinated initiators exhibited energy of activation lower than those of their non halogenated homologs (92 kJ/mol instead of 125 kJ/mol). Moreover, an extension of the chain-length makes the energy of activation greater. A Chinese team [79] performed an interesting survey about the thermal decomposition of fluorinated initiators prepared by direct addition of acid fluorides with hydrogen peroxide. These authors confirmed the previous work on low energy of activation and showed that the decomposition of the initiator led to mainly $R_F°$ radicals instead of $R_F-CO°$ ones. The particular reactivity of $(HCF_2-CF_2-COO)_2-$ is noteworthy, the extremely low decomposition of which was explained by the authors as a decomposition induced by transfer of $HCF_2-CF_2°$ radicals onto the $-CF_2H$ extremity of the initiator.

Starting in 1969, D.E. Rice used this method to prepare dicarboxylic oligomers which contained vinylidene fluoride and perfluoropropene. He used the way these fluorinated monomers fit the reaction of chain-termination by recombination but the basic difficulty lay in the preparation of the difunctional initiator [80–83]:

$$Cl-\overset{\overset{\displaystyle O}{\|}}{C}-R_F-\overset{\overset{\displaystyle O}{\|}}{C}-Cl \xrightarrow{ROH} Cl-\overset{\overset{\displaystyle O}{\|}}{C}-R_F-\overset{\overset{\displaystyle O}{\|}}{C}-OR$$

$$\xrightarrow{Na_2O_2;\ H_2O} RO-\overset{\overset{\displaystyle O}{\|}}{C}-R_F-\overset{\overset{\displaystyle O}{\|}}{C}-O-O-\overset{\overset{\displaystyle O}{\|}}{C}-R_F-\overset{\overset{\displaystyle O}{\|}}{C}-OR$$

The initiator was also prepared in situ, from the hemiester above according to the scheme described below [80]:

$$F_3C-CH_2O_2C-(CF_2)_3-\overset{\overset{\displaystyle O}{\|}}{C}-Cl \quad + \quad H_2O \quad + \quad H_2O_2 \quad + \quad NaOH$$

$$\xrightarrow[-5\,°C]{H_2C = CF_2/C_3F_6}$$

$$F_3C-CH_2O_2C-(CF_2)_3\left[(C_2H_2F_2)_{0.65}-(C_3F_6)_{0.35}\right]_n(CF_2)_3-\overset{\overset{\displaystyle O}{\|}}{C}-O_2-CH_2-CF_3$$

$$\bar{M}_n = 3990 \text{ (VPO)}$$

In the second patent [81], Rice obtained the initiator from anhydride:

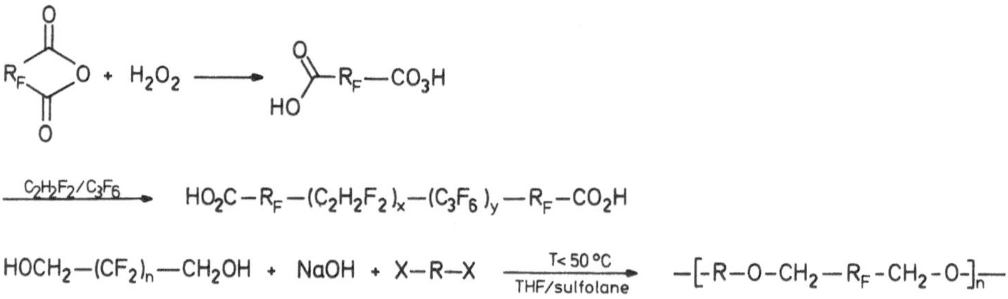

$$\xrightarrow{C_2H_2F_2/C_3F_6} \quad HO_2C-R_F-(C_2H_2F_2)_x-(C_3F_6)_y-R_F-CO_2H$$

$$HOCH_2-(CF_2)_n-CH_2OH \; + \; NaOH \; + \; X-R-X \quad \xrightarrow[\text{THF/sulfolane}]{T<50\,°C} \quad -[R-O-CH_2-R_F-CH_2-O-]_n-$$

The third patent is more orientated towards all the methods of syntheses of functional peresters [82] and we can notice the great decomposition-rates of these initiators since at 30 °C, the half life period is close to one hour [83]. Futhermore, the authors observed that this decomposition-rate increases when the monomer mixture is added.

Finally, these initiators were quicker than the corresponding fluorinated peresters such as $C_3F_7-CO_2-O_2C-C_3F_7$ (10% decomposition in one hour at 30 °C), however no explanation was given.

Rice and Sandberg [83] also noticed that hydrocarbonated initiators do not perform very well in the copolymerization of $C_2H_2F_2$ with C_3F_6. But the degrees of polymerization can be controled perfectly (initiator/monomer ratio) with fluorinated initiators and in good yields. On the other hand, all the experience showed the low reactivity of C_3F_6 since a $C_3F_6/C_2H_2F_2$ initial ratio of about 10 led only to 50% monomer in the copolymer.

3.8 Reaction Between a Dialcoholate with a Dihalogenide

Such a reaction was studied by Johncock et al. [84] particulary:

$$HOCH_2-(CF_2)_n-CH_2OH \quad + \quad NaH \quad + \quad X-R-X$$

$$\xrightarrow[\text{THF/sulfolane}]{T < 50\ °C} \quad \left(R-O-CH_2-R_F-CH_2-O\right)_n$$

Elastomers which resist temperatures up to about 300 °C were obtained.

In the following table, we sum up the main physical characteristics of these elastomers, according to the nature of R:

$R = -CH_2-$ (a)

$-CH_2-O-CH_2$ (b)

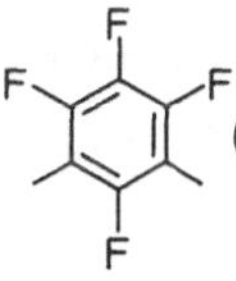

(c)   (d)

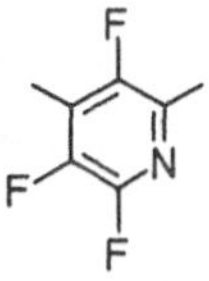

$-CH_2-C_6H_4-Z-C_6H_4-CH_2$ (e) when $Z = O, S, CO, CH_2, C(CH_3)_2$

n	3	3	3	3	3
R	a	b	c	d	e $(Z = -CH_2)$
dl/g η inh	0.38	0.59	0.06	0.50	0.47
Tg (°C)	-54	-57	crystalline	-5	-6

The possibility of obtaining of cyclic shapes which may be opened by catalysts such as PF_5 and lead to polyethers can be seen

For instance: 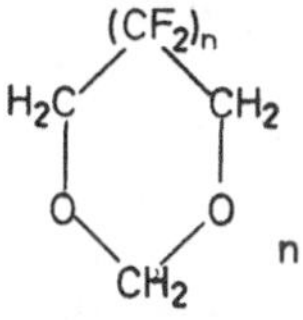   with $n = 3$ or 4

Other halogenides were used [85] such as $Cl-CH_2-O-CH_2-Cl$. Thus, Gosnell et al. [24] carried out the following reaction:

$$HOCH_2-(CF_2)_3-CH_2OH \quad + \quad (CH_2O)_3$$

$$H\left[O-CH_2-(CF_2)_3-CH_2-O-CH_2\right]_n O-CH_2-(CF_2)_3-CH_2OH$$

In the same way, Cook [87] obtained polyethers from perfluorinated dienes and fluorinated diols in an alkali medium according to the following scheme:

$$F_2C=CF-CF_2-CFCl-CF_2-CF=CF_2$$

$$+$$

$$\xrightarrow{\quad KOH \quad}$$

$$HOCH_2-CF_2-CF_2-CF_2-CH_2OH$$

$$\left[O-CH_2-(CF_2)_3-CH_2-O-CF_2-CFH-CF_2-CFCl-CF_2-CFH-CF_2\right]_n$$

Among these diols and dienes used, we can list:

$CF_2=CF-CF=CF_2$ $HO-CH_2-(CF_2)_4-CH_2OH$

$CF_2=CF-(CF_2)_4-CF=CF_2$ $HO-CH_2-(CF_2)_2-CH_2OH$

$CF_2=CF-(CF_2)_2-CF=CF_2$

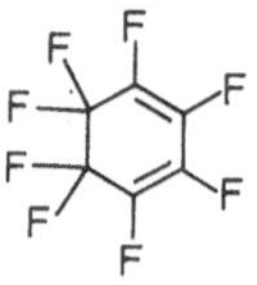

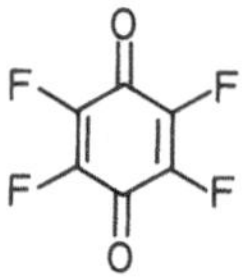

Such polymers exhibit glass transition temperatures lower than $-50\,°C$. The telomerization is the last method which allow to introduce fluorinated groups on the oligomeric chain.

3.9 The Telomerization with Functional Telogens

The telomerization is a well-known reaction for obtaining telechelic compounds and usually halogenated telogens are used. One of the forerunners, Csontos [88, 89] studied the telomerization of acrylates with $Br\,CCl_2-CH_2\,Br$ and obtained dibrominated polymers with molecular weights in the range of $10^4-2\times10^4$ in about 75% conversion-rate. After reacting with diamines, the oligomers led to elastomers with 100–200% strain. Other interesting telogens are available = CCl_4, CCl_3Br, CBr_4, CH_2Br_2, $Cl_3CCO_2CH_3$. Starks [90] produced an excellent review.

With reference to fluorinated compounds, hydrogen atoms must not be introduced in the telogen and furthermore, many investigations showed that the usable reactive groups are not numerous and in the following series:

$-CF_2-I$ or $-CF-I$ for iodides,
$\qquad\qquad\quad\ \ |$
$\qquad\qquad\quad CF_3$

$-CF_2Br$ or $-CFClBr, CCl_2Br$ for bromides,

$-CCl_3$ for chlorides.

These investigations were mainly carried out on monofunctional telogens in the 1970s and on the reactions with difunctional telogens are more recently and we give several examples with different fluorinated monomers. Most research has been performed with CF_4, as in the patent by Caporiccio et al. [91] where the telogens were:

$$I-(C_2F_4)_n-I \quad \text{and} \quad Z-(C_2F_4)_n-I \quad \text{where} \quad Z = Cl, Br.$$

Then, the products were brominated or chlorinated directly in the vessel. With C_3F_6, the same team [92] prepared the following telomers:

$$X\left(CF-CF_2\right)_a \left(C_2F_4\right)_b \left(CF_2CF\right)_c Y$$
$$\quad\quad |\quad\quad\quad\quad\quad\quad\quad\quad\quad |$$
$$\quad\quad CF_3\quad\quad\quad\quad\quad\quad\quad CF_3$$

$0 \leq a$ and $c \leq 2$; $1 \leq b \leq 3$; X and Y = I, Br or Cl

The interest and the originality of this work are that the authors isolated oligomers which exhibit sequences of two $-CF_2-CF(CF_3)$ units since such monomer were regarded as being unable to homopolymerize. From perfluoro-alkyle diiodides [93] block or statistic cotelomers were obtained and the authors used several mixtures with $C_2H_2F_2$; $C_2H_2F_2/CTFE$; $C_2H_2F_2/C_3F_6$; $C_2H_2F_2/C_2F_4/C_3F_6$. The applications of the elastomers obtained and crosslinked by several methods deals with products highly resistant to alcohols.

With $F_2C = CFCl$ as for C_3F_6, research is less developed and yet Dedek et al. [94] have carried out interesting investigations:

$$Br-CFCl-CF_2-Br \quad + \quad CF_2=CFCl$$

$$\xrightarrow{\text{UV, 20-40 °C}} \quad Br-(CF_2-CFCl)_n-(CFCl-CF_2)_m-Br$$

$n = m = 1$; $n = 1$ and $m = 2$; $n = m = 2$

Several adducts were isolated such as:

$$Br-CF_2-CFCl-CFCl-CF_2-Br$$
$$Br-CF_2-CFCl-CFCl-CF_2-CFCl-CF_2\ Br$$
$$Br-CF_2-CFCl-CF_2-CFCl-CFCl-CF_2-CFCl-CF_2-CF_2-Br$$

These structures show that both telomerization $-(CF_2CFClCF_2CFCl)-$ sequences and the coupling-reaction of the growing-chains under U.V. were in competition.

So far, no applications for the oligomers themselves have been found but they are excellent precursors for difunctional derivatives.

So far, we have described methods which lead to oligomers in which the fluorinated chain consists of a sequence of this oligomer. Now, we will detail methods which make it possible to prepare difunctional compounds with fluorinated grafts.

3.10 Polyethers with Fluorinated Grafts

The addition of fluorinated groups onto alcohols has been known for years and monoalcohols were prepared from such reactions. Thus:

$$CF_2-CFX \quad + \quad CH_3OH \xrightarrow{\text{peroxide}} \quad H-(CFX-CF_2)_n-CH_2OH$$

x = Cl or F

More recently, this reaction was performed again by Haszeldine [95] with trifluoropropene as the monomer:

$$H-CR^1R^2-OH \quad + \quad C_3F_6$$

$$\xrightarrow{\text{peroxide/hv}} \quad H-CF-CF_2-CR^1R^2OH \quad + \quad F_2HC-CF-CR^1R^2OH$$

$$\underset{CF_3}{\mid} \qquad\qquad\qquad\qquad \underset{CF_3}{\mid}$$

with R^1R^2 = H, Me, Et, n Pr and even CF_3

Obviously, the expected and normal adduct is predominant but we can obtain 45% of the false adduct. Furthermore, CF_3-CH_2OH is a telogen whereas $(CF_3)_2$ $CH-OH$ is not. Haszeldine proposed a special mechanism:

$$C_3F_6 \xrightarrow{\text{UV}} (C_3F_6)^* \xrightarrow{\text{RH}} R\bullet \quad + \quad (C_3HF_6)\bullet$$

$$\xrightarrow{\text{RH}} \quad F_3C-CHF-CHF_2 \quad + \quad R\bullet$$

$$\xrightarrow{C_3F_6} \quad RCF_2-\overset{\bullet}{C}F-CF_3 \quad + \quad F_3C-CRF-CF_2\bullet$$

$$\text{RH} \downarrow \qquad\qquad\qquad\qquad \downarrow \text{RH}$$

$$R\bullet \quad + \quad RCF_2-CFH-CF_3 \qquad F_3C-CFR-CHF_2 \quad + \quad R\bullet$$

In the same way, Chambers et al. [96] studied the addition of THF onto the perfluoropropene:

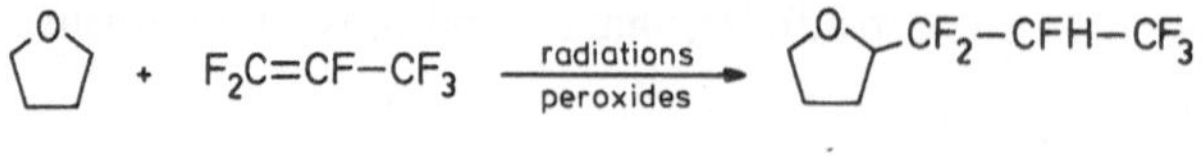

$+ \quad F_2C=CF-CF_3 \xrightarrow[\text{peroxides}]{\text{radiations}}$ $-CF_2-CFH-CF_3$

These authors showed that the methylene group in the α position of an ether group is also able to initiate the addition of the perfluoropropene:

$$1\text{--}3; \; 1\text{--}4 \text{ and } 1\text{--}3 \; + \; 1\text{--}4$$

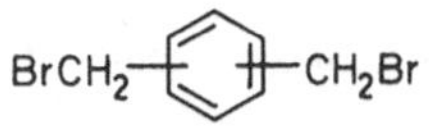

BrCH₂—⟨ ⟩—CH₂Br   ClCH₂—⟨ ⟩—O—⟨ ⟩—CH₂Cl

(f) 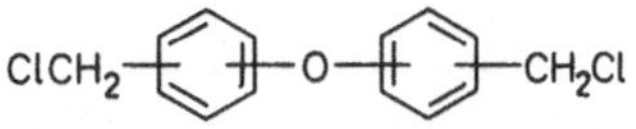   (g)

with the same diol $HOCH_2-(CF_2)_n-CH_2OH$.

In the following table, we sum up as previously, the main characteristics related to the corresponding Tg's and viscosities:

n	3	3	3	3	3
R	e	f 1–3	f 1–4	f 1–3 + 1–4	g
dl/g η inh	0.87	0.65	0.48	0.65	0.55
T_g (°C)	−43	−35	−26	−32	6

Halogenated perfluorinated aromatics were used [86] but the obtained polyethers exhibit Tg's greater than the previous ones:

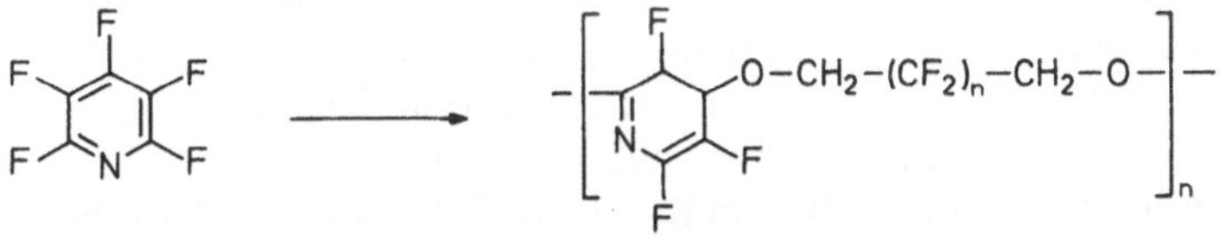

$$T_g = -5 \text{ to } -12 \text{ °C}$$

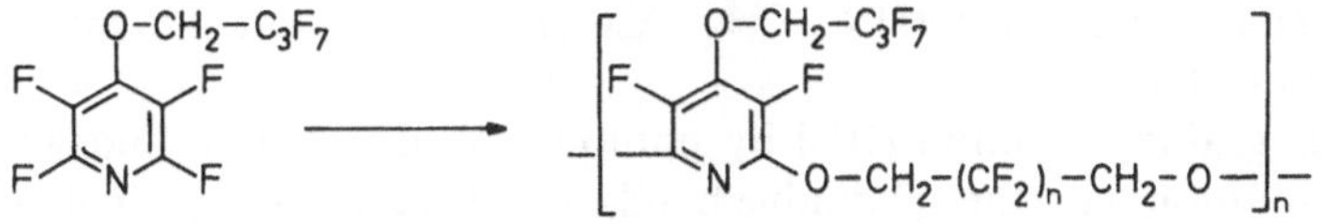

$$T_g = 2 \text{ °C}$$

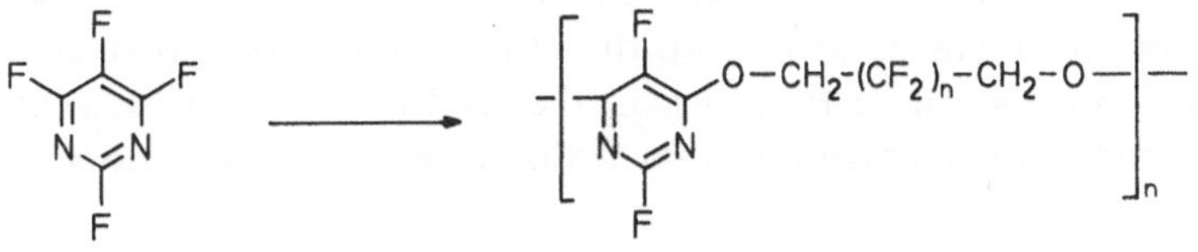

$$T_g = 0 \text{ °C}$$

In order to complete the synthesis of fluorinated polyethers, we can mention Gosnell et al. [24] and Cook's [87] research.

The application of telechelic compounds was carried out by ICI [97] by achieving the above reaction on poly THF hydroxytelechelic. The industrial application consists of the coating of optical fibers obtained after substitution of the hydroxyl groups by acrylic ones.

The reaction on poly THF 650 led to the following product:

$$HO-(C_4H_8O)_x-(C_4H_7O)_y-H \quad \text{with} \quad x \approx y \approx 4.5$$
$$\underset{C_3F_6H}{|}$$

The acrylic groups were added in two steps: first by addition of an excess of diisocyanate and then reaction with hydroxy-2 ethylacrylate.

The formulations with N-vinyl pyrrolidone as reactive diluent show an important decrease of water absorption about non fluorinated polymers

3.11 Polybutadienes with Fluorinated Grafts

Thiokol Chem. Corp. [98] and Ciba Geigy [99] are keen on adding fluorinated compounds onto polydienes and among them there are the polybutadiene hydroxytelechelics (PBHT).

In the first case, the authors graft perfluoroalkyliodides $C_nF_{2n+1}-I$ onto PBHT (Acro), PBCT (Goodrich) and PB phenyltelechelic by radical initiation. The reaction was carried out with benzoyl peroxide at temperatures in the range 60–110 °C in the dioxane and led to an amount of grafted fluorine of 7.0–21.5%.

With ADIB at 55 °C in acetone, such amount was 5.0% and with the U.V. initiator, it was 5.4%.

From these results, we think that the grafting occurs on the 1,2-double bonds only but the data are not conclusive enough.

More recently, the second survey used the results obtained from the Company Ciba Geigy on the addition of thiols onto insaturated compounds. The authors showed that the addition of thiols onto dienes by ADIB led to highly grafted products since they may contain 45% by weight of fluorine (70% grafting), but such products are only soluble in freon or hexafluoroxylene.

Thus, $RF-I$ can be grafted more easily than $R_F-C_2H_4SH-$. Furthermore, both fluorinated thiols and functional ones ($HS-CH_2-CO_2H$) may be grafted to make them emulsifiable in water.

We repeated this grafting reaction [100] by photochemical initiation and we showed that if the addition onto the 1,2 double bands was favored, the fluorinated thiols could be grafted onto all double bands of PBHT and diols with a fluorine amount of 55% could be obtained. Moreover, this study shows an improvement of the dispersion (observed in G.P.C.) when the grafting was realized by photochemical initiation in comparison to that realized by classical initiators.

4 Conclusion

The synthesis of fluorinated molecular or macromolecular telechelic compounds has been developed over many years and in numerous studies. Actually, the first surveys were carried out in the 1970s but after a 10 year-stop, renewed interest has been shown. This is due to the particularly interesting properties of fluorinated products e.g. thermal resistance and also their surface properties. However, several difficulties may appear: non solubility, special reactivity of functional groups in positions about perfluorinated groups and we consider that the compromise between the expected properties and the difficulties of synthesis has not been reached yet. The most recent work seems to isolate these two kinds of properties and is orientated towards two kinds of polymers: those which exhibit good high temperature properties and those with improved surface properties. However, we hope that increased knowledge will allow us to produce oligomers with a low enough molecular weight to have better solubility so that they can be used industrially.

5 References

1. Simons JH (1950) Fluorine chemistry vol 1, Academic, New York, p 40
2. Lovelace AM, Rausch DA, Postelnek W (1958) J Am Chem Soc, Monograph Series, Reinhold, New Yorker
3. Evans DEM, Tatlow JC (1954) J Chem Soc, 377a
4. Fear EPJ, Thrower J, Veitch J (1955) J Appl Chem 5: 589
5. Knunyants IL, Shoshrina VV, Li Chih-Yuan (1959) Proceed Acad USSR, p 971
6. Haszeldine RN (1951) Nature, 167: 139
7. Barnhart WS, Wade RH (1957) US Patent 2,806,865
8. Boutevin B, Pietrasanta Y (1975) European Polym J 12: 219
9. Boutevin B, Cals J, Pietrasanta Y (1974) Tetrahedron Letters 12: 939
10. Boutevin B, Pietrasanta Y (1975) European Polym J 12: 231
11. Hauptschein M, Braid M (1961) J Am Chem Soc 83: 2500
12. Ameduri B, Boutevin B, Lecrom C, Pietrasanta Y, Parsy R (1988) Makromol Chem 189: 2545
13. Boutevin B, Pietrasanta Y (1988) French Patent 8,801,882
14. Smeltz K (1969) US Patent 3,478,116 (Du Pont De Nemours)
15. Smeltz K (1969) US Patent 3,504,016 (Du Pont De Nemours)
16. Boutevin B (unpublished results)
17. Foulletier, L. Lalu JP (1970) Ger Patent 2,004,1956, Chem Abst 71: 110486c
18. Bouvet P, Lalu JP (1970) French Patent 2,034,142 (Atochem)
19. Anello L (1970) French Patent 1,959,703
20. Battais A, Boutevin B, Hugon JP, Pietrasanta Y (1980) J of Fluorine Chem 16: 397
21. Foulletier L, Lalu JP (1970) French Patent 1,588,865 (Atochem)
22. Boutevin B, Hugon JP, Pietrasanta Y (1981) J. of Fluorine Chem 17: 357
23. Filler R, O'Brien JF, Fenner JV, Hauptschein M (1953) J Am Chem Soc 75: 966
24. Gosnell RB, Hollander J (1967) J Macromol Sci Phys B14: 831
25. Boutevin B, Dongala FB, Pietrasanta Y (1981) J of Fluorine Chem 17: 113
26. Evans FW (1968) US Patent 3,388,078
27. O'Brace N (1964) US Patent 3,145,222
28. Hollander J, Trischler FD (1967) J Am Chem Soc, Div Polym Chem, Preprints 8: 491
29. Knunyants IL (1972) USSR Patent 352,890
30. O'Rear JG, Griffith JR (1972) J Am Chem Soc, Div Org Coatings Plast Chem 32(1): 417–9
31. Pittman AG, Wasley WL, Roitman J (1970) J Polym Sci, Part B 8(12): 873
32. O'Rear JG, Griffith JR, Reines SA (1971) J Paint Technol 43 (552): 113
33. Griffith JR (1974) Can Patent 956,398

34. Damont FR, Sharpe LH, Schonhorn HJ (1965) Polymer Sci B3 12: 1021
35. Griffith JR, Quick JE (1970) Advance Chem Ser 92: 8
36. Cupples AL, Lee, HL, Stoffy DG (1970) Advance Chem Ser 92: 173
37. Moore EP (1966) US Patent 3,250,802
38. Moore EP (1967) US Patent 3,322,826
39. Hill JT (1974) J Macromol Sci Chem 8: 499
40. Eleuterio HS (1972) J Macromol Sci Chem 6: 1027
41. Tanesi D, Pasetti A, Corti C (1969) US Patent 3,442,942
42. Caporiccio G, Viola G, Corti C (1983) Europ Patent 89,820
43. Sianesi D, Pasetti A, Belardinelu G (1984) US Patent 4,451,646
44. Sovolov SV (1975) Zhwin Org 11: 303
45. Mitsch RA, Zollinger JL (1970) Ger Patent 2,011,774
46. Warnell JL (1966) US Patent 3,250,805
47. Johncock P, Hewins MAH (1975) J of Polym Sci 13: 807
48. Johncock P, Hewins MAH (1976) J of Polym Sci 14: 365
49. Sokolov SV (1975) Zhwin Org Kai 11(3): 552
50. Caporiccio G (1977) Ger Patent 2,633,736
51. Riley MD, Kim YK, Pierce OR (1977) J of Fluorine Chem 10: 85
52. Harris JF (1965) J Org Chem 30: 2182
53. Zollinger JL (1969) J Macromol Sci Chem 3: 1443
54. Caporiccio G (1986) J of Fluorine Chem 33: 314
55. Mitsch RA, Zollinger JL (1976) US Patent 3,972,856
56. Stump EC, Rochow SE (1972) US Patent 3,671,497
57. Stump EC, Rochow SE (1973) US Patent 3,755,265
58. Pilati F, Bonora V, Nanaresi P, Munari A, Toselli M (1989) J of Polym Sci, Part A, Polym Chem 27: 951
59. Rosser RW, Parker JA, De Pasquale RJ, Stump EC (1975) ACS Sym Ser 6, 185–198 (Polyethers Symp. 1974)
60. Caporicci G, Bargigia G (1974) Ger Patent 2,502,505
61. Grindahl GA, Pierce OR (1967) J Org Chem 603
62. Webster JA, Butler JM, Morrow TJ (1972) Nuovo Chim 48: 51–4
63. De Pasquale RJ (unpublished results) Symposium Gainsville, Florida 1975
64. Evers RC (1978) J of Polym Sci Polym Chem Ed 16: 2817
65. Evers RC (1974) Polym Prepr Am Chem Soc Div Polym Chem 15: 685
66. Gerhardt GE, Lagow RJ (1978) J Org Chem 43: 45.05
67. Gerhardt GE, Lagow RJ (1981) J Chem Soc Perkin Trans 1: 1321
68. Persico DF, Gerhardt GE, Lagow RJ (1985) J Am Chem Soc 107: 1197
69. Persico DF, Lagow RJ (1985) Macromolecules 18: 1383
70. Persico DF, Lagow RJ (1987) International Patent WO 87/00538
71. Modena, S, Calini P, Gregorio G, Hoggi G (1988) J of Fluorine Chem 40: 349
72. Weinmayr V (1963) J Org. Chem 28: 492
73. Yohnosuke O, Takashi T, Shoji T (1984) Eur Patent Appl 0,148,482 (Daikin) Chem. Abst 104: 069315
74. Thiokol Corp. (1951) Brit Patent 957,652
75. Boutevin B, Advances in Polymer Science (in press)
76. Bulitt O (1951) US Patent 2,559,630 (Du Pont de Nemours)
77. Miller WT (1951) US Patent 2,580,358
78. Novikov VA, Sass VP, Ivanova LS, Sokolov LF, Sokolov SV (1975) Polym Sci USSR, Ser. A9: 1414
79. Zhao Cheng Yue, Zhou Renmo, Pan Heqi, Jin Xiangshan, Qu Yangling, Wu Chengjiu, Jiang Xikui (1982) J Org Chem 47(11): 2009
80. Rice DE (1969) US Patent 3,438,953 (3 M)
81. Rice DE (1969) US Patent 3,457,245 (3 M)
82. Rice DE (1969) US Patent 3,461,155 (3 M)
83. Rice DE, Sanberg CL (1971) Polym Prep Am Chem Soc Div Polym Sci 12(1): 396
84. Johncock P (1970) Ger Patent 1,954,999

85. Johncock P, Hewins MAH (1975) J of Polym Sci 13: 807
86. Johncock P, Hewins MAH (1976) J of Polym Sci 14: 365
87. Cook EW (1968) US Patent 3,391,118
88. Csontos AA (1973) US Patent 3,730,862
89. Csontos AA (1973) US Patent 3,775,276
90. Starks CM (1974) Free radicals telomerization, Academic New York
91. Caporiccio G, Bargigia G, Tonelli C, Tortelli V (1988) US Patent 4,731,170
92. Caporiccio G, Bargigia G, Tonelli C, Tortelli V (1986) Eur Patent Appl 0,200,988
93. Furakawa Y (1981) Eur Patent Appl 0,045070 (Daikin Kogyo Co.)
94. Dedek V, Chatal Z (1986) J of Fluorine Chem 31: 363
95. Haszeldine RN, Rowland R, Sheppard RP, Tipping AE (1985) J of Fluorine Chem 28: 291
96. Chambers Rd, Rievson BG, Drakesmith FG, Powell RL (1985) J of Fluorine Chem 29: 323
97. Head RA, Johnson S (1987) Eur Patent Appl 260,842 (I.C.I.)
98. Villa JL, Iserson H (1974) Ger Patent 2,325,561 (Thiokol Chem Corp)
99. Mueller KF (1984) Eur Patent Appl 115,253 (Ciba Geigy) Chem Abst 102: 47471u
100. Boutevin B, Hervaud Y, Nouiri M (1990) Eur Polym J (to be Published)

Editor H.-J. Cantow
Received July 24, 1990

Synthesis and Applications of Fluorinated Telechelic Monodispersed Compounds

B. Améduri and B. Boutevin
Ecole Nationale Supérieure de Chimie de Montpellier, URA CNRS D 11930,
8 Rue Ecole Normale, 34053 Montpellier Cedex, France

Different methods of synthesis involved in the preparation of fluorinated telechelic monodispersed compounds are described. First, a bibliographical approach develops the molecules in which the fluorine atoms are located on the main chain: obtaining by chemical change of fluorinated molecules (e.g. diacids, telomers), by bistelomerization and by direct or special syntheses. In a second part, investigations concerning the different preparations and uses of fluorinated telechelic molecules in which the fluorinated groups are branched on the back-bone are mentioned. Synthetic methods of obtaining 1,x-diols where $1 \leqq x \leqq 4$ are described: diols prepared from diamines, by condensation, by simple addition of fluorinated telogens onto difunctional olefines. Furthermore, other novels syntheses such as the ring opening polymerization of fluorinated oxetanes lead to these compounds. In both parts, several chemical and physical parameters are supplied and the applications described.

1 Introduction 135

2 Fluorinated Difunctional Products with a Fluorinated Main Chain 135
  2.1 Obtaining Difunctional Compounds by Chemical Change
     of Fluorinated Molecules 136
  2.2 Chemical Change of Telomers 137
     2.2.1 Synthesis of Telogens 137
     2.2.2 Chemical Change of Telomers 139
     2.2.3 Bistelomerization 140
  2.3 Chemical Change of Diacids 141
  2.4 Synthesis of Fluorinated Telechelic Diisocyanates 141
  2.5 Special Difunctional Molecules Obtained by Direct Synthesis . . . 144
     2.5.1 2,2-bis(4-Hydroxyphenyl)Hexafluoropropane 144
     2.5.2 1,3-bis(2-Hydroxyhexafluoro-2-propyl)Benzene 147
     2.5.3 Diaromatic Difunctional Compounds Linked by Fluorinated
        Chains 150
  2.6 Other Special Difunctional Products 152

3 Difunctional Products with Fluorinated Lateral Chain 153
  3.1 Nature of the Fluorinated Chains 153
     3.1.1 Linear Perfluorinated Chains 153
     3.1.2 Branched Perfluorinated Chains 153
     3.1.3 Chlorofluorinated Chains 154
     3.1.4 Hydrogenofluorinated Chains 155

Advances in Polymer Science, Vol. 102
© Springer-Verlag Berlin Heidelberg 1992

3.2 Synthesis of Diols from Diamines 155
3.3 Synthesis of Diols by Condensation. 157
   3.3.1 Condensation of Fluorinated Iodides with Thiols-Diols 157
   3.3.2 Condensation of Thiol Halogenide with Glycidol 157
   3.3.3 Condensation of Fluorinated Alcohols with Glycidol 157
3.4 Addition of Fluorinated Telogens onto Difunctional Olefines 158
   3.4.1 With Fluorinated Iodides as Telogen. 158
   3.4.2 With a Fluorinated Thiol as Telogen. 158
   3.4.3 With Epoxies as Monomers 159
3.5 Radical Telomerization of a Fluorinated Monomer with a Diol-Thiol 160
3.6 Synthesis of Fluorinated Propane-1,3-diol from Alkyl Malonates . . 160
3.7 Ring Opening Polymerization of Fluorinated Oxetanes 160
3.8 Miscellaneous 162
3.9 Synthesis of Fluorinated Aromatic Diisocyanates 163

4 Conclusion 164

5 References 164

1 Introduction

The fluorinated telechelic compounds are prepared for the synthesis of poly-condensates which have specific properties such as high thermal resistance, very good surface properties and important chemical resistance. Such properties lead them to particular applications and to a very high "added value". Fluorinated telechelic monodispersed products are numerous and diacids or diesters derivatives can be found or prepared.

We can mention diols but also less abundant ones such as diepoxies [1], diamines [2, 3], diisocyanates [4, 5], dialdehydes [6] and diacetylenic compounds [7].

Usually the adjustment of properties is carried out by using mixtures of high molecular weight polydispersed telechelic prepolymers and simple low molecular weight molecules. Recently, we produced a review about the prepolymers mentioned above [8] which we will not describe in this paper. This work concerns the synthesis of low molecular weight monodispersed molecules, their properties and eventually their applications.

Two main kinds of fluorinated difunctional products exist whatever the reactive functions: those in which the fluorinated group is located in the main chain and those in which the fluorinated group is lateral about the main chain. This split is important since it induces differences on both the syntheses and the properties. In this review, these two kinds of compounds are studied respectively.

2 Fluorinated Difunctional Products With a Fluorinated Main Chain

The most common and the oldest simple difunctional compounds are obviously the molecules which have a $-CF_2-$ group on the back bone. Usually they are prepared by electrochemical fluorination of the corresponding hydrocarbonated acid chlorides, and, perfluorinated acid difluorides:

$$F-\overset{\overset{\displaystyle O}{\|}}{C}-(CF_2)_n-\overset{\overset{\displaystyle O}{\|}}{C}-F$$

are obtained in poor yields [9].

Most difunctional compounds can be prepared from these products: diesters of course, or by reduction, diols which have the following formula: $HOCH_2-(CF_2)_n-CH_2OH$ [9–11].

The alcohols $R-CF_2OH$ are known to be unstable and produce acid fluoride directly.

Furthermore, fluorinated diamines $H_2NCH_2-(CF_2)_n-CH_2NH_2$ and diisocyanates $O=C=N-(CF_2)_n-N=C=O$ have also been obtained. As in the case of the alcohols, all the diamines have a methylene group between the difluoromethylene and the amine groups [9].

Other branched chains can be prepared but they will not be mentioned in this paper since their syntheses are rather old and well known; however this article deals with more particular difunctional compounds.

Several methods can be proposed for preparing these kinds of molecules and they can only be classified with difficulty. Either they can be obtained by direct synthesis, or by modification of available functional compounds, or from oligomers which can be changed.

2.1 Obtaining Difunctional Compounds by Chemical Modification of Fluorinated Molecules

The most classical synthesis of the diacids concerns the oxidation of double bonds which are obtained by reactions of dehydrohalogenation or dehalogenation [12–16].

Two methods can be used:

- permanganic oxidation [17, 18],
- catalytic oxidation by ruthenium dioxide [19].

In the case of the oxidation by $KMnO_4$, the reactions are usually performed in a hot medium, either in a basic system (KOH) by monitoring the pH or in buffer medium (sodium hydrogenocarbonate [20]). It has been shown that when such a reaction is carried out in an acid medium the yield decreases drastically [21].

The usually accepted reaction mechanism [22–23] involves an α-diketonated intermediate which, in an alkali medium, leads to the salts of both corresponding perfluorinated acids as shown in the following scheme:

$$\longrightarrow R_F CO_2^- + R'_F CO_2^-$$

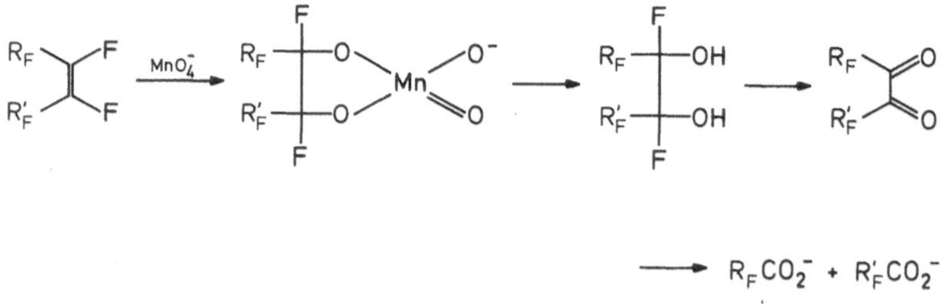

At room temperature, the oxidation by RuO_2 requires an oxidant such as peracetic acid, periodic acid or sodium hypochlorite.

This method has been used with perhalogenated unsaturated cyclic olefines to lead to diacids as described in the following scheme:

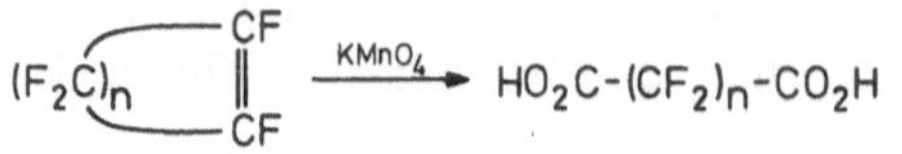

However, such a cleavage only occurs if fluorine atoms are branched onto the double bond; in the contrary case, the diol is obtained as Chambers [24] and Husain [25] showed (Table 1).

Table 1. Fluorinated telechelic diols obtained from the permanganic oxidation of fluorinated alkenes

Alkene	Diol	Yield	Ref.
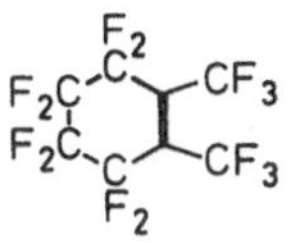	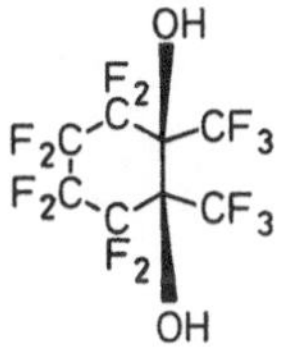	74.4	24
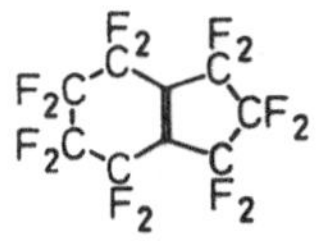	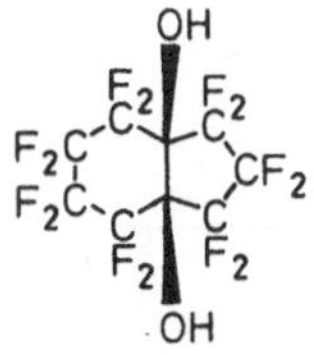	24.0	24
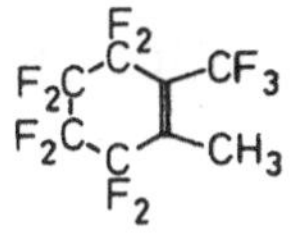	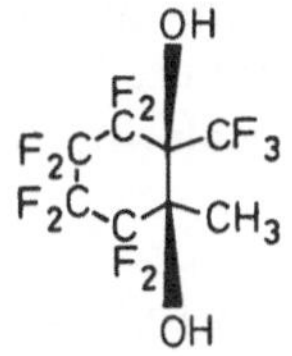	28.8	24
F_3C / C_2F_5 $\diagup\!\!\!\!\diagdown$ CF_3 / C_2F_5 (alkene)	$F_3C-\underset{\underset{C_2F_5}{|}}{\overset{\overset{OH}{|}}{C}}-\underset{\underset{C_2F_5}{|}}{\overset{\overset{OH}{|}}{C}}-CF_3$	59.0	25

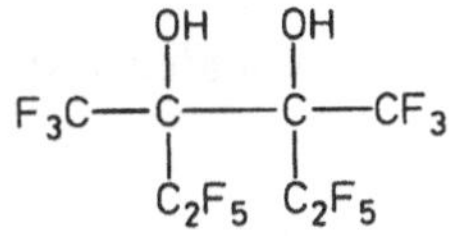

2.2 Chemical Change of Telomers

2.2.1 Synthesis of Telogens

Telomerization is an excellent means of obtaining fluorinated or chlorofluorinated chains which exhibit well defined end-groups.

Table 2 lists several exemples of syntheses described in the following scheme:

$$CCl_2XY + F_2C = CFZ \xrightarrow{\text{catalyst}} X-CCl_2-(CF_2-CFZ)_n-Y$$

where X or Y = Cl or Br

and   Z = F, Cl, Br.

Table 2. Telomerization of fluorinated taxogens with halogenated telogens

Taxogen	Telogen	Initiator or Catalyst	Ref.
	Cl_3CBr	peroxide	26−28
		U.V.	29−30
		$AlCl_3$	31
		redox	32
$F_2C=CFCl$	CCl_4	redox	33−35
	SO_2Cl_2	peroxide	36
	PCl_5	peroxide	37
	Br_2	redox	38
	$BrCF_2CFCLBr$	U.V.	39
$F_2C=CF_2$	$I-C_2F_4-I$	thermal	40−42
C_3F_6	$I-C_2F_4-I$	thermal	43

From these compounds, the end-groups must be activated in order to perform the functionalizations. Two methods are used:

− The chemical change in α, ω bis(trichloromethyled) end-groups by $AlCl_3$ [34, 35, 44, 45]:

$$Cl_3C-(CF_2CFCl)_n-Cl \xrightarrow[80\,°C]{AlCl_3/CCl_4} Cl_3C-(CF_2CFCl)_{n-1}-CF_2-CCl_3 \,.$$

− The coupling of brominated or iodinated with zinc [12] or with UV-radiation [46]:

$$Cl_3C-(CF_2CFCl)_n-Br \xrightarrow[Ac_2O]{Zn} Cl_3C-(CF_2CFCl)_n-(CFClCF_2)_n-CCl_3 \,.$$

In the case of iodinated compounds, the synthesis was carried out in two different ways: the telomerization and the decarboxylation of silver salts.

Concerning the telomerization, Bedford et al. [47] studied the telomerization of tetrafluoroethylene with IC_2F_4I and I_2. The results are listed in the following table:

Reactant	C_2F_4/I_2	Yield (%)	IC_2F_4I	IC_4F_8I
I_2	2	49	64.2	28.5
I_2	4	47	65.5	29.5
IC_2F_4I	1	50	66.6	29.1
IC_2F_4I	2	52	50.0	33.3
IC_2F_4I	4	67	34.1	41.4

The reaction was carried out at $200-220\ ^\circ C$, under a pressure of $20-27$ bar and without any catalyst.

Recently Caporiccio et al. [48] pursued such a synthesis to obtain the compounds $Br-(CF_2)_n-Br$ by halogenation with Cl_2 or with Br_2.

These authors used perfluoropropene instead of C_2F_4 [49] and prepared $I-(C_3F_6)_x-(C_2F_4)_x-(C_3F_6)_y-I$ (x and y worth 1 mainly) but they showed that two C_3F_6 groups were in the role. This study, the mechanism of telomerization and the ^{19}F NMR characteristics have been published recently [50].

Concerning the decarboxylation of silver salts in presence of iodine, the reaction is the following:

$$AgO_2C-(CF_2)_n-CO_2Ag \xrightarrow{2I_2,\ \triangle} I-C_nF_{2n}-I\ .$$

This is an old method, often applied to monofunctional compounds that Rice [51] used from the diacid $O(C_2F_4CO_2H)_2$.

Finally, aliphatic [52] and aromatic [53] diamines were obtained by McLoughlin. Thus, from $ClCO(CF_2)_nCOCl$, reacting with KI at $200\ ^\circ C$, in anhydrous medium, he prepared a mixture of $I(CF_2)_nCOCl$ and $I-(CF_2)_n-I$. The author described diiodides with $n = 6, 7, 8$ and 9. With regards to aromatic ones, the synthesis is as follows:

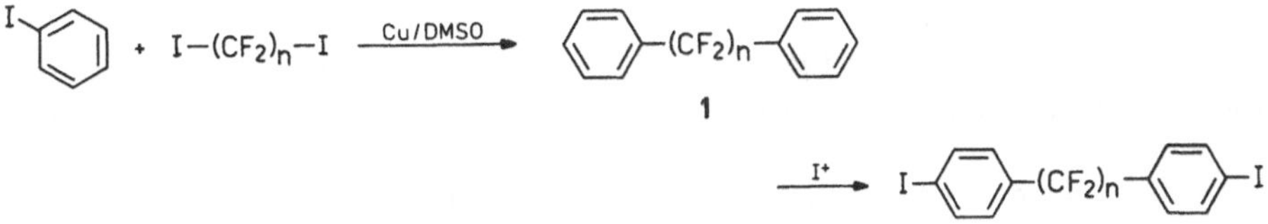

2.2.2 Chemical Change of Telomers

The α, ω bis(trichloromethyled)telomers can be used for the preparation of diacids after reaction with oleum [54, 55].

$$Cl_3C-(CF_2CFCl)_n-CF_2CCl_3 \xrightarrow[SO_3]{H_2SO_4} HO_2C-(CF_2CFCl)_n-CF_2CO_2H\ .$$

We can mention in our recent study [56] concerning the intermediates obtained during the hydrolysis that when $n = 0, 2$ and higher base-units number, acid dichlorides were prepared, and yet the compound $n = 1$ produced the acid anhydride:

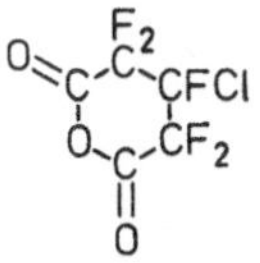

The hydrolysis with oleum can be performed for diiodinated compounds [57] but the reaction was more difficult.

In the same series of chlorofluorinated telomers, Wujclak et al. [58] prepared $HOCH_2(CF_2CFCl)_nCF_2CH_2OH$ (n = 0–7) from the reduction of the corresponding acid dichlorides.

2.2.3 Bistelomerization

From α, ω bis(trichloromethyled) telogens, we carried out various telomerizations with methyl undecylenate [59], ethylene [60], vinyl acetate [61] allyl acetate [62–63], and with acrylates [61, 64]. We can notice that the allyl and acrylic derivatives lead to a mixture made of monoacetate and diadduct [63]:

$$Cl_3C-(CF_2CFCl)_n-CF_2-CCl_3 + H_2C=CH-\text{\textcircled{G}}$$

$$\downarrow \text{redox}$$

$$Cl_3C-(CF_2CFCl)_n-CF_2-CCl_2-CH_2CHCl-\text{\textcircled{G}}$$

$$+$$

$$\text{\textcircled{G}}-CHClCH_2-CCl_2-(CF_2CFCl)_n-CF_2-CCl_2-CH_2CHCl-\text{\textcircled{G}}.$$

The amounts of both products depend upon the reaction-time, the temperature, the quantity and the nature of the catalyst (copper or iron salts and ruthenium complex).

With ethylene and methyl undecylenate, the reactions were nearly quantitative, especially when ethylene was used with an adequate pressure. Thus, at 40 bar at 150 °C with $CuCl_2$ as catalyst, 90% of diadducts was obtained.

With diiodinated compounds, the reaction leads to the following telomer [40]:

$$I-(CH_2-CH_2)_m-C_nF_{2n}-(CH_2-CH_2)_m-I$$

where n = 3–20, m is lower than 10 (usually m = 1).

Allyl olefines are mainly used and, from these reactants, Brace prepared diols and diacids by reacting silver acetate and by permanganic oxidation, respectively [65].

From diiodides, Piccardi et al. [66] obtained non-conjugated dienes, $H_2C=CH-(CF_2)_n-CH=CH_2$ and studied the addition of CCl_4 by redox catalysis. They isolated the diadducts:

$$Cl_3C-CH_2-CHCl-(CF_2)_n-CHCl-CH_2-CCl_3 .$$

Finally, we can mention interesting surveys about α,ω-dibrominated compounds $Br-C_2H_4-(CF_2)_n-C_2H_4Br$ prepared by Smeltz [67].

2.3 Chemical Change of Diacids

The chemical change of diacids is carried out in the classic manner and yet the high degree of activity brought about by the difluoromethylene groups modify their reactivity.

The reduction into diols is performed by using KBH_4 [68], $AlLiH_4$ [69, 70] and also by catalytic reduction (copper and chromium oxide) [68] or Ru/C [71].

The dialdehydes are obtained from diesters [6] and $NaAlH_2(OCH_2CH_2OCH_3)_2$ which leads to the dihydrate. This latter is dehydrated to yield the dialdehyde. However, because of their reactivity, Chemical Companies use a spacer between the fluorinated chain and the functional group [4].

$$IC_2H_4-(CF_2)_n-C_2H_4I \longrightarrow HOC_2H_4-(CF_2)_n-C_2H_4OH$$

$$\textbf{2}$$

$$\textbf{2} + CrO_3 \longrightarrow HO_2C-CH_2-(CF_2)_n-CH_2-CO_2H \longrightarrow \text{diisocyanate}.$$

In previous research [32], we chose two methods for introducing alcohols from diesters [72], described as follows:

$$MeO_2C-(CF_2CFCl)_n-CF_2CO_2Me$$

$$\xrightarrow{HOC_2H_4OH} HOC_2H_4-\overset{\overset{\displaystyle O}{\|}}{C}-(CF_2CFCl)_n-CF_2-\overset{\overset{\displaystyle O}{\|}}{C}-OC_2H_4OH$$

$$\xrightarrow{H_2NC_2H_4OH} HOC_2H_4NH-\overset{\overset{\displaystyle O}{\|}}{C}-(CF_2CFCl)_n-CF_2-\overset{\overset{\displaystyle O}{\|}}{C}-NHC_2H_4OH.$$

2.4 Synthesis of Fluorinated Telechelic Diisocyanates

The synthesis of fluorinated diisocyanates are described in references [3, 73]. Perfluorotrimethylene diisocyanate was prepared by the Curtius degradation of perfluoroglutaryl diazide:

$$Cl-\overset{\overset{\displaystyle O}{\|}}{C}-(CF_2)_3-\overset{\overset{\displaystyle O}{\|}}{C}-Cl \xrightarrow{NaN_3} N_3-\overset{\overset{\displaystyle O}{\|}}{C}-(CF_2)_3-\overset{\overset{\displaystyle O}{\|}}{C}-N_3 \xrightarrow{\Delta}$$

$$\xrightarrow{\Delta} OCN-(CF_2)_3-NCO.$$

In aromatic series, the same pattern is also used:

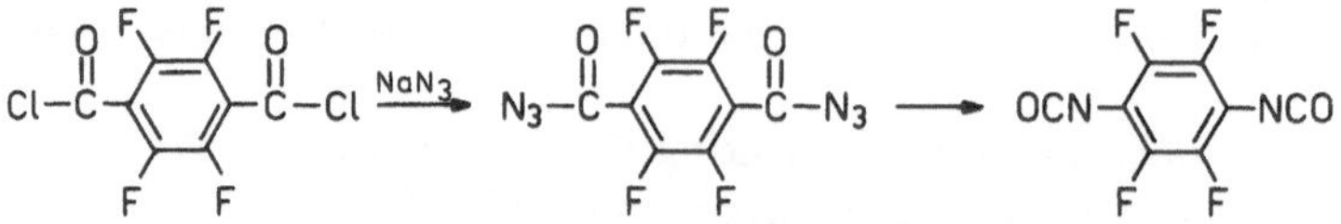

This latter telechelic compound **3** can be prepared from the fluorinated aromatic diamine according to both the following schemes:

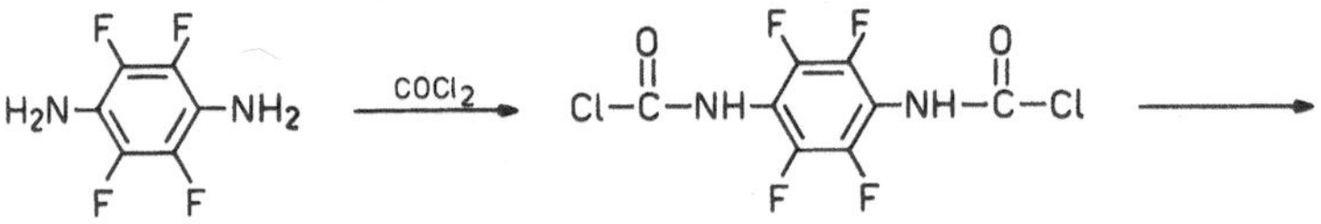

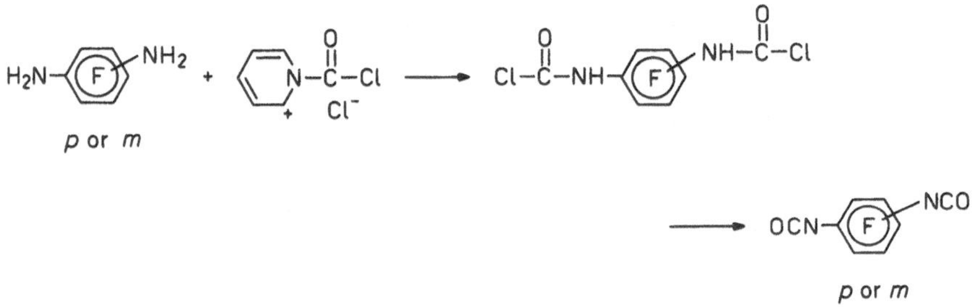

Furthermore a chlorofluorinated aromatic diisocyanate can be prepared from the same chlorocarbonyl pyridinium chloride:

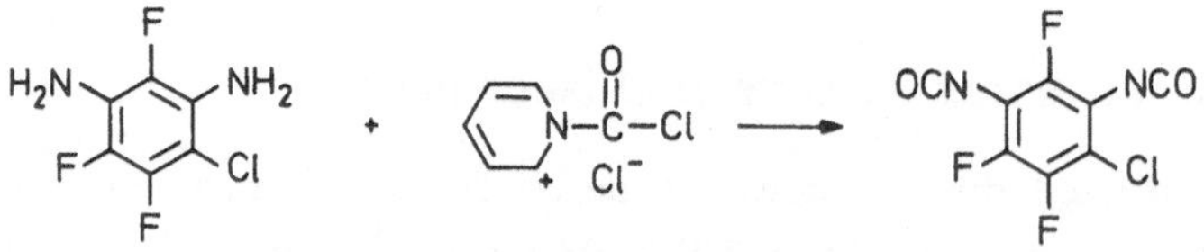

Such a scheme can also be performed on aliphatic compounds:

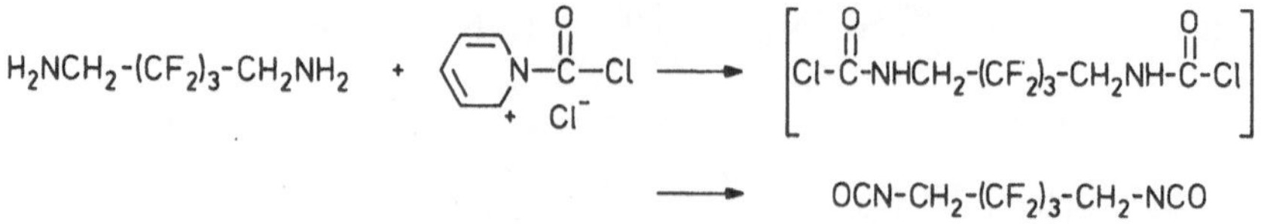

4,4′-Diisocyanatooctafluorobiphenyl was prepared by the chlorocarbonylation method and also by normal phosgenation:

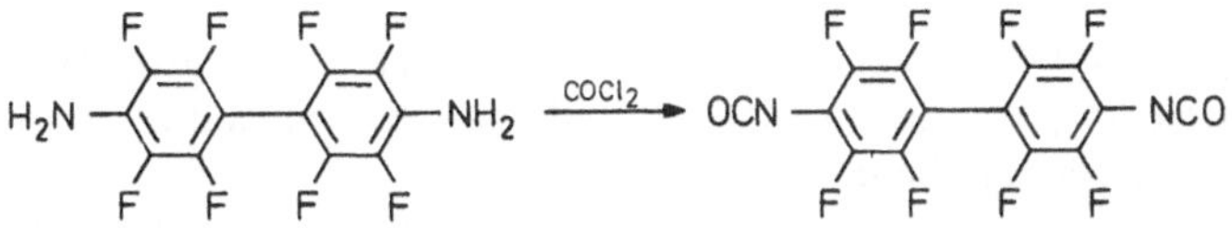

Such diisocyanates can lead to polyurethanes. The first fluorinated polyurethane was patented in 1958 [74]. An interesting survey [75] details the comparison of the reactivity of fluorinated diols about that of non halogenated ones for the preparation of such polymers. The first one which contained fluorine was synthesized by reaction of hexafluoropentanediol and hexamethylene diisocyanate [76]:

$$OCN-(CH_2)_6-NCO + HOCH_2-(CF_2)_3-CH_2OH \longrightarrow$$

$$\longrightarrow \left[\overset{\displaystyle O}{\overset{\|}{C}}-NH-(CH_2)_6-NH-\overset{\displaystyle O}{\overset{\|}{C}}-OCH_2-(CF_2)_3-CH_2O \right].$$

Then, polyurethanes based on aromatic diisocyanates were prepared. The same fluorinated diol was involved in the synthesis of such polymers with 3,5-tolylene diisocyanate and with a mixture of 2,4-tolylene diisocyanate and 3,3′-bitolylene-4,4′-diisocyanate [77].

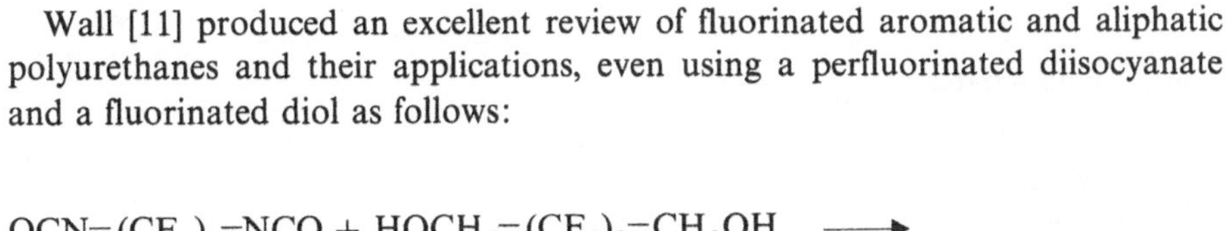

Wall [11] produced an excellent review of fluorinated aromatic and aliphatic polyurethanes and their applications, even using a perfluorinated diisocyanate and a fluorinated diol as follows:

$$OCN-(CF_2)_3-NCO + HOCH_2-(CF_2)_3-CH_2OH \longrightarrow$$

$$\longrightarrow \left[\overset{\displaystyle O}{\overset{\|}{C}}-NH-(CF_2)_3-NH-\overset{\displaystyle O}{\overset{\|}{C}}-OCH_2-(CF_2)_3-CH_2O \right]_n.$$

However, such a polyurethane is hydrolytically unstable [78].

Furthermore, Wall reviewed the fluorinated polyurethanes prepared from polyesters of fluorinated diols [79], from fluorinated polyethers [80] and detailed their properties [81].

Takakura et al. [82] investigated on the structure and the mechanical properties of fluorinated segmented poly(urethaneurea)s, showed that their structures were highly ordered and such polymers behaved as elastomers.

Yoon [83] studied by X-ray photoelectron spectroscopy the surface structure of segmented poly(ether urethane)s and poly(ether urethane urea)s with various perfluorinated chain extenders and noticed that the surface topography of such polymers depended strongly on the extent of phase separation.

2.5 Special Difunctional Molecules Obtained By Direct Synthesis

2.5.1 2,2-bis(4-Hydroxyphenyl)Hexafluoropropane

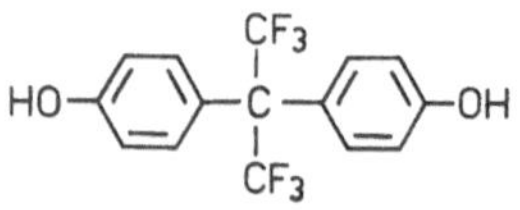

is one of the most important fluorinated diols and certainly the most readily available industrially.

It has been used for years to synthesize polyesters [84]:

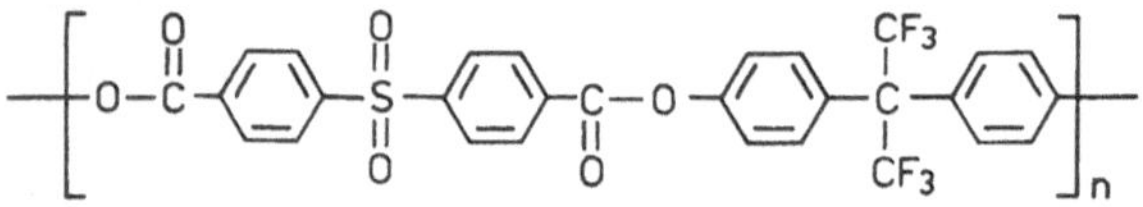

and also recently [85] to prepare:

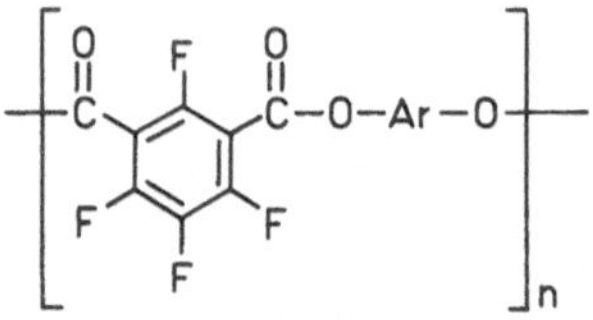

where Ar designates:

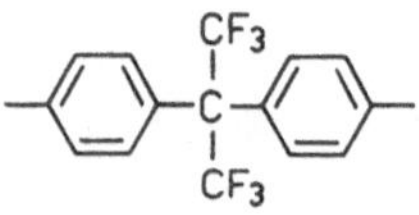

In both cases, the authors noticed these polyesters exhibited excellent thermal resistances (higher than 420 °C under air) and $T_g \geqq 150$ °C.

Furthermore, the $-C(CF_3)_2-$ group gives them exceptional properties of solubility (even in chloroform).

These diols are also used for the synthesis of polycarbonates. Morgan [86] prepared the following polymers:

$$HO-\!\!\!\bigcirc\!\!\!-\overset{\overset{\displaystyle CF_3}{|}}{\underset{\underset{\displaystyle CF_3}{|}}{C}}-\!\!\!\bigcirc\!\!\!-OH \;+\; 2\,COCl_2 \longrightarrow Cl-\overset{O}{\overset{||}{C}}-O-\!\!\!\bigcirc\!\!\!-\overset{\overset{\displaystyle CF_3}{|}}{\underset{\underset{\displaystyle CF_3}{|}}{C}}-\!\!\!\bigcirc\!\!\!-O-\overset{O}{\overset{||}{C}}-Cl$$

4

$$\mathbf{4} \xrightarrow{\;\underset{HN\underset{}{\frown}NH}{\overset{H_3C\quad}{}}\;} \left[CO_2-\!\!\!\bigcirc\!\!\!-\overset{\overset{\displaystyle CF_3}{|}}{\underset{\underset{\displaystyle CF_3}{|}}{C}}-\!\!\!\bigcirc\!\!\!-O-\overset{O}{\overset{||}{C}}-N\underset{}{\frown}N\right]$$

An interesting method performed by Saegusa et al. [87] should be mentioned:

$$HO-\!\!\!\bigcirc\!\!\!-\overset{\overset{\displaystyle CF_3}{|}}{\underset{\underset{\displaystyle CF_3}{|}}{C}}-\!\!\!\bigcirc\!\!\!-OH \;+\; Cl-\overset{O}{\overset{||}{C}}-OCCl_3 \xrightarrow[\underset{CH_2Cl_2/H_2O}{catalysis}]{Phase\ transfer} polycarbonate$$

As previously mentioned, these polymers exhibit a good solubility, a high T_g ($>150\,°C$) and a decomposition temperature higher than $400\,°C$.

Finally, aromatic diols are also forerunners of diepoxides [88–90]. In this latter patent, we can mention the twinned use of both a perfluorinated chained-monoepoxide with these diepoxides.

Numerous derivatives of such a hexafluorinated bisphenol have been prepared:

$$ClOC-\!\!\!\bigcirc\!\!\!-\overset{\overset{\displaystyle CF_3}{|}}{\underset{\underset{\displaystyle CF_3}{|}}{C}}-\!\!\!\bigcirc\!\!\!-COCl\;[91], \qquad H_2N-\!\!\!\bigcirc\!\!\!-\overset{\overset{\displaystyle CF_3}{|}}{\underset{\underset{\displaystyle CF_3}{|}}{C}}-\!\!\!\bigcirc\!\!\!-NH_2\;[92]$$

(both compounds can be used alone or in mixture for the synthesis of fluorinated polyamides) and:

$$H_2N-\!\!\!\bigcirc\!\!\!-O-\!\!\!\bigcirc\!\!\!-\overset{\overset{\displaystyle CF_3}{|}}{\underset{\underset{\displaystyle CF_3}{|}}{C}}-\!\!\!\bigcirc\!\!\!-O-\!\!\!\bigcirc\!\!\!-NH_2$$

which is attracting increasing interest.

First referred to in a publication in 1983, this latter monomer, has been involved in much research: first, American Aerospace prepared composites [93–96] e.g. Fusaro et al. prepared polyimides with pyromellitic anhydride ($Tg = 390\,°C$);

second, N.A.S.A. carried out investigations about thermostable transparent materials from such a diamine and the corresponding anhydride:

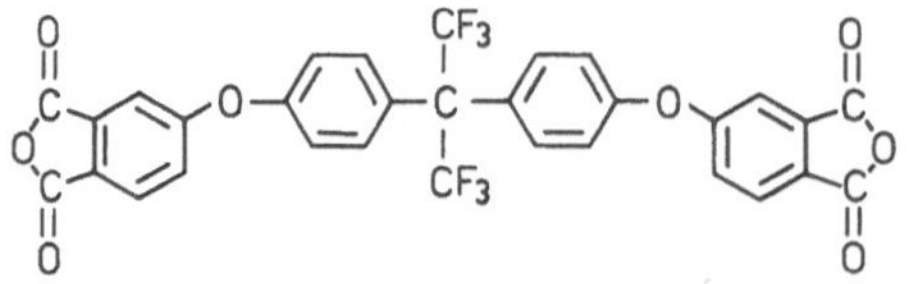

[86-87]

third, the American Company TRW Inc. [99–102] which was the first to obtain both monomers as follows:

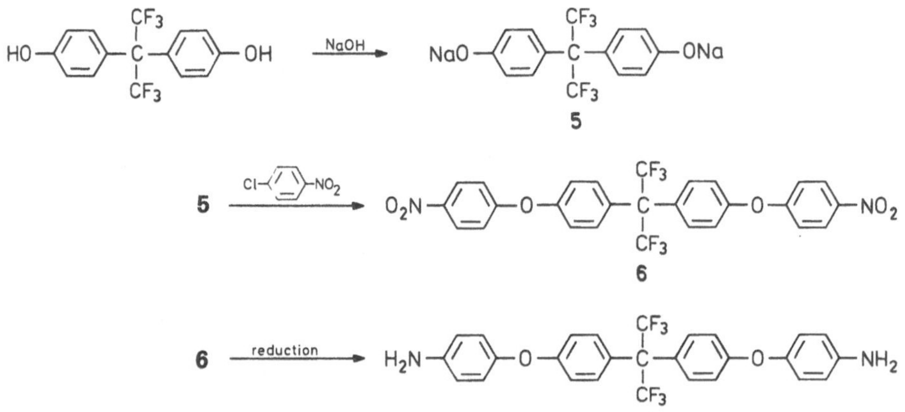

5

6

In addition, the Japanese Companies, the Hitachi Chemical Co and the Mitsui Toatsu Chemical Inc. have developed interesting applications with polyimides: semi-conductors, liquid crystals, poorly colorable thermostable films [103–107].

Finally, let us mention monomers obtained from diamines; first the synthesis of a diacetylenic compound [7]:

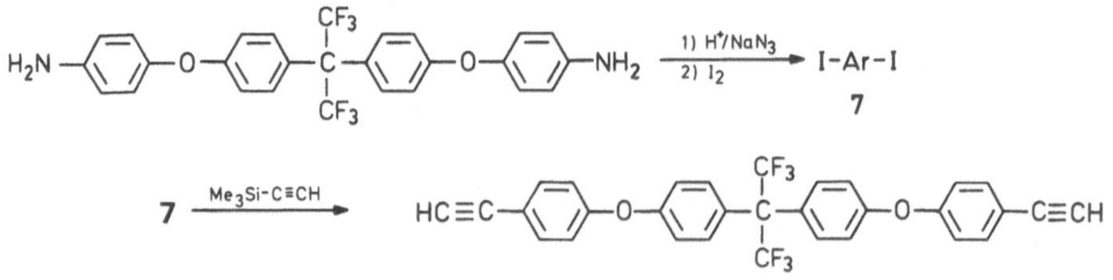

7

The trimerization at 300 °C leads to a polymer (T_g = 333 °C) which exhibits a decomposition temperature higher than 500 °C.

Similarly, Union Carbide prepares tetraglycidyl ether from such an amine [108].

2.5.2 1,3-bis(2-Hydroxyhexafluoro-2-propyl)Benzene

Other commonly used monomers have the formula:

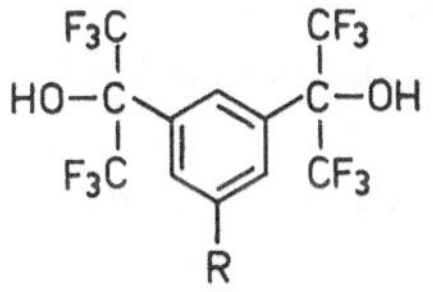

where R = H or a fluorinated group. They are prepared according to the following scheme [109]:

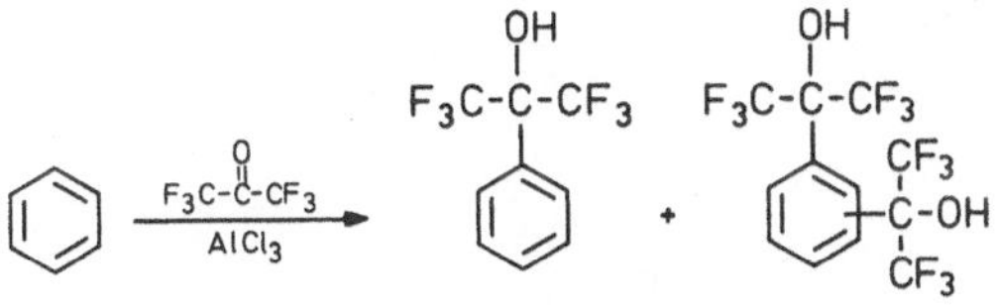

O'Rear and Griffith [110] developed such a product and its derivatives as diglycidyl ether of 1,3-bis(2-hydroxy-hexafluoro-2-propyl)-5-(perfluoro-n-alkyl)-benzene.

Introducing the fluorinated group in the 5-position represents the main difficulty [111].

Actually, two pathways are used from the *meta* isomer:

— the first one concerns the nitration, the reduction, the diazotation, the iodination, and finally the expected aromatic iodide is coupled with a iodoperfluoroalkane in a copper/DMSO system.

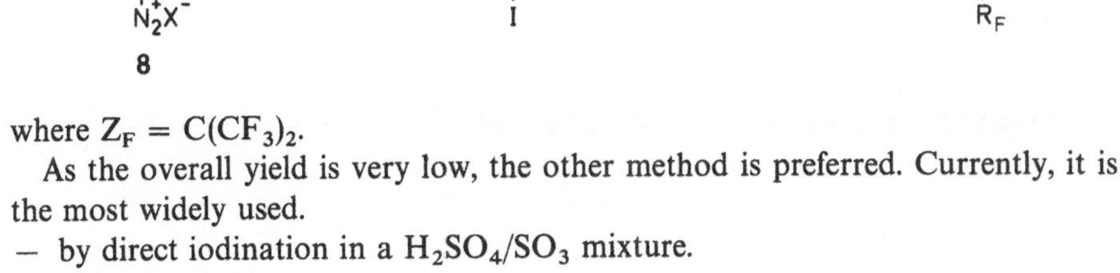

where $Z_F = C(CF_3)_2$.

As the overall yield is very low, the other method is preferred. Currently, it is the most widely used.

— by direct iodination in a H_2SO_4/SO_3 mixture.

Such a route leads to much better results especially with the improvement of the coupling reaction (78–91% yield) because of the complex between the DMSO and the diol [112].

Numerous polymers require such a precursor: for instance, by polycondensation with hexamethylene diisocyanate, Keller [113] developed interesting polyurethanes which are amorphous, insoluble and transparent when they are prepared at temperatures higher than 75 °C, whereas they are brittle when the reaction is performed at a temperature lower than 75 °C.

Griffith's investigations of fluorinated epoxy resins are many and it is difficult to summarize them. An interesting patent [89] concerns a large variety of diepoxies (below) where Ⓖ represents the $H_2C-CH-CH_2-O-$ group:

$$\underset{\diagdown O \diagup}{H_2C-CH-CH_2-O-}$$

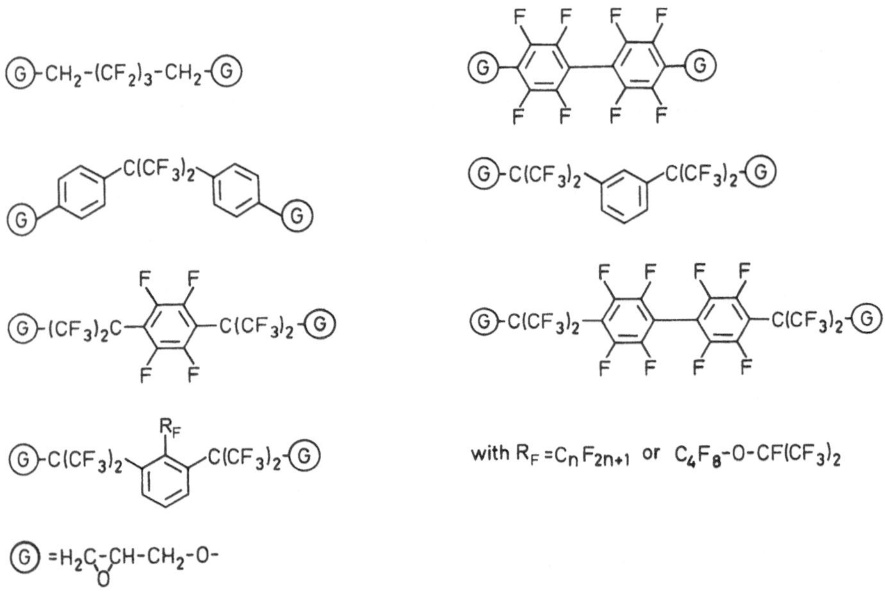

In the same way, anhydrides were prepared for the curing of epoxies:

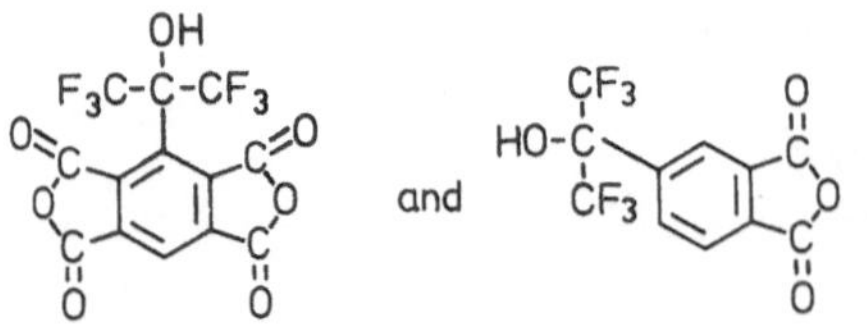

Griffith and O'Rear also obtained triepoxies [113]:

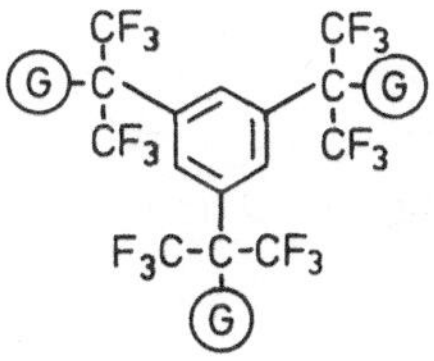

Other investigations performed by these authors [115–116] dealt with the acrylic derivatives of

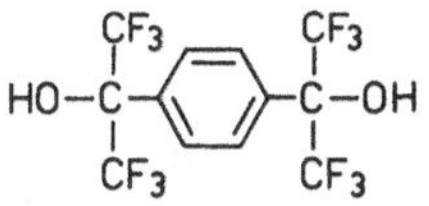

the diacrylates:

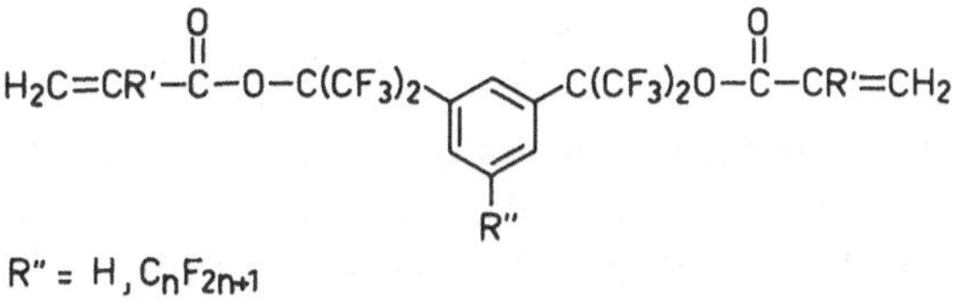

[115]

$R'' = H, C_nF_{2n+1}$

where $R'' = H$ or C_nF_{2n+1}
and the monoacrylates [116]:

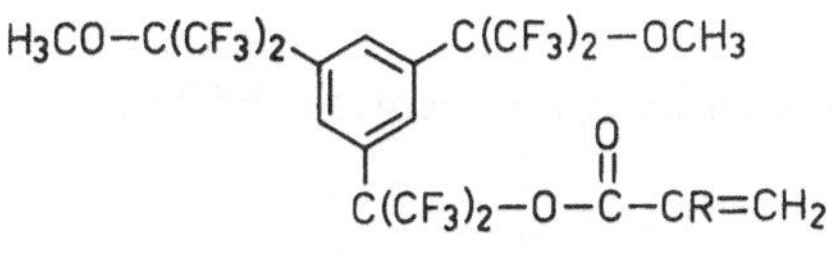

where $R = H$ or CH_3.

The advantage of these monoacrylates about the classical fluorinated acrylates [117] concerns the *gem* trifluoromethyl groups which are "water repellant umbrella".

Even if the diacrylates were easily obtained (simple by direct esterification of diol), the monoacrylates were prepared in several steps:

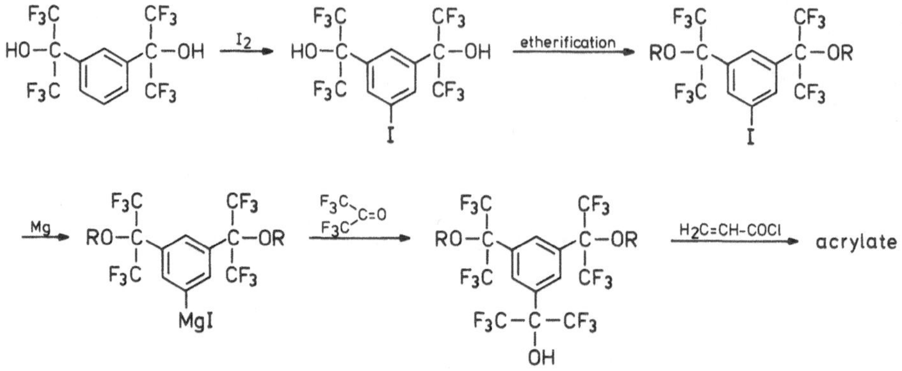

2.5.3 Diaromatic Difunctional Compounds Linked to Fluorinated Chains

Mc Loughlin et al. [118] carried out extensive research into the synthesis of such molecules according to the following reaction:

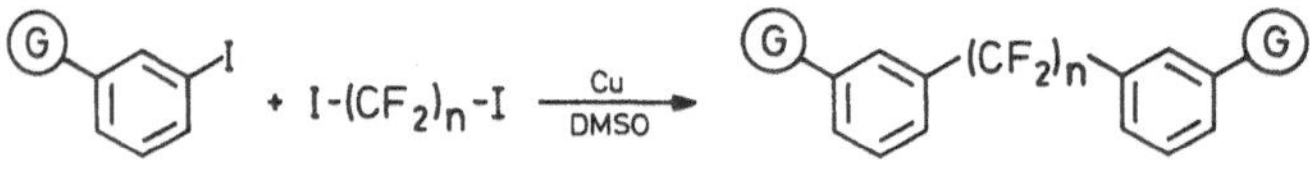

where Ⓖ represents a functional group such as hydroxy(phenol), carboxylate, nitro (precursor of amine), in para or meta positions about the fluorinated chain. This team also branched several functional groups onto each aromatic ring [120] such as tetracarboxylates:

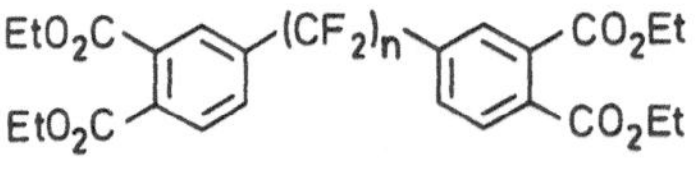

From these compounds, novel polymers were prepared: polyesters [121], silicones [53], polyimides [53], etc . . .

Webster et al. [122] obtained the same kind of molecules as mentioned below:

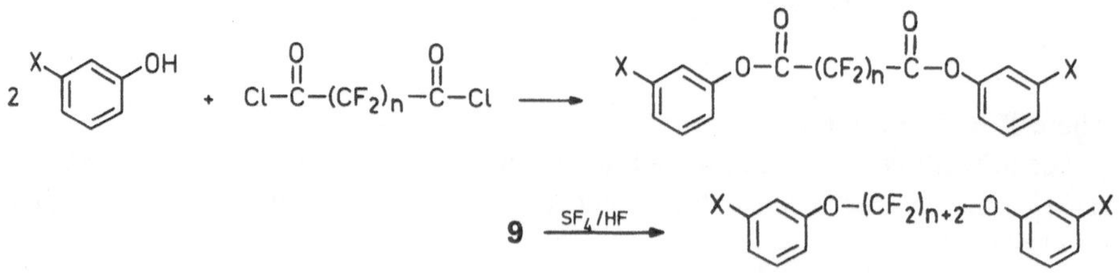

This team [123] performed the synthesis of diisocyanates by both methods to yield:

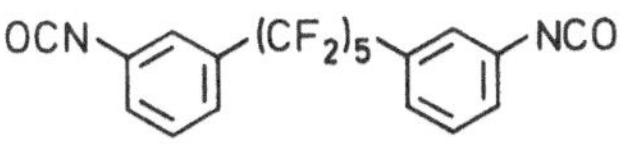

$$OCN - C_6H_4 - (CF_2)_5 - C_6H_4 - NCO \quad \text{and} \quad OCN - C_6H_4 - O-(CF_2)_5-O - C_6H_4 - NCO$$

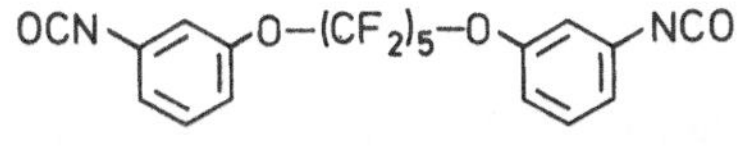

which produced polyisocyanurates. These latter ones are compared each other and exhibited identical thermal properties; however, Webster showed that the stability to the hydrolysis of the polymers which had the $C_6H_4-OCF_2$ group was better than that of those which contained the $C_6H_4-CF_2$ group.

Finally, this team prepared diamines:

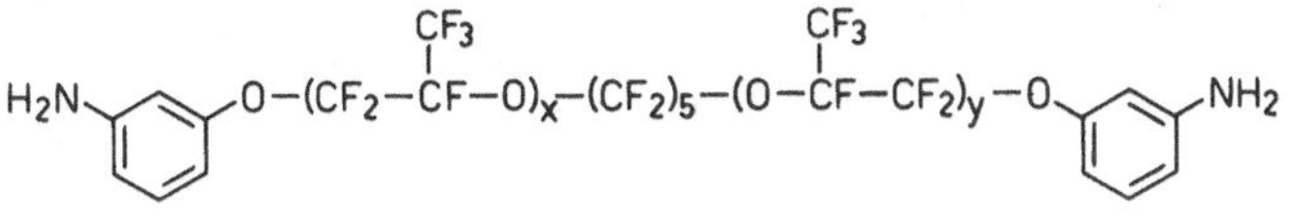

$$H_2N - C_6H_4 - O-(CF_2-\overset{CF_3}{\underset{|}{CF}}-O)_x-(CF_2)_5-(O-\overset{CF_3}{\underset{|}{CF}}-CF_2)_y-O - C_6H_4 - NH_2$$

where $x + y = 0, 1, 2$ and 3.

The monomers for bisbenzoxazoles can also be mentioned. Evers has been investigating this topic for several years [124]. The reaction is as follows:

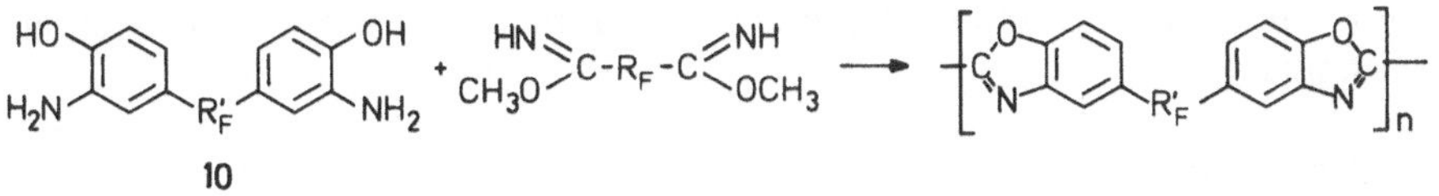

$$\underset{\textbf{10}}{HO-C_6H_3(NH_2)-R_F'-C_6H_3(NH_2)-OH} + \underset{}{CH_3O-\overset{HN}{\underset{}{C}}-R_F-\overset{NH}{\underset{}{C}}-OCH_3} \longrightarrow \left[\text{benzoxazole}-R_F'-\text{benzoxazole} \right]_n$$

and the synthesis of the forerunners is:

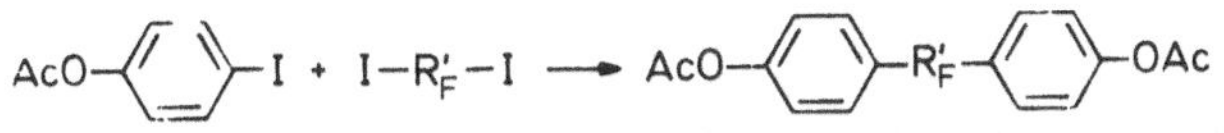

$$AcO-C_6H_4-I \; + \; I-R_F'-I \longrightarrow \underset{\textbf{11}}{AcO-C_6H_4-R_F'-C_6H_4-OAc}$$

$$\textbf{11} \xrightarrow[\text{HNO}_3]{\text{HOAc}} HO-C_6H_3(O_2N)-R_F'-C_6H_3(NO_2)-OH \xrightarrow[\text{2) NaHCO}_3]{\text{1) 10\% Pd/C, HCl}} \textbf{10}$$

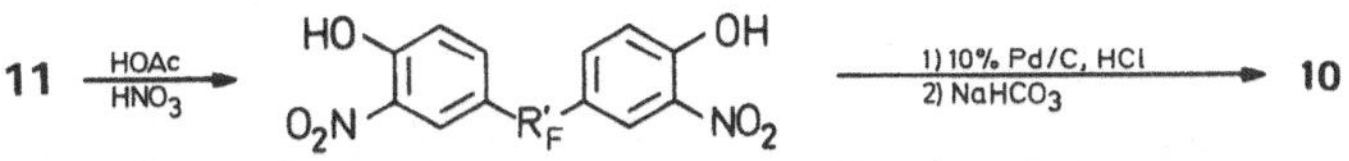

where R_F' represents $(CF_2)_n$, with $n = 3, 8$, or $(CF_2)_2-O-(CF_2)_5-O-(CF_2)_2$.

The dimethyl perfluorodiimidate was obtained by addition of methanol onto the corresponding dinitrile, under nitrogen, with a tertiary amine as catalyst.

Evers published numerous investigations on the outstanding thermal properties of these polybenzoxazoles.

Other interesting fluorinated aromatic telechelic compounds were prepared by Ohsaka et al. [125]:

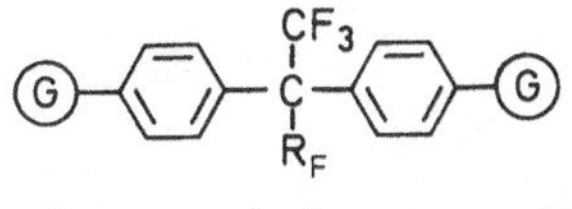

$$\textcircled{G}-C_6H_4-\overset{CF_3}{\underset{R_F}{\overset{|}{\underset{|}{C}}}}-C_6H_4-\textcircled{G}$$

where $\textcircled{G}$ designates $-OH$, $-CO_2H$ or $-NH_2$ and R_F represents the C_8F_{17},

C_8F_{17}, C_4F_9,   or $F-C\overset{CF_3}{\underset{OC_3F_7}{<}}$ groups.

For instance, the telechelic diols were prepared from phenol, a fluorinated ketone and HF according to the following reaction:

$$R_F C_2H_4-\overset{\overset{\displaystyle O}{\|}}{C}-CF_3 \; + \; \langle\text{phenol}\rangle-OH \; + \; HF \; \longrightarrow \; HO-\langle\text{ }\rangle-\overset{\overset{\displaystyle CF_3}{|}}{\underset{\underset{\displaystyle C_2H_4R_F}{|}}{C}}-\langle\text{ }\rangle-OH$$

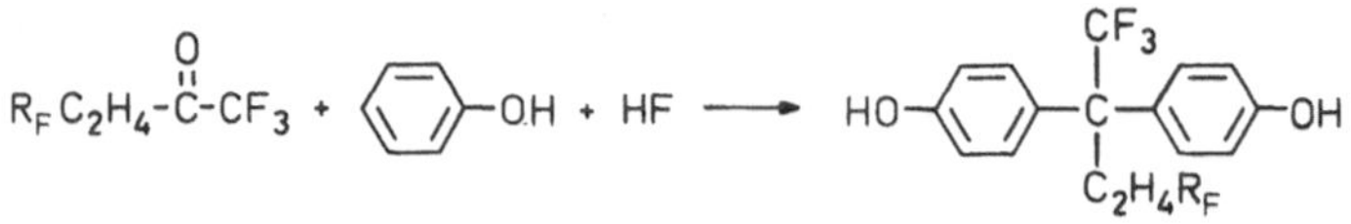

2.6 Other Special Difunctional Products

Tarrant et al. [125] obtained 30% yield of:

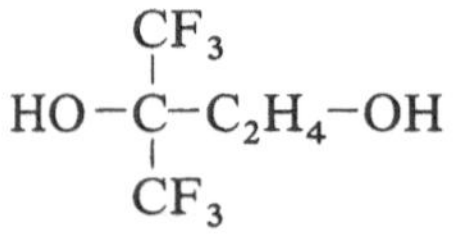

$$HO-\overset{\overset{\displaystyle CF_3}{|}}{\underset{\underset{\displaystyle CF_3}{|}}{C}}-C_2H_4-OH$$

from fluorinated oxetane.

Other difunctional compounds were prepared mainly from hexafluoroacetone; for example:

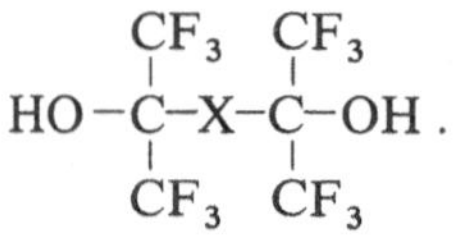

$$HO-\overset{\overset{\displaystyle CF_3}{|}}{\underset{\underset{\displaystyle CF_3}{|}}{C}}-X-\overset{\overset{\displaystyle CF_3}{|}}{\underset{\underset{\displaystyle CF_3}{|}}{C}}-OH \; .$$

The first diol of the series is the perfluoropinacol which is obtained by the following reaction [127]:

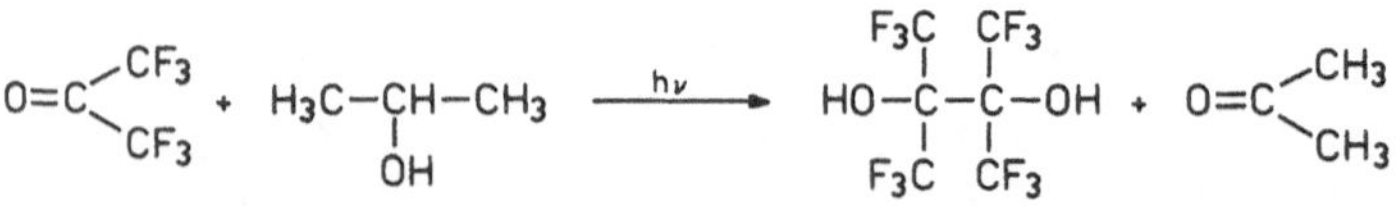

$$O=C\overset{\displaystyle CF_3}{\underset{\displaystyle CF_3}{\big\langle}} \; + \; H_3C-\underset{\underset{\displaystyle OH}{|}}{C}H-CH_3 \; \xrightarrow{h\nu} \; HO-\overset{\overset{\displaystyle F_3C}{|}}{\underset{\underset{\displaystyle F_3C}{|}}{C}}-\overset{\overset{\displaystyle CF_3}{|}}{\underset{\underset{\displaystyle CF_3}{|}}{C}}-OH \; + \; O=C\overset{\displaystyle CH_3}{\underset{\displaystyle CH_3}{\big\langle}}$$

Other authors obtained such a diol by reductive dimerization using sodium [128].

The internal chain can be longer; thus Loeb et al. [129] developed the condensation of acetone and hexafluoroacetone to produce:

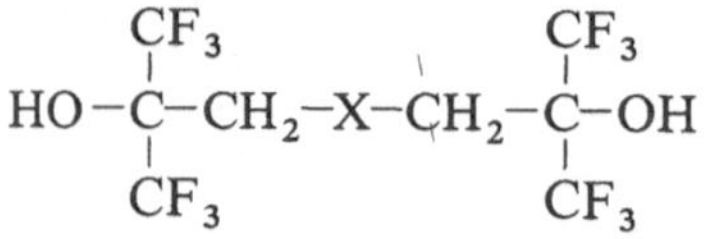

$$HO-\overset{\overset{\displaystyle CF_3}{|}}{\underset{\underset{\displaystyle CF_3}{|}}{C}}-CH_2-X-CH_2-\overset{\overset{\displaystyle CF_3}{|}}{\underset{\underset{\displaystyle CF_3}{|}}{C}}-OH$$

where X designates CO or CHOH. Such a compound is used as coordination-agent called "tridental" of metals such as Ni^{2+} or Cu^{2+}.

The products for which X is an alkyl group [130, 131] can be prepared from the reaction of ligands onto non conjugated dienes (butadiene derivatives).

However, X can be an ether group, as shown by Tarrant [126] who obtained:

$$\underset{\overset{|}{CF_3}}{\overset{\overset{|}{CF_3}}{HO-C-C_2H_4-O-C_2H_4-C-OH}}$$

from a fluorinated oxetane.

Another example of diol which exhibits the hexafluoropropanol group was given by Langkammerer [132].

In conclusion hexafluoroacetone is a key reactant for the preparation of fluorinated diols and was involved in many investigations. The other interesting method concerns the oligomerization of fluorinated monomers; the end-groups of such obtained telomers are changed by different pathways to produce difunctional compounds.

3 Difunctional Products With a Fluorinated Lateral Chain

The monofunctional fluorinated compounds can be obtained more easily than their difunctional homologues and this explains many articles published on these products.

First, we mention the main methods of synthesis of these fluorinated monofunctional chain.

3.1 Nature of the Fluorinated Chains

The fluorinated chains can be linear or branched according to their mode of synthesis. They may be perfluorinated or not, i.e. they contain hydrogen or chlorine atoms, either on the end-group or in a regular way in the chain, or at random.

3.1.1 Linear Perfluorinated Chains

The only pathway for obtaining a perfectly linear chain as Lichtenberger [133] showed, concerns the telomerization of tetrafluoroethylene with perfluoroalkyl iodides such as C_2F_5I:

$$C_2F_5I + C_2F_4 \xrightarrow{\text{catalyst}} C_nF_{2n+1}-I.$$

A mixture of telomers which contain 4 to 20 carbon atoms with a high majority of C_6 and C_8 is obtained and exhibits a very narrow molecular distribution. These methods of oligomerization were mainly reviewed by Allied, Daikin, I.C.I., Hoechst, Dupont and Atochem.

3.1.2 Branched Perfluorinated Chains

The most commonly used pathway is an electrochemical fluorination as shown by Simons [134]. From an acid chloride or a hydrocarbonated acid sulfochloride, the corresponding acid fluorides were obtained in 10% yield. During the reaction, the cleavages of the chain led to $\dot{C}F_3$ radicals which create short branching. The 3M Company developed this process.

The ionic oligomerization of C_2F_4 allows us to obtain branched chains like Fielding et al. [135] did according to the reaction:

$$C_2F_4 + CsF \longrightarrow C_nF_{2n}$$

The most abundant oligomer is the pentamer.

These perfluorinated olefines are not very reactive [136, 137] since the double bond is internal in the chain. However, the ICI Company [135] functionalizes these alkenes by adding phenol onto the perfluorinated olefines:

$$C_{10}F_{20} + C_6H_5OH \xrightarrow[-HF]{} C_{10}F_{19}OC_6H_5$$

and further reactions can be carried out on the aromatic ring.

Finally, branched compounds are prepared from perfluoroacetone.

Thus, Hollander et al. [138] obtained hexafluoroisopropanol and Boerner et al. [139] synthesized a novel derivative according to the chemical scheme:

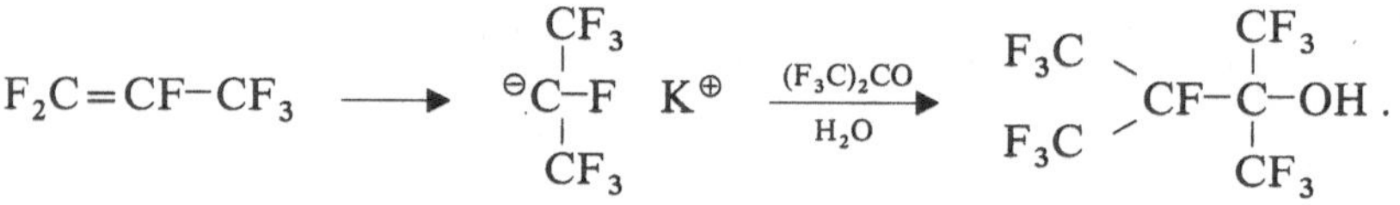

$$F_2C=CF-CF_3 \longrightarrow \overset{\overset{\displaystyle CF_3}{|}}{\underset{\underset{\displaystyle CF_3}{|}}{\ominus C-F}} K^{\oplus} \xrightarrow[H_2O]{(F_3C)_2CO} \begin{matrix} F_3C \\ F_3C \end{matrix} CF-\overset{\overset{\displaystyle CF_3}{|}}{\underset{\underset{\displaystyle CF_3}{|}}{C}}-OH .$$

3.1.3 Chlorofluorinated Chains

As we mentioned previously (Sect. 2.2.3), the telomerization producing oligomers and chlorotrifluoroethylene is mainly used.

The introduction of chlorine atoms can be performed either from chlorofluoro-acetone [140]:

$$\begin{matrix} ClF_2C \\ ClF_2C \end{matrix} C=O \longrightarrow \begin{matrix} ClF_2C \\ ClF_2C \end{matrix} CH-OH$$

or by U.V. irradiation, developed by Brace [141]:

$$H-C_4F_8-CH_2OH \xrightarrow[70\,°C]{Cl_2/U.V.} Cl-C_4F_8-CH_2OH$$

or from a chlorofluoro monomer [142] involving a monosodium derivative of malonic ester:

$$F_3C-CCl=CClCF_3 \longrightarrow F_3C-CCl=CCF_3-CH(CO_2H)_2$$

by hydrolysis of 4-(2-chloro-1,1,2-trifluoroethyl)-2,2-dimethyl-1,3-dioxolane [143]:

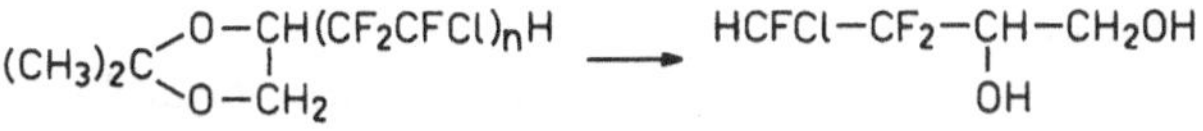

$$(CH_3)_2C \overset{O-CH(CF_2CFCl)_nH}{\underset{O-CH_2}{\big\langle}} \quad \longrightarrow \quad HCFCl-CF_2-\underset{\underset{OH}{|}}{CH}-CH_2OH$$

3.1.4 Hydrogenofluorinated Chains

The telomerization of fluorinated olefines with alcohols as telogens leads to ω-hydroperfluoro alcohols:

$$nF_2C=CFX + R-CH_2OH \quad \longrightarrow \quad H-(CFX-CF_2)_n-\overset{\overset{R}{|}}{CH}-OH$$

with X = F, Cl, CF_3 and R = H or alkyl.
    The reactivity series of the alcohols is the following [144]:

$$CH_3OH > C_2H_5OH > (CH_3)_2CHOH.$$

Peroxides [145] or γ irradiation [146] can be used.
    There are numerous varieties of monofunctional fluorinated compounds which exhibit a wide range of linear or branched, hydrogenated or chlorinated structures.
    From these compounds, many investigations were attempted to perform the synthesis of difunctional products.

3.2 Synthesis of Diols from Diamines

These diols were prepared in two steps.
    The first one was based on the synthesis of a diamine and the second one dealt with the introduction of hydroxy functions [2]:

$$R_F-C_2H_4I + HN-(CH_2)_x-NH \quad \longrightarrow \quad R_F-C_2H_4-N-(CH_2)_x-NH.$$
$$\qquad\qquad\qquad R \qquad\quad R' \qquad\qquad\qquad\qquad R \qquad\quad R'$$

For R = H, the synthesis can be continued.
    Thus, with propiolactone and with propanesultone, Bloechl [2,147] obtained:

$$R_F-C_2H_4-N-(CH_2)_x-N-R' \quad \text{and} \quad R_F-C_2H_4-N-(CH_2)_x-N-R' \quad \text{respectively}$$
$$\qquad\quad C=O \qquad\quad C=O \qquad\qquad\qquad\quad SO_2 \qquad\quad SO_2$$
$$\qquad\quad C_2H_4OH \quad C_2H_4OH \qquad\qquad\quad C_3H_6OH \quad C_3H_6OH$$

In the same way triazines derivatives were prepared.
    These authors observed the dehydrofluorination of the compounds during reactions with amines:

$$R'_F-CF=CH-CH_2-N-(CH_2)_x-NH.$$
$$\qquad\qquad\qquad\qquad R \qquad\quad R'$$

They also prepared polyurethanes with the diols, and polyamides with the intermediate diamines. These polymers are used for treating textiles and papers and as expansion agents of plastics.

At the same time, the Ugine Kuhlmann Compagny [148] performed the addition of $R_F C_2 H_4 I$ onto functional amines such as $H_2 N C_2 H_4 OH$. The formation of $(R_F C_2 H_4)_2 N C_2 H_4 OH$ was observed as a by-product. Furthermore, the acrylation of the latter amino-alcohol led to a mixture of $R_F - C_2 H_4 - NH - C_2 H_4 - O - CO - CH = CH_2$ and $R_F - C_2 H_4 - N - CO - CH = CH_2$.

$$\underset{C_2 H_4 OH}{|}$$

Some time later, Foulletier et al. [149–150] extended this research to the obtaining of the following diols:

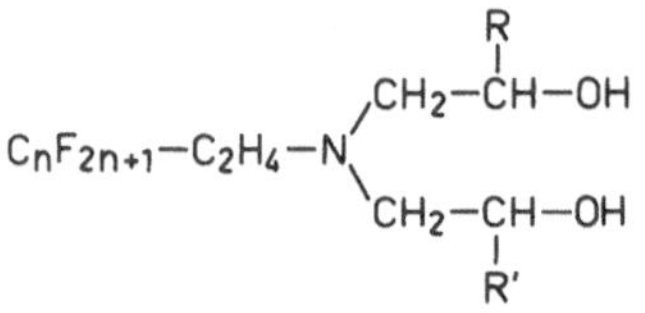

$$C_n F_{2n+1} - C_2 H_4 - N \underset{CH_2 - \underset{R'}{\overset{|}{CH}} - OH}{\overset{CH_2 - \overset{R}{\overset{|}{CH}} - OH}{\big<}} \qquad \text{with R} = CH_3 \text{ and R}' = H \text{ or } CH_3$$

This synthesis involved the latter secondary amino-alcohols reacting with epoxides such as:

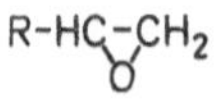

$$R - HC - CH_2 \atop O$$

However, we can see that the obtained diols are primary-secondary rather than primary-primary.

From sulfonamides and amides, varied diols were prepared.

— primary-primary diols [151, 152]

$$C_n F_{2n+1} - SO_2 - N(C_2 H_4 OH)_2$$

$$C_n F_{2n+1} - SO_2 - N - (C_2 H_4 O)_x - H \qquad \text{with } x + y = 3-30$$
$$\overset{|}{(C_2 H_4 O)_y - H}$$

$$C_n F_{2n+1} - CO - N - C_2 H_4 OH$$
$$\overset{|}{C_2 H_4 OH}$$

— primary-secondary diols [153]

$$C_n F_{2n+1} - SO_2 - N - CH_2 - CH - CH_2 OH$$
$$\overset{|}{CH_3} \qquad \overset{|}{OH}$$

— secondary-secondary diols [151]

$$C_n F_{2n+1} - SO_2 - N(CH_2 - CH - CH_3)_2$$
$$\overset{|}{OH}$$

In this case too, polyurethanes were produced in order to obtain water and oil repellant products, which have good abrasion resistance and very low surface energy.

3.3 Synthesis of Diols by Condensation

Several methods of condensation were used and each of them consisted of binding the perfluorinated chain to a molecule which exhibited two hydroxy groups or a forerunner of the diol (e.g. epoxide-alcohol or diepoxide).

3.3.1 Condensation of Fluorinated Iodides with Thiol Diols [154–156]

$$R_F-C_2H_4-I + HS-CH_2-\underset{\underset{OH}{|}}{CH}-CH_2OH \xrightarrow[\text{EtOH}]{\text{MeOH}}$$

$$\xrightarrow[\text{EtOH}]{\text{MeOH}} R_F-C_2H_4-S-CH_2-\underset{\underset{OH}{|}}{CH}-CH_2OH .$$

3.3.2 Condensation of Thiol Halogenide or Thiol Alcohol Systems

In the latter system, Bloechl [147] carried out the condensation of a thiol with glycidol and obtained, after basic hydrolysis, the fluorinated diol:

$$R_F-C_2H_4-S-CH_2-\underset{\underset{OH}{|}}{CH}-CH_2OH .$$

The author oxidized this product with hydrogen peroxide.

Such a reaction is used by several Companies, e.g. Pennwalt [157], whatever the glycidol or epichlorhydrin [158].

3.3.3 Condensation of Fluorinated Alcohols with Glycidol

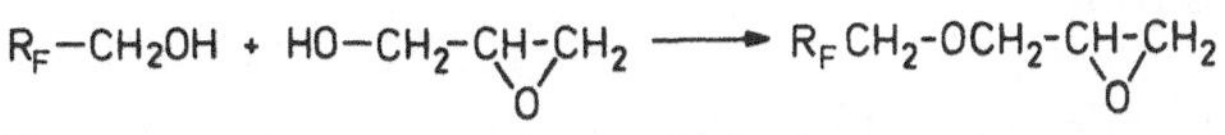

$$R_F-CH_2OH + HO-CH_2-CH-CH_2 \longrightarrow R_F CH_2-OCH_2-CH-CH_2$$

with $R_F = C_nF_{2n+1}$ [159], $C_nF_{2n}H$ [160], $(CF_3)_3C$ [161–163].

In the last case, the synthesis of the alcohol is described by the following scheme:

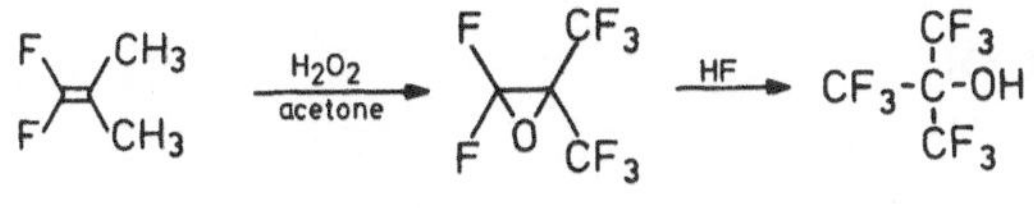

3.4 Addition of Fluorinated Telogens onto Difunctional Olefines

The literature shows that the most commonly used pathway is from fluorinated iodides or thiols.

3.4.1 With Fluorinated Iodides, the Reaction is the Following

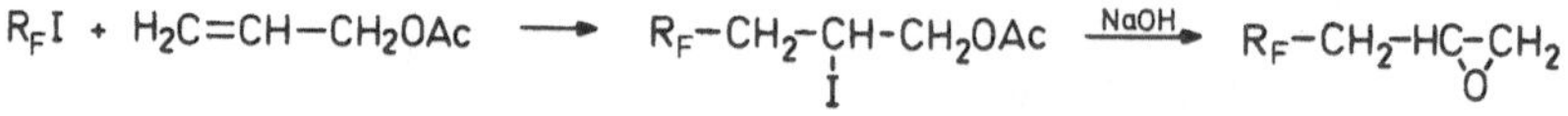

R_FI + $H_2C=CH-CH_2OAc$ ⟶ $R_F-CH_2-\underset{I}{CH}-CH_2OAc$ $\xrightarrow{NaOH}$ $R_F-CH_2-HC\underset{O}{-}CH_2$

Many types of fluorinated groups are possible:

$(CF_3)_2CF-OC_2F_4-$ [164], $C_nF_{2n+1}-$ [165–166],

$(CF_3)_2CF-(CF_2)_4-$ [167–168].

However, using Zn, Blancou [169] observed, in the first step, the formation of the following difunctional fluorinated compound:

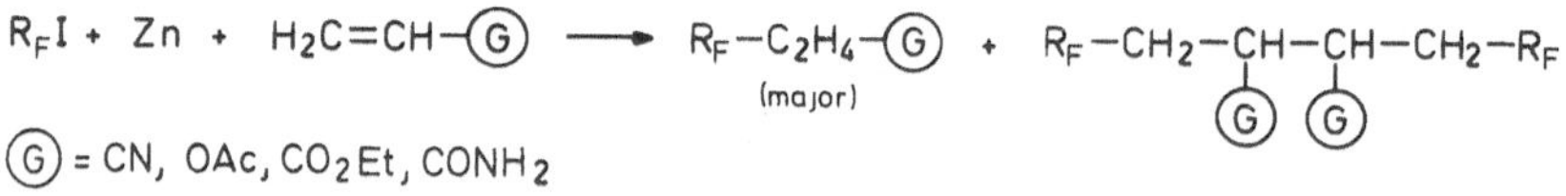

R_FI + Zn + $H_2C=CH-$Ⓖ ⟶ $R_F-C_2H_4-$Ⓖ + $R_F-CH_2-\underset{Ⓖ}{CH}-\underset{Ⓖ}{CH}-CH_2-R_F$
                                                        (major)

Ⓖ = CN, OAc, CO$_2$Et, CONH$_2$

Furthermore, different methods of initiation are interesting: with classical initiators [164, 167, 168], also with mercury or manganese [165] and by electrochemistry [166, 170].

3.4.2 Other Authors used Fluorinated Thiols as Telogens

Gresham [171] studied the following monoaddition:

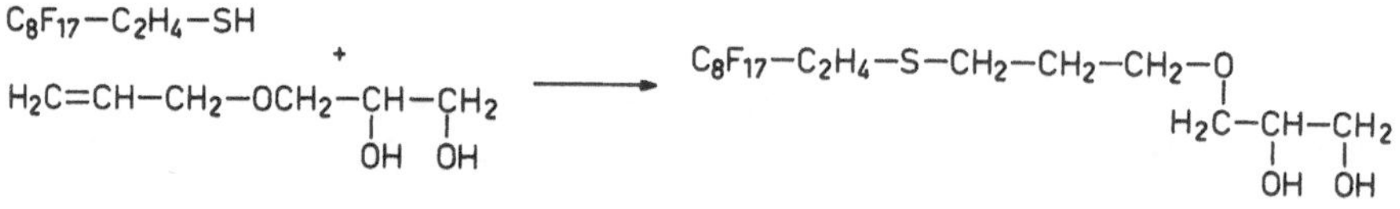

$C_8F_{17}-C_2H_4-SH$

+

$H_2C=CH-CH_2-OCH_2-\underset{OH}{CH}-\underset{OH}{CH_2}$ ⟶ $C_8F_{17}-C_2H_4-S-CH_2-CH_2-CH_2-O$
                                                                    $H_2C-CH-CH_2$
                                                                         $\underset{OH}{\quad}\underset{OH}{\quad}$

Other olefines can be involved such as [172]:

$H_2C=CH-CH_2-CH(CO_2Et)_2$.

Furthermore, Gresham [171] performed the addition of fluorinated thiols onto telechelic alkenes and obtained a primary-primary diol as follows:

$HOCH_2-CH=CH-CH_2OH$ + $R_FC_2H_4SH$ ⟶

⟶ $R_FC_2H_4S-\underset{CH_2OH}{CH}-CH_2-CH_2OH$.

Dear et al. [173] tried the telechelic alkyne $HOCH_2-C\equiv C-CH_2OH$ and prepared the disubstituted diol:

$$R_FC_2H_4-S-\underset{\underset{CH_2OH}{|}}{CH}-\!\!-\!\!-\underset{\underset{CH_2OH}{|}}{CH}-S-C_2H_4R_F \, .$$

3.4.3 With Epoxies as Monomers

Allyl glycidyl ether is used the most since this monomer is commercially available. However, the synthesis requires a further step: the chemical change of the oxirane into the diol.

Usually a strong acid is used [174]. Two original methods can be mentioned: the first one leads to a semiacetylated intermediate in the presence of triethylamine [175].

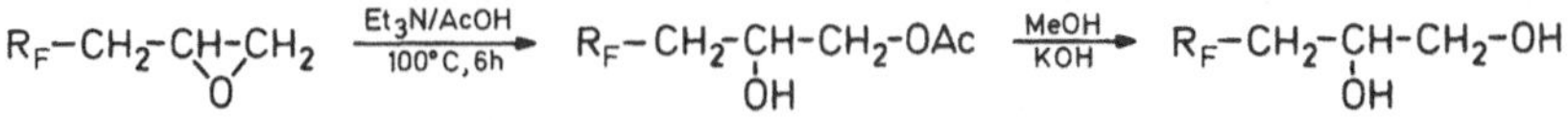

$$R_F-CH_2-\underset{O}{CH-CH_2} \xrightarrow[100^\circ C,\,6h]{Et_3N/AcOH} R_F-CH_2-\underset{\underset{OH}{|}}{CH}-CH_2-OAc \xrightarrow[KOH]{MeOH} R_F-CH_2-\underset{\underset{OH}{|}}{CH}-CH_2-OH$$

The second method [176] consists of hydrolizing the epoxide into the diol over a resin of sulfonated polystyrene. Cambon [177] performed the same reactions and showed that the longer the length of the perfluorinated chain, the greater the yields of the diols: 60, 64 and 75% for C_4F_9, C_6F_{13} and C_8F_{17}, respectively.

However, such an acidic hydrolysis can be carried out at rather high temperature (100–140 °C) for $R_F = CF_3$, C_2F_5 and C_3F_7 [178].

Furthermore, 2-fluoroalkyl ethane-1,2-diols:

$$R_F-\underset{\overset{|}{OH}}{CH}-CH_2-OH$$
12

were prepared either from the corresponding epoxides ($R_F = CF_3$ [179], C_3F_7 [180]) or from the corresponding monophosphates of perfluoroalkyl ethane diols ($R_F = CF_3(CF_2)_{p=3-1}$, $(CF_3)_2CF(CF_2)_{q=1-14}$, $(CF_3CF_2\underset{\overset{|}{CF_3}}{CF})_{r=1-4}$ and $(CF_3)_2CFCF_2\underset{\overset{|}{CF_3}}{CF})_{s=1-4}$ [181].

The perfluoroalkyl diols **12** exhibit good hydrophobic and oleophobic properties. For instance, they are used for treating porous (paper, wood, leather, fibers, textiles) or non porous materials (surfaces of glass, metals or synthetic products).

In addition, Ayari et al. [177] developed three methods by the oxidation of fluoroalkyl olefine with $KMnO_4$ and were able to improve significantly the yield of the synthesis of the fluorinated diols **12**.

3.5 Radical Telomerization of a Fluorinated Monomer With a Diol Thiol

In contrast to the case described in Sect. 3.4.2, the addition of functional thiols onto fluorinated alkenes is also possible:

$$C_8F_{17}-CH=CH_2 + HS-CH_2-\underset{OH}{CH}-\underset{OH}{CH_2} \longrightarrow$$

$$\longrightarrow C_8F_{17}-CH_2-CH_2-S-CH_2-\underset{OH}{CH}-\underset{OH}{CH_2} \quad [171].$$

3.6 Synthesis of Fluorinated Propane-1,3-Diol from Alkyl Malonates

The first investigations were performed by McBee et al. [182]:

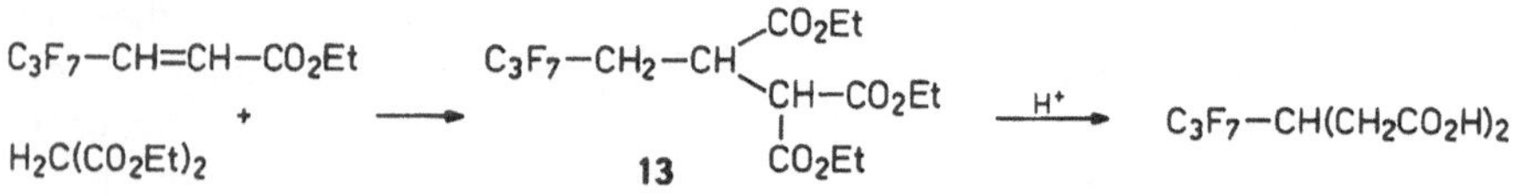

$$C_3F_7-CH=CH-CO_2Et$$

$$+$$

$$H_2C(CO_2Et)_2$$

$$\longrightarrow \quad C_3F_7-CH_2-CH\underset{CH-CO_2Et}{\overset{CO_2Et}{\diagup}} \quad \overset{H^+}{\longrightarrow} \quad C_3F_7-CH(CH_2CO_2H)_2$$

$$\mathbf{13} \quad CO_2Et$$

Then, Smeltz [183] for the Dupont Compagny, developed such a method as following:

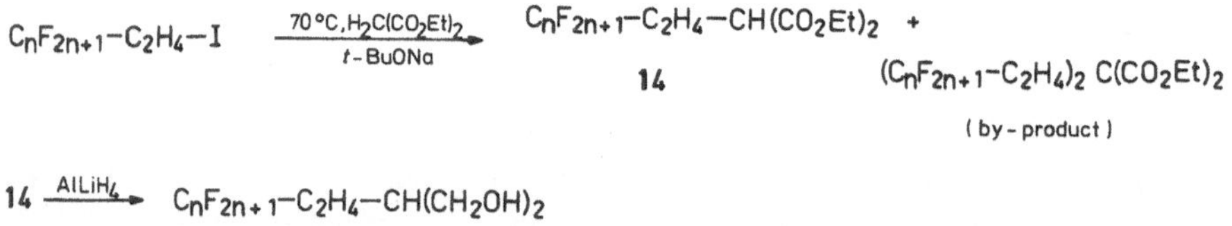

$$C_nF_{2n+1}-C_2H_4-I \quad \xrightarrow[t-BuONa]{70\,^\circ C, H_2C(CO_2Et)_2} \quad C_nF_{2n+1}-C_2H_4-CH(CO_2Et)_2 \;+$$

$$\mathbf{14} \qquad (C_nF_{2n+1}-C_2H_4)_2\,C(CO_2Et)_2$$

$$(\text{by-product})$$

$$\mathbf{14} \xrightarrow{AlLiH_4} C_nF_{2n+1}-C_2H_4-CH(CH_2OH)_2$$

Several polyesters were prepared from these diols [184] but also polycarbonates [185] and polyurethanes [186–187].

3.7 Ring Opening Polymerization of Fluorinated Oxetanes

This a very interesting way of obtaining a fluorinated diol and yet the synthesis is long [188]:

$$C(CH_2OH)_4 \xrightarrow{\text{SOCl}_2} C(CH_2Cl)_x(CH_2OH)_{4-x}$$
15

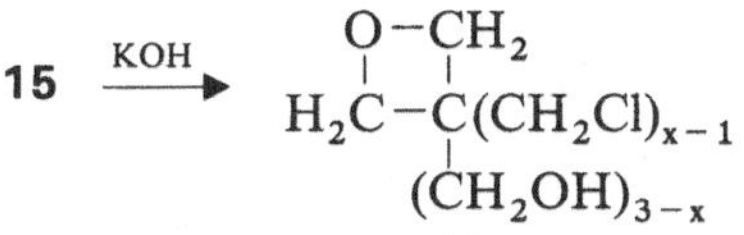

$$\textbf{15} \xrightarrow{\text{KOH}} \begin{array}{c} O-CH_2 \\ | \quad | \\ H_2C-C(CH_2Cl)_{x-1} \\ | \\ (CH_2OH)_{3-x} \end{array}$$
16

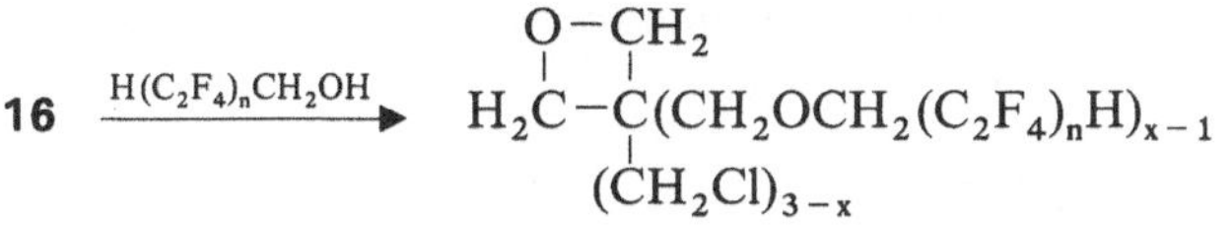

$$\textbf{16} \xrightarrow{\text{H(C}_2\text{F}_4)_n\text{CH}_2\text{OH}} \begin{array}{c} O-CH_2 \\ | \quad | \\ H_2C-C(CH_2OCH_2(C_2F_4)_nH)_{x-1} \\ | \\ (CH_2Cl)_{3-x} \end{array}$$
17

$$\textbf{17} \xrightarrow{\text{H}_2\text{SO}_4} \begin{array}{c} (CH_2OH)_{3-x} \\ | \\ HOCH_2-C-(CH_2OCH_2(C_2F_4)_nH)_{x-1} \\ | \\ CH_2OH \end{array}$$

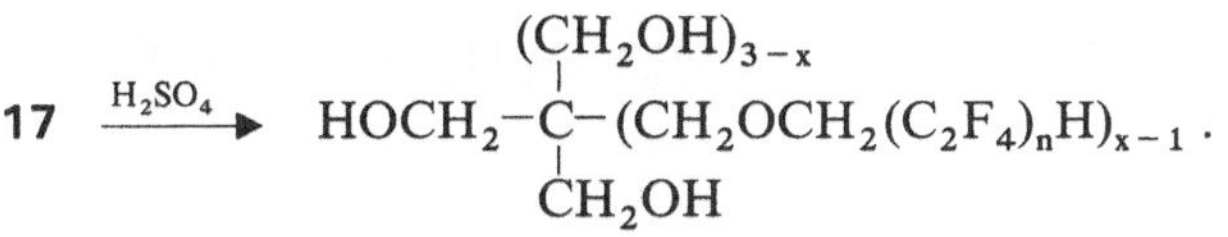

Fluorinated oxetanes, oxolanes and oxanes were prepared by several authors [189–194].

Harris and Coffman [189] obtained fluorinated oxetanes from aldehydes, ketones or acid fluorides according to the reaction:

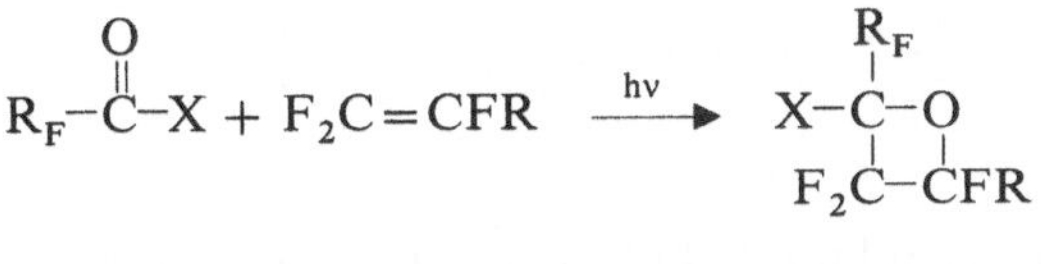

$$\begin{array}{c} O \\ \| \\ R_F-C-X \end{array} + F_2C=CFR \xrightarrow{h\nu} \begin{array}{c} R_F \\ | \\ X-C-O \\ | \quad | \\ F_2C-CFR \end{array}$$

where X = H, F or R_F, R = R_F or Cl.

Cook described the cycloaddition of hexafluoroacetone to ethylene, vinyl fluoride and vinylidene fluoride [190] and he noticed the formation of isomeric oxetanes:

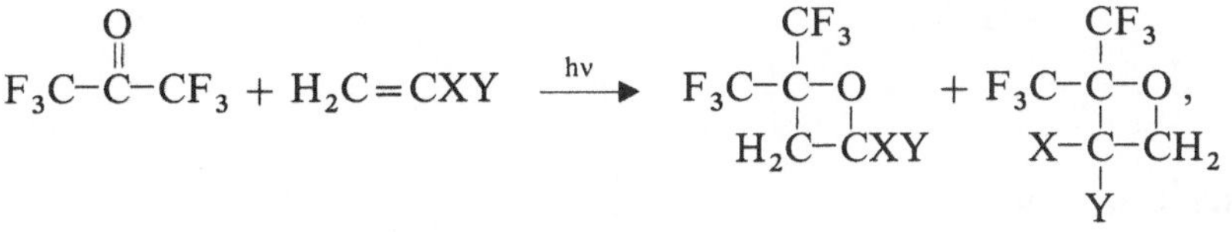

$$\begin{array}{c} O \\ \| \\ F_3C-C-CF_3 \end{array} + H_2C=CXY \xrightarrow{h\nu} \begin{array}{c} CF_3 \\ | \\ F_3C-C-O \\ | \quad | \\ H_2C-CXY \end{array} + \begin{array}{c} CF_3 \\ | \\ F_3C-C-O \\ | \quad | \\ X-C-CH_2 \\ | \\ Y \end{array},$$

X = H, F; Y = H, F

The polyfluorooxetanes were reported to exhibit exceptional thermal stability [189].

Numerous patents were published and the most pertinent one was applied for in 1984 [195]. In this patent, the authors studied the polymerization of $CF_2CF_2CH_2O$ and its copolymerization with:

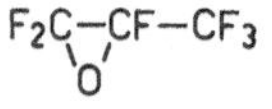

$$F_2C-CF-CF_3$$
$$\underset{O}{\diagdown\diagup}$$

in order to obtain block copolymers according to a catalytic process.

The following examples of the synthesis of monofunctional and difunctional polymers are significant:

a) CsF + $C_3F_7O-\underset{\underset{CF_3}{|}}{CF}-C\overset{O}{\underset{F}{\diagdown\!\!\!\diagup}}$ $\longrightarrow$ $C_3F_7O-C_3F_6O^- Cs^+$

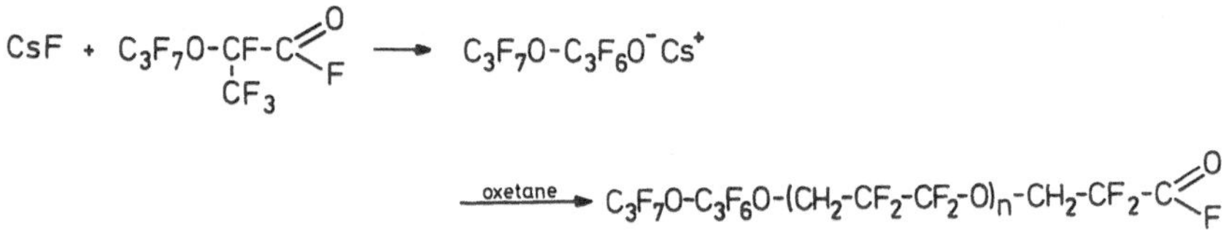

$\xrightarrow{\text{oxetane}}$ $C_3F_7O-C_3F_6O-(CH_2-CF_2-CF_2-O)_n-CH_2-CF_2-C\overset{O}{\underset{F}{\diagdown\!\!\!\diagup}}$

b) $\underset{F}{\overset{O}{\diagdown}}C-C\overset{O}{\underset{F}{\diagdown\!\!\!\diagup}}$ + CsF $\longrightarrow$ $Cs^+\ {}^-O-CF_2-CF_2-O^- Cs^+$

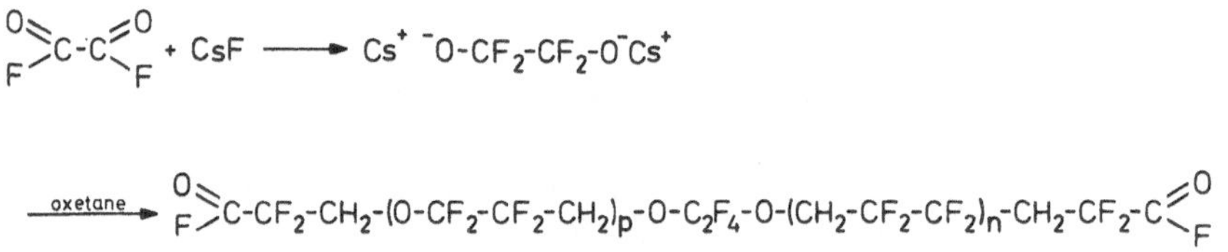

$\xrightarrow{\text{oxetane}}$ $\underset{F}{\overset{O}{\diagdown}}C-CF_2-CH_2-(O-CF_2-CF_2-CH_2)_p-O-C_2F_4-O-(CH_2-CF_2-CF_2)_n-CH_2-CF_2-C\overset{O}{\underset{F}{\diagdown\!\!\!\diagup}}$

c) KI + $oxetane$ $\xrightarrow{\text{MeOH}}$ $I-(CH_2-CF_2-CF_2-O)_n-CH_2-C\overset{O}{\underset{OCH_3}{\diagdown\!\!\!\diagup}}$

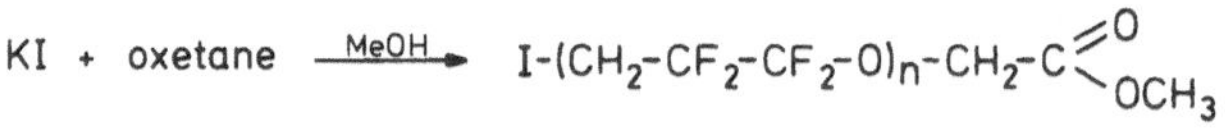

Furthermore, these authors performed fluorinations and even chlorinations to increase the thermal stabilities of these products and they mentioned the block copolymers:

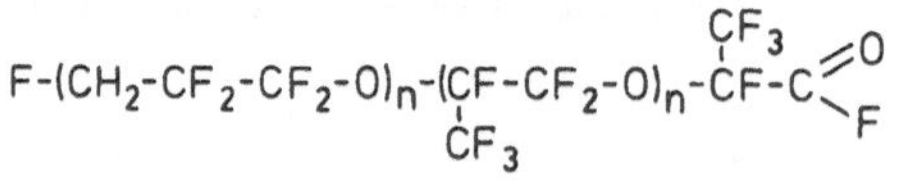

$$F-(CH_2-CF_2-CF_2-O)_n-(\underset{\underset{CF_3}{|}}{\overset{\overset{CF_3}{|}}{CF}}-CF_2-O)_n-CF-C\overset{O}{\underset{F}{\diagdown\!\!\!\diagup}}$$

3.8 Miscellaneous

We previously mentioned that Webster [122] changed an ester group into the OCF_2 group thanks to sulfur tetrafluoride and so did Persico with polyesters [196]. Actually, the use of such a fluorinating agent was pioneered by the Du Pont Company [197, 198] more than thirty years ago.

From this reactant Sheppard [199] obtained α-fluorinated ethers whereas Hudlicky [200] achieved the synthesis of fluorinated diols:

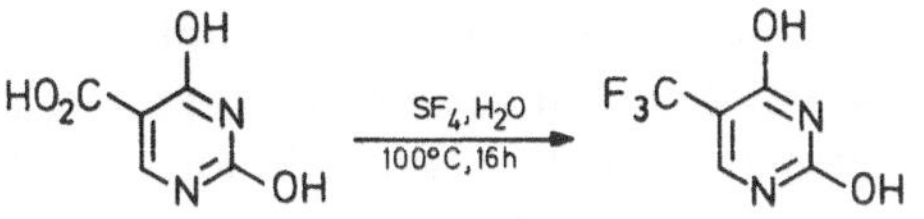

Furthermore Dmowski [192] prepared telechelic diols as follows:

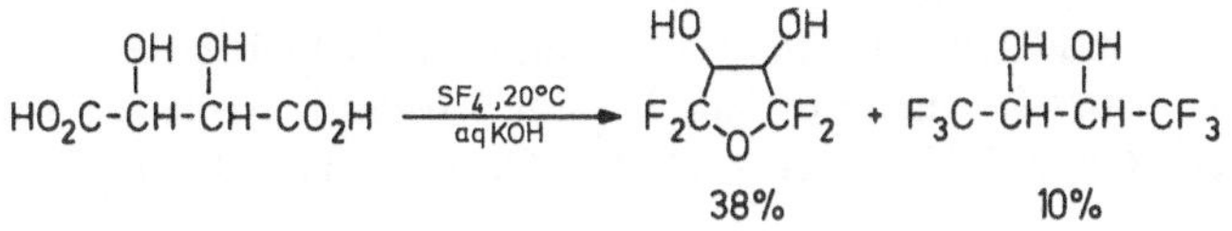

Moreover, the fluorinated *gem*-diols can be prepared by reduction of the corresponding perfluorinated acids in spite of the formation of the monoalcohol. This was illustrated by Sokoya's investigations [201]:

$$\mathrm{n\,C_7F_{15}CO_2H \xrightarrow[Et_2O]{AlLiH_4} n\,C_7F_{15}CH_2OH + n\,C_7F_{15}CH(OH)_2\,.}$$

Also, the reaction between a perfluoroiodide and acetone in presence of calcium led to 30% fluorinated diol after acidic hydrolysis [202]:

$$\mathrm{H_3C{-}\overset{\overset{\displaystyle O}{\|}}{C}{-}CH_3 + Ca + C_6F_{13}I \xrightarrow[2)\ H^+/H_2O]{1)\ THF}}$$

$$\mathrm{\xrightarrow[2)\ H^+/H_2O]{1)\ THF} (CH_3)_2\underset{\underset{\displaystyle OH}{|}}{C}{-}C_6F_{13} + C_6F_{13}{-}\underset{\underset{\displaystyle OH}{|}}{\overset{\overset{\displaystyle CH_3}{|}}{C}}{-}\underset{\underset{\displaystyle OH}{|}}{\overset{\overset{\displaystyle CH_3}{|}}{C}}{-}CH_3\,.}$$

3.9 Synthesis of Fluorinated Aromatic Diisocyanates

There are very important investigations into the synthesis of polyurethanes done by Malichenko et al. [75]. These authors carried out the synthesis of numerous diisocyanates:

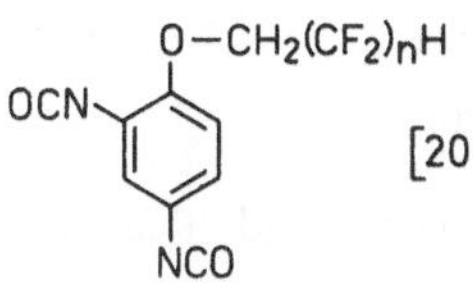

[203]

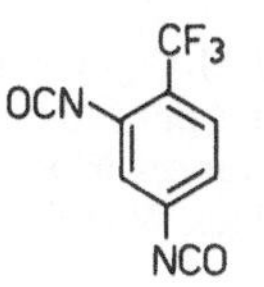

[204]

4 Conclusion

This bibliographical survey shows the wide range of syntheses of the fluorinated difunctional compounds, whatever the position of the fluorinated substituent in the molecule. However, major efforts were developed, for example, to protect a group from hydrolysis in the case of esters protected by *gem*-trifluoromethyl groups. Actually, because of the electroattractive effects of the difluoromethyl groups in α position about the functions, the fluorinated polyesters and polyurethanes exhibit weak points which affect the applications of such polymers.

Another typical example concerns the introduction of aromatic rings between the fluorinated chain and the chemical function in order to improve the thermal properties. The same behavior occurs for the fluorinated chains/aromatic nucleus links ($C_6H_5-CF_2$, $C_6H_5-O-CF_2$, $C_6H_5-CH_2-OCF_2$, $C_6H_5-CH_2-OCH_2CF_2$ groups).

When the molecules have a lateral chain, surface properties are investigated and then little attention is shown to the thermal resistances. In these conditions, the fluorinated chain/functional group junction is not critical and the authors look for the simplest method which processes the most accessible products in fluorine chemistry.

Furthermore, properties of chemical and thermal resistances are expected when the fluorinated groups are located in the main chain. Thus the authors investigated the fluorinated chain/functional group stability. So far, aromatic groups are usually used but the syntheses require many steps and therefore the yields are still poor. However, novel and interesting polymers have been prepared. They have glass transition temperatures higher than 200 °C, decomposition-temperatures which reach 400 °C and a satisfactory softness which represents the weak point of the non fluorinated thermostable products.

We must be aware of the large amount of research still to be performed since the fluorinated groups modify the functions-reactivity and cannot be used for certain syntheses.

However, introducing a spacer or protective groups gives hope that such syntheses may be successfully performed.

5 References

1. Field DE (1974) US Patent 3,852,222 (12/03/74) Chem Abst 82: 99139e
2. Bloechl W (1969) Ger Patent 1,925,555 (05/20/69) Chem Abst 74: 87347t
3. Gosnell R, Hollander J (1967) J Macromol Sci Phys B, 1: 831
4. Takakura T, Yamabz M, Kato M (1985) Nippon Kagaku Kaishi 11: 2208 (Chem Abst 105: 175822k)
5. Hollander J, Trischler FD, Edward ES (1967) Amer Chem Soc Div Polym Chem Prepints 8: 1149 (Chem Abst 70: 97293d)
6. Greenwald RB, Evans DH (1976) J Org Chem 41: 1470
7. Lau KS, Kelleghan WJ PCT Int Appl No 8, 601, 504 (03/13/86)
8. Boutevin B, Robin JJ (1992) Advances in Polym Sci 102: 111
9. Lovelace AM, Rausch DA, Postelnek W (1958) Aliphatic fluorine compounds, Reinhold, New York, p 141

10. Mc Bee ET, Marzluff WF, Pierce OR (1952) J Am Chem Soc 74: 444
11. Wall LA (1972) Fluoropolymers, Wiley Intersc New York, p 201
12. Haszeldine RN (1955) J Chem Soc 4291
13. Haszeldine RN (1951) Nature 167: 139
14. Evans DEM, Tatlow JC (1954) J Chem Soc 3779
15. Fear EP, Thrower J, Veitch J (1955) J Appl Chem 5: 589
16. Knunyants IL, Shoshkina VV, Chich-Yuan L (1959) Proceed Acad USSR 971
17. Haszeldine RN (1951) J Chem Soc 586
18. Stacey M, Tatlow JC, Sharpe AG (1965) Advances in Fluorine Chem 4: 93
19. Guizard C, Cheradame H (1979) J Fluor Chem 13: 173
20. Haszeldine RN, Osborne JE (1955) J Chem Soc 3880
21. Mc Genty, British Patent 642 459 (Chem Abst 1951: 3866)
22. Jenkins AD (1967) Advances in Free Rad Chem 2: 139
23. Burdon J, Tatlow JC (1958) J Appl Chem 8: 293
24. Chambers RD, Lindley AA, Philpot PD, Fielding HC, Hutchinson J, Whittaker G (1978) J Chem Soc Perkin I 216
25. Husain SZ, Plevey RG, Tatlow JC (1986) Bull Soc Chim Fr 6: 891
26. Haszeldine RN, Steek BR (1953) J Chem Soc 1592
27. Miller WT, Howald J. (1952) ACS Meeting
28. Barnhart WS, Wade RM, US Patent 2,806,865 (1957)
29. Ehrenfeld RL US Patent 2,788,375 (1957)
30. Henne AL, Krauss DW (1951) J Amer Chem Soc 73: 1791
31. Henne AL, Krauss DW (1951) J Amer Chem Soc 73: 5503
32. Boutevin B, Cals J, Pietrasanta Y (1975) Eur Polym J 12: 225
33. Boutevin B, Pietrasanta Y (1973) Tetrahedron Lett 12: 219
34. Boutevin B, Pietrasanta Y (1975) Europ Polym J 12: 219
35. Boutevin B, Pietrasanta Y (1989) In: Allen G, Bevington JC, Eastmond AL (eds) Comprehensive Polymer Sciences, vol 3, Telomerization, Pergamon, Oxford
36. Barnhart WS, Granford NJ (Kellog Corp.), US Patent 2,770,659 (11/13/56)
37. Hugh LR (ICI) Ger Patent 1,233,375 (1967)
38. Dannels BF, Fifolt MJ, Tang DY, Amherst E, US Patent 4,808,760 (02/28/89)
39. Dedek V, Chvatal Z (1986) J Fluor Chem 31: 363
40. Brace NO, US Patent 3,145,222 (1961) Chem Abst 61: 10589g
41. Rondestvedt CS (Dupont) French Patent 1,521,774 (1968) Chem Abst 71: 2955
42. Iserson H, Magazzu J, Osburne S, Ger Patent 2,130,378 (1972)
43. Caporiccio G, Bargigia G, Tonelli C, Tortelli N Eur Patent Appl 194,781 (1986) Chem Abst 106: 69110k
44. Paleta O, Posta A (1966) Coll Czech Chem 31: 2389
45. Barnhart WS, Wade RH, US Patent 2,938,888 (1960)
46. Kim YK, Pierce OR (1968) J Org Chem 33 (1): 442
47. Bedford CD, Baum K (1980) J Org Chem 45: 347
48. Caporiccio G, Bargigia G, Tonelli C, Tortelli V, US Patent 4,731,170 (03/15/88)
49. Caporiccio G, Bargigia G, Tonelli C Eur Patent Appl 200,908 (03/27/88)
50. Tortelli V, Tonelli C (1990) J Fluor Chem 47: 199
51. Rice DE (1968) Polymer Lett 6: 335
52. McLoughlin VCR (1968) Tetrahedron Lett 46: 4761
53. Critchley JP, McLoughlin VCR, Thrower J, White IM (1970) Br Polym J 2: 288
54. Barnhart WS, Wade RH, US Patent 2,904,567 (1959)
55. Paleta O, Posta A (1968) Coll Czech Chem Commun 33: 2970
56. Boutevin B, Ranjalahy L, Rousseau A (1990) J Fluor Chem (submitted for publication)
57. Hauptschein M, Braid M (1961) J Amer Chem Soc 83: 2500
58. Wujclak DW, Wade RH, Barnhart WS, US Patent 2,824,897 (02/25/58)
59. Battais A, Boutevin B, Pietrasanta Y, Sarraf T (1982) Makromol Chem 183: 2359
60. Boutevin B, Pietrasanta Y (ATOCHEM) French Patent No 8,801,882 (1988)
61. Battais A, Boutevin B, Hugon JP, Pietrasanta Y (1980) J Fluor Chem 16: 397

62. Ameduri B, Boutevin B, Lecrom C, Pietrasanta Y, Parsy R (1988) Makromol Chem 189: 2545
63. Ameduri B, Boutevin B, J Poly Sci, Polym Chem (in press)
64. Alfiguigui C, Ameduri B, Boutevin B, (unpublished results)
65. Brace NO, US Patent 3,016,407 (01/09/62)
66. Piccardi P, Massardo P, Modena M, Santoro E (1973) J Chem Soc Perkin I: 982
67. Smeltz KL (Dupont) US Patent 3,055,953 (1962)
68. Dolgopol'Skii IM, Dobina KA, Sinaiskaya MI, Konshin AI, Kamysheva SA, Balashova LG Russ Patent 282,307 (09/28/70) Chem Abst 74: 87374z
69. Haszeldine RN (1954) Ann Reports 51: 279
70. Husted DR, Ahlbrecht AH (1953) J Amer Chem Soc 75: 1605
71. Fisher H, Ger Patent 1,944,381 (1971)
72. Boutevin B, Dongala EB, Pietrasanta Y (1981) J Fluor Chem 17: 113
73. Ref 11 p 199
74. British Patent 797,795 (07/09/58)
75. Buist JM, Gudgeon H (1970) Advances in polyurethane technology, Elsevier, Amsterdam
76. Malichenko BF, Yazlovitskii AV, Nesteron AE (1970) Vysokomol Svedin Ser A 12 (8): 1700 (Chem Abst 73: 120958x)
77. Bonanni AP, US Naval Air Engineering Center Report NO NAEC AML 1636 (03/13/67)
78. Hollander J, Trischler FD, Gosnell RB (1967) J Poly Sc A, 1 5: 2757
79. Ref 11, p 213
80. Ref 11, p 215
81. Ref 11, p 218
82. Takakura T, Kato M, Yamabe M (1990) Makromol Chem 191: 625
83. Yoon SC, Ratner BD (1986) Macromolecules 19: 1068
84. Pietrusza EW, Pederson JR (Allied Chem) US Patent 3,573,250 (03/30/71) Chem Abst 75: 37022h
85. Kakimoto M, Harada S, Oishi Y, Imai Y (1987) J Polym Sc (Polym Chem) 25: 2747
86. Morgan PW (Dupont) US Patent 3,373,139 (03/12/68) Chem Abst 68: 96752z
87. Saegusa Y, Kukiki M (in press)
88. O'Rear JG, Griffith JR, Reines SA (1971) J Paint Techno 43 (552): 113
89. Griffith JR, O'Rear JG, US Patent 4,045,408 (08/30/77)
90. Ohmori A (Daikin Kogyo Co) British Patent 7,924,300 (09/28/79)
91. Kwolek SL (Dupont) US Patent 3,328,352 (06/27/67) Chem Abst 68: 60466v
92. Knunyants IL, Vinogadova SV, Livshits BR, Russ Patent 226,845 (09/12/68) Chem Abst 70: 48054t
93. Fusaro R (1983) Nasa Nemo 21 (12) Chem Abst 99: 160971q
94. Fusaro R (1984) Asle Trans 27 (3): 189 (Chem Abst 101: 133539h)
95. Fusaro R, Hady W (1984) Nasa Tech Memo (1984), Sci Tech Aerosp Rep 22 (15) (Chem Abst 102: 27793x)
96. Delvigs P (1986) Nasa Tech Nemo 1985 Sci Tech Aerosp Rep 24 (3) (Chem Abst 104: 208176k)
97. Saint Clair AK, Waynz S, Sample J (1985) 21 (4): 28 (Chem Abst 104: 150860k)
98. Saint Clair AK, Saint Clair TL, US Patent Appl 643,589 (07/10/89) Chem Abst 106: 197320r
99. Green HE, Jones RJ, O'Rell MK (TWR Inc) US Patent 4,173,700 (11/06/79) Chem Abst 92: 95022v
100. Jones RJ, O'Rell MK, Hom JM (TWR Inc) US Patent 4,111,906 (09/05/78)
101. Jones RJ, O'Rell MK, Hom JM (TWR Inc) US Patent 4,203,922 (05/20/80)
102. Jones RJ, Chang GE, Powell SH, Green HE (TWR Inc) 1982, Polyimides Synth Charact Appl 2: 1117
103 Shoji T, Masahiro O (Mitsui Toatsu Chem) Europ Patent Appl EP 234,882 (09/02/87) Chem Abst 108: 95747j

104 Masahiro O, Saburo K (Mitsui Toatsu Chem) Jap Patent 62,185,715 (08/14/87) Chem Abst 108: 76724h.
105. Shumichi N, Koji F (Hitachi Ltd) Europ Patent Appl 142,149 (05/22/85) Chem Abst 103: 125168n
106. Hitachi Ltd Jap Patent 5,976,451 (05/01/84) Chem Abst 102: 88541k
107. Hitachi Ltd Jap Patent 5,891,430 (05/31/83) Chem Abst 120594x
108. Hill Newman Evans R, (Union Carbide Corp) Europ Patent Appl 239,804 (10/07/87) Chem Abst 108: 39088h
109. Farah BS, Gilbert EE, Sibilia JP (1965) J Org Chem 30: 998
110. O'Rear JG, Griffith JR, US Patent 3,879,430 (1975) Chem Abst 83: 58638
111. N'Guyen TT, Amey RL, Martin JC (1982) J Org Chem 47: 1024
112. Sepiol J, Soulen RL (1984) J Fluor Chem 24: 61
113. Keller TM (1985) J Polym Sc, Polym Chem Ed 23 (9): 2557
114. Griffith JR, O'Rear JG (1987) Poly Sci Techno 35: 63
115. Griffith JR, O'Rear JG, US Patent 4,452,998 (06/05/84)
116. Griffith JR, O'Rear JG, US Patent 4,578,508 (03/25/86)
117. Boutevin B, Pietrasanta Y, (1988) Les Acrylates et Polyacrylates Fluorés; Dérivés et Applications EREC Ed France
118. Mc Loughlin VCR, Thrower J. (1969) Tetrahedron Lett 25: 5921
119. Mc Loughlin VCR, Thrower J, US Patent 3,408,411 (10/29/68)
120. McLoughlin VCR, Thrower J, British Patent 1,208,451 (05/24/67)
121. Mc Loughlin VCR, Thrower J, Hewtins MAH, Pipett JS, White MA (Royal Aircraft Establish) Technical Report 70, 160 Sept 1970 UDC 678, 674
122. Webster JA, Butler JM, Morrow TJ (1972) Polym Preprint Amer Chem Soc Div Polym Chem 13 (1): 612
123. Webster JA (1972) Nuova Chim 48 (10): 51
124. Evers RC (1974) Polym Preprint Amer Chem Soc Div Polym Chem 15 (1): 685 (Chem Abst 84: 17775p)
125. Ohsaka Y, Kobayashi T, Kubo M (Daïkin Ind Ltd) Eur Patent Appl 285,160 (03/31/88)
126. Tarrant P, Bull RN (1988) J Fluor Chem 40: 201
127. Middleton WJ, Lindsey Jr RV (1964) J Amer Chem Soc 86: 4948
128. Allan M, Janzen AF, Willis CJ (1968) Can J Chem 46: 3671
129. Loeb SJ, Martin JW, Christopher CJ (1978) Can J Chem 56 (17): 2369
130. Green M, Lewis B (1973) J Chem Soc Chem Commun 4: 114
131. Green M, Lewis B (1975) J Chem Soc Chem Commun 12: 1137
132. Langkammerer CM (Dupont) US Patent 3,627,847 (12/14/71) Chem Abst 76: 591189
133. Lichtenberger L (1971) Chimie et Ind 104 (7): 815
134. Simons JH (1950) Fluorine chemistry, Academic, New York, vol 1, p 401
135. Fielding HC, British Patent 1,130,822 (Chem Abst 70: 11364t)
136. Battais A, Boutevin B, Moreau P (1979) J Fluor Chem 13 (1): 39
137. Battais A, Boutevin B, Pietrasanta Y, Sierra P (1981) J Fluor Chem 19: 35
138. Hollander J, Woolf C (Allied Chemical Co) Belg Patent 634,368 (1963)
139. Boerner D, Koerner G, Rossmy G, Ger Patent 2,062,816 (1972)
140. Hollander J, Woolf C (Allied Chemical Co) US Patent 3,177,187 (1965) Chem Abst 63: 500
141. Brace NO, Cornack WB (Dupont) US Patent 3,047,610 (1962) Chem Abst 61: 7834a
142. Henne AL, Latif KA (1953) J Indian Chem Soc 80 (12): 809
143. Dedek V, Hemer I (1985) Coll Czech Chem Commun 50: 2743
144. Lin WT, Chen CH, Hung HC, Chou HY, Hsun TF, Huang WY (1964) Ko Fen Tzu T'ung Hsun 6 (5): 363 Chem Abst 63: 16147
145. Joyce RN (Du Pont) US Patent 2,559,628 (1951)
146. Chutny B, Frantisek L, Dedek V (1969) Ustav Jad Vyzk 2237 (Chem Abst 74: 17968)
147. Bloechl W, Ger Patent 2,052,579 (05/04/72) Chem Abst 77: 100756j
148. Ugine Kuhlmann, French Patent 1,532,284 (01/02/67)
149. Foulletier L, Lalu JP, French Patent 2,031,650 (02/03/69)

150. Foulletier L, Lalu JP, Ger Patent 2,004,156 (08/06/70) Chem Abst 73: 110486c
151. Koemm U, Getsler (Bayer) Ger Patent 3,319,368 (11/29/84) Chem Abst 102: 1985
152. Jap Patent 61,252,220 (11/10/86)
153. Lazerte JD, Guenthner RA (3M) Ger Patent 1,620,965 (05/17/73) Chem Abst 79: 67291v
154. Toukan SS, Hauptschein M, Ger Patent 2,239,709 (02/22/73) Chem Abst 79: 20652y
155. Toukan SS, Hauptschein M, US Patent 3,948,887 (04/06/76)
156. Toukan SS, Hauptschein M, US Patent 3,906,049 (09/16/77)
157. Hager BR, Toukan SS, Walter GJ (Pennwalt) Ger Patent 2,342,888 (03/07/72) Chem Abst 81: 153076b
158. Hager BR, US Patent 3,893,984 (07/08/75) Chem Abst 84: 5826t
159. Gervits LL, Makarov KN, Pletnev M Yu (1985) Zh Vses Khim 30 (6): 578 (Chem Abst 104: 131990m)
160. Tesero GC, US Patent 3,470,258 (09/30/69) Chem Abst 71: 123, 558m
161. Pavlik FJ (3M) US Patent 3,385,904 (05/28/68) Chem Abst 69: 26753x
162. Pavlik FJ (3M) US Patent 3,981,928 (09/21/76) Chem Abst 86: 16294k
163. Pavlik FJ (3M) US Patent 4,010,212 (03/01/77) Chem Abst 86: 189226t
164. Evans FW, Morton H (Allied Chem Co) US Patent 3,388,078 (06/11/68) Chem Abst 69: 28654q
165. Commeyras A, Calas P (PCUK) British Patent 2,486,526 (07/08/80)
166. Commeyras A, Calas P (PCUK) British Patent 2,486,522 (07/08/80)
167. Daikin Kogyo Co French Patent 1,535,485 (04/17/67) Chem Abst 71: 12585y
168. Daikin Kogyo Co French Patent 1,475,237 (04/07/66)
169. Blancou H, Commeyras A (1990) 1st French Symposium in Fluorine Chemistry (Eveux, France)
170. Abe T, Nagase S, Baba H (1973) Bull Chem Soc Jpn 46: 2524
171. Gresham JT (FMC Co) Ger Patent 2,245,722 (04/05/73)
172. Gresham JT, Skillman (FMC Co) US Patent 3,759,874 (09/18/73)
173. Dear REA, Falk RA, Mueller KL (Ciba Geigy) Ger Patent No 2,503,872 (10/30/75) Chem Abst 84: 45224b
174. Boutevin B, Hugon JP, Pietrasanta Y (1981) J Fluor Chem 17: 357
175. Amimoto Y, Daimon S, Okamoto M, Kodaya Y, Okamura K (Daikon Kogyo Co) Jap Patent 7,884,909 (08/26/78) Chem Abst 89: 214891c
176. Konrad W (Hoechst AG) Ger Patent 3,525,494 (07/17/85)
177. Ayari A, Szonyi S, Rouvier E, Cambon A (1990) J Fluor Chem 50: 67
178. Park JD, Rogers FE, Lacher JR (1961) J Amer Chem Soc 26: 2089
179. Mc Bee ET, Burton TM (1952) J Am Chem Soc 74: 3022
180. Rausch DA, Lovelace AM, Coleman Jr LE (1956) J Org Chem 21: 1328
181. French Patent (Ciba Geigy) 2, 235, 105 (1974)
182. Mc Bee ET, Roberts CW, Wilson jr G (1957) J Amer Chem Soc p 2323
183. Smeltz KC (Dupont) US Patent 3,478,116 (11/11/69) Chem Abst 72: 33199u
184. Smeltz KC (Dupont) US Patent 3,504,016 (03/31/70) Chem Abst 72: 122867v
185. Thayer Jr GL (Dupont) US Patent 3,510,458 (05/05/70) Chem Abst 73: 4744a
186. Smeltz KC (Dupont) US Patent 3,547,894 (02/28/69) Chem Abst 74: 113167f
187. Smeltz KC (Dupont) US Patent 3,578,701 (05/11/71) Chem Abst 75: 64665b
188. Vakhlamova LSH, Kashkin AV, Krylov AI, Kashkina GS, Sukhinin VS (1978) Zh Vses Khim 23 (3): 357 (Chem Abst 89: 110440p)
189. Harris JF, Coffman DD (1962) J Amer Chem Soc 84: 1553
190. Cook EW, Landrum J (1965) J Heterocyclic Chem 2: 327
191. Chambers RD, Grievson B (1985) J Chem Soc Perkin Trans I: 2215
192. Dmowski W (1986) J Fluor Chem 32: 255
193. Bargigia G, Tonelli C, Tato Marco (1987) J Fluor Chem 36: 449
194. Weinmayr J, (1963) J Org Chem 28: 492
195. Yohnosuke O, Takashi T, Shoji T, Eur Patent Appl (DAIKIN) 148,482 (12/26/83) Chem Abst 104: 69315

196. Persico DF, Gerhardt GE, Lagow RJ (1985) J Amer Chem Soc 107: 1197
197. Smith WC, US Patent 2,859,245 (1958)
198. Hasek WR, Smith WC, Engelhardt VA (1960) J Amer Chem Soc 82: 543
199. Sheppard WA (1964) J Org Chem 29 (1): 1
200. Hudlicky M (1976) Chemistry of organic fluorine compounds, Wiley
201. Sokoya A, Hiromaka S (1986) Sekiyu Gakkaishi 29 (2): 183
202. Santini G, Le Blanc M, Riess JG (1975) J Chem Soc Commun 16: 678
203. Yazlovitskii AV, Malichenko BF (1971) Vysokomol Svedin Ser A 13 (3): 734 (Chem Abst 74: 142711j)
204. Malichenko BF, Tsypina ON, Nestorov AE (1969) Vysokomol Svedin Ser B 11 (1): 67 (Chem Abst 70: 97296g)

Editor H.-J. Cantow
Received April 15, 1991

Synthesis and Metal Complexation of Poly(ethyleneimine) and Derivatives

Bernabé L. Rivas[a]) and Kurt E. Geckeler
Institute of Organic Chemistry, University of Tübingen, Auf der Morgenstelle 18, 7400 Tübingen, FRG

This article surveys the research work on the synthesis and modification reactions of poly(ethyleneimine) as well as its applications to metal complexation processes. Poly-(ethyleneimine), one of the most simple heterochain polymers exists in the form of two different chemical structures: one of them is branched, which is a commercially available and the other one linear which is synthesized by cationic polymerization of oxazoline monomers and subsequent hydrolysis of poly[(N-acylimino)ethylene]. The most salient feature of poly(ethyleneimine) is the simultaneous presence of primary, secondary, and tertiary amino groups in the polymer chain which explains its basic properties and gives access to various modification reactions. A great number of synthetic routes to branched and linear poly(ethyleneimine)s and polymer-analogous reactions are described. In addition, the complexation of poly(ethyleneimine) and its derivatives with metal ions is investigated. Homogeneous and heterogeneous metal separation and enrichment processes are reviewed.

1 Introduction 173

2 General Properties of Poly(ethyleneimine) (PEI) 173
  2.1 Chemical and Physical Properties 173
  2.2 Analysis and Applications 174

3 Synthesis of Poly(ethyleneimine) and Derivatives 174
  3.1 Linear Poly(ethyleneimine) (LPEI) and Derivatives 174
    3.1.1 LPEI Homopolymers 174
    3.1.2 LPEI Copolymers 175
  3.2 Synthesis of Branched Poly(ethyleneimine) (BPEI) 176
  3.3 Quaternized Derivatives 176
    3.3.1 Polystyrene-Grafted Copolymers 176
    3.3.2 Quaternized LPEI 176

4 Homogeneous Phase Complexation 177
  4.1 PEI-Metal Interaction 177
  4.2 Metal Interaction with PEI Derivatives 179

[a] Permanent address: Department of Chemistry, Faculty of Science, University of Concepcion, Casilla 3-C, Concepción, Chile

5 Heterogeneous Phase Complexation 180
  5.1 PEI-Resins 180
    5.1.1 Alkylated Resins 181
    5.1.2 Resins by Cross-Linking and *N*-Methylation of BPEI and
         LPEI 181
    5.1.3 Resins by Zwitterion Copolymerization of Ethyleneimine . . 185
  5.2 Graft PEI Resins and Other Derivatives 185

6 References 186

1 Introduction

In the series of the poly(alkyleneimine) polymers poly(ethyleneimine)* (PEI) is the most important polymer due to its salient properties and commercial availability. The linear polymer has the empirical formula $(C_2H_5N)_n$ and the molecular mass of its repeating unit is 43.07. This unit, however, does not express the real chemical structure and reactivity. More appropriate and closer to reality is the following structure:

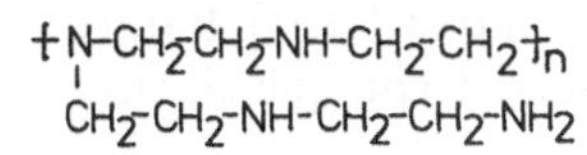

$$+N-CH_2CH_2NH-CH_2CH_2+_n$$
$$CH_2-CH_2-NH-CH_2-CH_2-NH_2$$

Due to its versatile industrial use there are a number of trade names. The most frequent examples are: Polyaziridine, Corcat, Montrek, Polyamin, Polymin P.

The complexation chemistry of the low-molecular analogs shows that the stability of the chelates with metal ions is strongly dependent on the structure of the amine moiety. As poly(ethyleneimine) contains three different types of amino groups the complexation properties are correspondingly versatile and interesting.

2 General Properties of Poly(ethyleneimine) (PEI)

2.1 Chemical and Physical Properties

Two different states are characteristic of PEI, the non-ionic and the polyelectrolyte state. However, the situation is more complicated because many commercial types of PEI are not linear but branched to different degrees. That means that the macromolecule contains three different types of amino groups: in the main chain secondary and tertiary groups and in the side-chain secondary and primary amino groups. The ratios are between $1:1:1$ and $1:2:1$ referred to primary, secondary, and tertiary, but variable in principle, depending on the degree of branching [1]. Practically, the polymer should be regarded as a polymeric amine due to its basic amine characteristics.

PEI is commercialized in a broad range of molecular masses, most frequently between 600 and 100000. The commercial PEI is amorphous but true linear PEI is highly crystalline. The polymer is very well water-soluble and, depending on the molecular mass, partially or fully soluble in lower alcohols such as methanol and ethanol but insoluble in most other organic solvents. Usually the hygroscopic PEI is available in the form of a viscous, aqueous solution (40–60%), which is strongly alkaline (pH 9–10). The thermal stability can be considered as relatively good; it decomposes very little up to 200 °C and up 500 °C in nitrogen. Contamination from the atmosphere can stem from the absorption of carbon dioxide. Therefore, protection during storage and use is required for analytically pure applications.

---

* IUPAC nomenclature: Poly(imino-1,2-ethanediyl)

2.2 Analysis and Applications

Analysis of the imine polymer can be performed by a colorimetric method which is based on the complexation reaction with copper [2]. Elemental analysis serves for the determination of the nitrogen content. Potentiometric and conductometric titration is also used in analysis [3]. For quantitative routine determination, preferably in the form of the hydrochloride, refractometry is recommended.

Detection using a fluorescent agent is also applicable [4]. Infrared spectrometry using potassium bromide requires a modified technique because of the highly viscous state of branched PEI [5].

A great variety of applications makes PEI an important technical polymer. It is used as an adhesive, for coating, flocculation, textiles, ion exchange, plastics, and many other purposes.

3 Synthesis of Poly(ethyleneimine) and Derivatives

The availability of different types of amino groups allows a variety of reactions typical for amines which are suitable for the derivatization of PEI. Reaction partners can be aldehydes, ketones, alkyl halides, isocyanates and thioisocyanates, epoxides, cyanamides, guanides, ureas, acids, and anhydrides. The synthesis and chemical modification reactions of PEI will be discussed in the next sections.

3.1 Linear Poly(ethyleneimine) (LPEI) and Derivatives

3.1.1 LPEI Homopolymers

LPEI is obtained by cationic polymerization of cyclic iminoethers as oxazoline and oxazine derivatives producing a linear low-molecular compound with a high crystallinity [6–20].

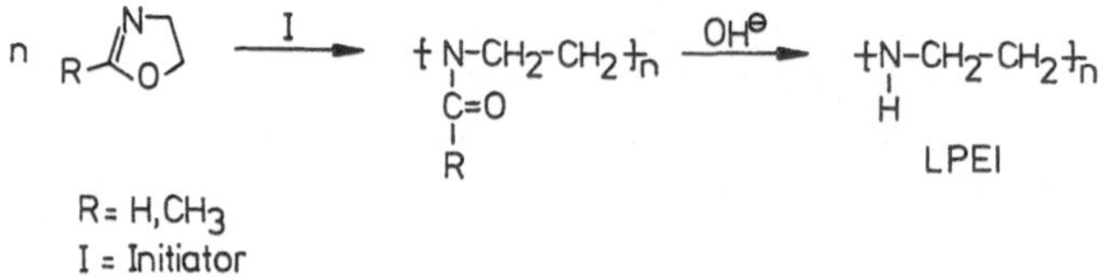

$$R = H, CH_3$$
$$I = Initiator$$

LPEI contains only secondary amino groups in the main chain in contrast to the commercial poly(ethyleneimine) with a branched structure.

3.1.2 LPEI Copolymers

Saegusa and Ikeda prepared copolymers of LPEI with polybutadiene (M = 7500) with the following structure [21]:

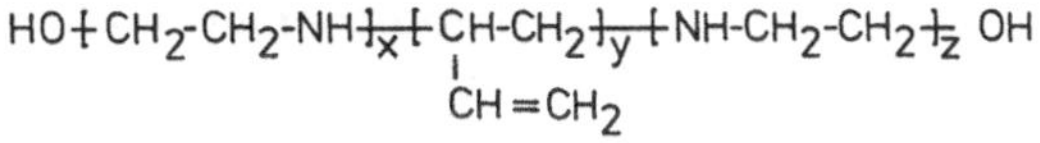

$$HO \text{-} (CH_2\text{-}CH_2\text{-}NH)_x \text{-} (CH\text{-}CH_2)_y \text{-} (NH\text{-}CH_2\text{-}CH_2)_z \text{ OH}$$
$$\underset{CH=CH_2}{|}$$

They started from dihydroxy poly(butadiene), converted it to the ditosylate and allowed this compound to react with 2-oxazoline. The block copolymer obtained was hydrolyzed in aqueous medium to yield the above copolymer which is soluble in chloroform. According to NMR investigations the LPEI unit content in the copolymer was 63 mol-% and the LPEI blocks had an average molecular mass of 4700.

Tanaka et al. described the application of a LPEI lithium aluminium hydride complex to reduction reactions [22]. They assigned the following structure to the product obtained:

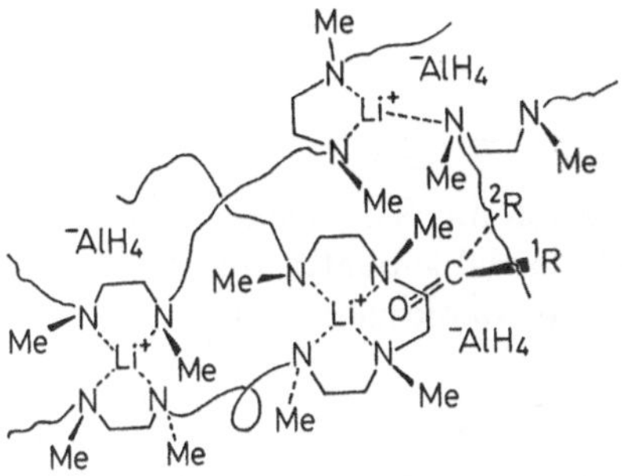

Fig. 1. Complex of poly(N-methylethyleneimine) with lithium aluminium hydride [22]

The basis polymer of the complex, which was suspended in hydrocarbon media, was prepared from poly(N-benzoyl ethyleneimine) by acid-catalyzed hydrolysis and had a molecular mass of 10^5. The methyl derivative of LPEI was obtained by reductive N-methylation and used for the preparation of the complex. The reduction procedure was shown to be successful for aldehydes, ketones, and esters. For several reductions, e.g. diphenylmethanol, benzyl alcohol, 2-methyl-1-propanol, they reported degrees of conversion of 100% according to NMR assay.

LPEI was also used to prepare interpolymers with poly(epichlorohydrin) for membrane applications [23].

The simple preparation procedure was to heat a mixture of the two polymers in DMF and to cure the material after casting. A typical section of the structure is the following:

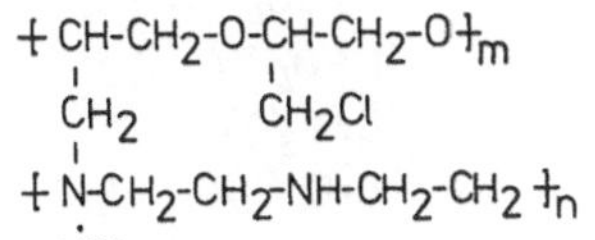

$$\text{-} (CH\text{-}CH_2\text{-}O\text{-}CH\text{-}CH_2\text{-}O)_m$$
$$\underset{CH_2}{|} \quad \underset{CH_2Cl}{|}$$
$$\text{-} (N\text{-}CH_2\text{-}CH_2\text{-}NH\text{-}CH_2\text{-}CH_2)_n$$
$$\underset{HCl}{|}$$

3.2 Synthesis of Branched Poly(ethyleneimine) (BPEI)

BPEI is produced by cationic polymerization of ethyleneimine. This reaction is performed with different catalysts [24–31]. In all cases, a polymer with a low molecular mass is obtained which is branched and contains up to 38% primary amino groups.

3.3 Quaternized Derivatives

3.3.1 Polystyrene-Grafted Copolymers

Polystyrene-grafted PEI was converted to an effective phase transfer catalyst by quaternization with butyl bromide [32]:

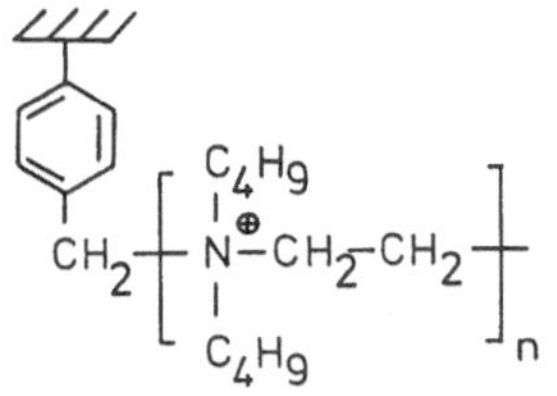

This derivative is more effective as a catalyst for S_N2 reactions than the corresponding nongrafted compound. The water-soluble quaternized LPEI showed even higher activity almost comparable to low-molecular analogs.

3.3.2 Quaternized LPEI

LPEI was converted to a cationic polyelectrolyte by polymer-analogous methylation [33]. The soluble anion exchange polymer had trimethyl ammonium moieties in the side chain:

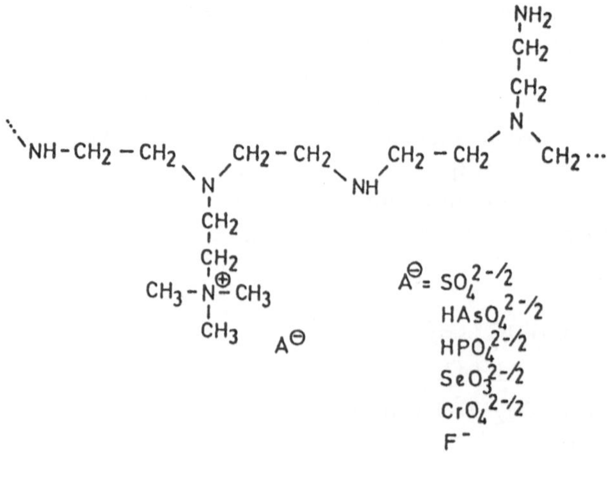

The polymer derivative was shown to be applicable to anion exchange separation of a number of anions.

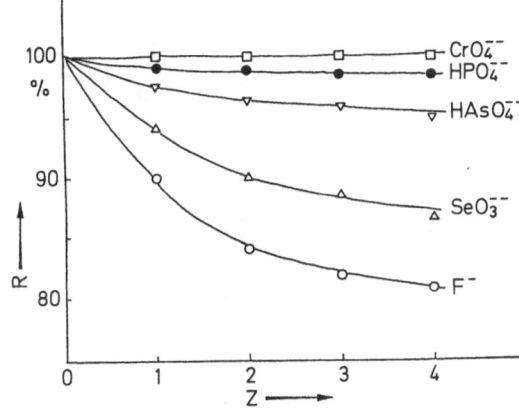

Fig. 2. Anion exchange separation of trimethylammonium poly(ethyleneimine): Retention as a function of the filtration factor Z (ratio of filtrate volume to cell volume) (4% w/w, 0.01 M sodium salts, pH 8.5) [33]

The homogeneous phase separation process is influenced by pH and concentration of polymer reagent. Cleavage of the anionic species from the polymer was performed with more highly concentrated sodium salt solutions.

4 Homogeneous Phase Complexation

Polymeric soluble supports have been developed for the complexation of various metals [34–43]. In particular, the preparation of soluble polymeric chelating agents has received attention by the application to metal recovery from dilute solutions. In developing systems in conjunction with membrane filtration for the application of such polymeric materials the first step has been achieved towards technological exploitation. Membrane filtration easily allows to separate metals bound to soluble chelating polymers from non-chelated metals. These chelating polymers, termed polychelatogenes, have been prepared by functionalizing various polymers. One of the most capable polymers for this application is poly(ethyleneimine) [41–52].

4.1 PEI-Metal Interaction

The complexation properties of PEI were investigated in the aqueous phase for Co, Ni, Zn, Cd, and Cu ions using membrane filtration [35, 39, 40]. According to the elution behavior, which was measured for each metal at different pH values, PEI is an effective polymeric complexing agent suitable for the retention and separation of metals in aqueous diluted solutions.

Table 1. Metal content of some soluble complexes of poly(ethyleneimine) (PEI) and derivatives

PEI Derivative	Metal ion	pH	Capacity	
			[mg g^{-1}]	[mmol g^{-1}]
—	Co^{2+}	4.0	105	1.8
	Ni^{2+}	4.0	135	2.3
	Cu^{2+}	4.0	180	2.8
Iminodiacetic acid	Cu^{2+}	4.0	130	2.0
	Pd^{2+}	2.5	80	0.8
	Ag^{+}	2.5	40	0.4
N-Methyl thiourea	Au^{3+}	2.5	180	0.9
	Pt^{4+}	2.5	135	0.7
	Hg^{2+}	4.0	100	0.5

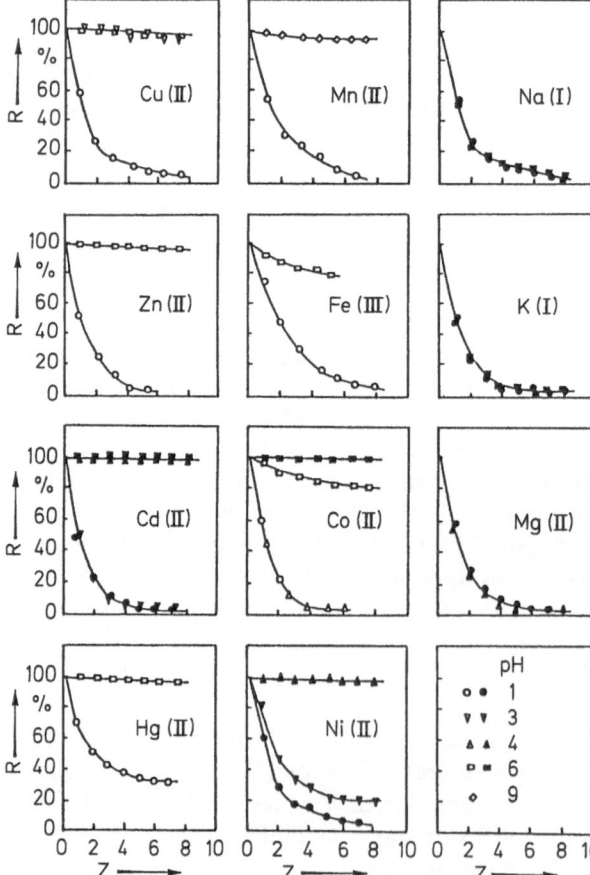

Fig. 3. Complexation of metal ions by poly(ethyleneimine) (1% w/w) at different pH values in the absence (filled symbols) and presence (open symbols) of 0.15 M NaNO$_3$ as a function of the ratio of filtrate volume to cell solution volume (Z) [43]

Another study showed that PEI can also serve as a useful polymeric agent for separation and preconcentration of elements [43].

Due to its relatively weak complexing ability it can be used for complexation in neutral solutions. As shown in Fig. 3, this polymer quantitatively retains several divalent metals (Cu, Zn, Cd, Hg, Mn, Co, Ni) at pH 6. Interestingly, Mn(II) is also completely retained at pH > 8, that is important because its extraction from aqueous solutions is generally a great problem.

Copper(II) forms the most stable complexes with PEI. It is separable from other elements at lower pH values (≤ 3). In contrast to the elements mentioned above, alkali and alkaline earth metals are not retained at any pH value by PEI. Neutral salts (e.g. sodium chloride or nitrate) influence the complexation of PEI (Fig. 4) [53].

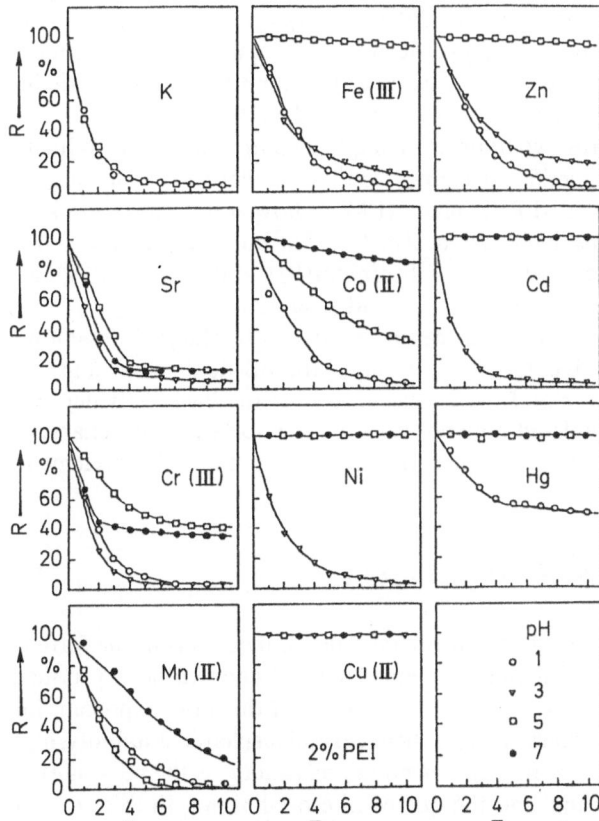

Fig. 4. Dependence of the retention of metal ions by poly(ethyleneimine) at several pH values in the presence of 0.15 M NaNO$_3$ + HNO$_3$ on the filtration factor [53]

4.2 Metal Interaction with PEI Derivatives

A series of different PEI derivatives has been synthesized to introduce selective ligands (Table 2).

Table 2. Poly(ethyleneimine) and functional derivatives for complexation of metal ions (R = PEI backbone)

Functional Group	Name	Acronym	Ref.
$R-NH_2$	Amine	PEI	[35]
$R-NH-C(=S)-NH-CH_3$	N-Methyl thiourea	PTU	[43]
$R-N^+(CH_3)_3$	Trimethylammonium	PMP	[44]
$R-NH-SO_2-oxine$	8-Hydroxyquinoline	POX	[53]
$R-N(CH_2-COOH)_2$	Iminodiacetic acid	PDA	[35]

The PEI methyl thiourea derivative is an effective reagent for the separation of mercury and noble metals from other elements [35, 43]. For example, Hg(II), Ag(I), Au(III), Pd(II), and Pt(IV) are quantitatively complexed at pH 1, whereas alkali and non-ferreous metal ions such as Co(II), Zn(II), Cd(II), and Ni(II) do not interact with this sulfur-containing polymer derivative. Remarkable are the kinetic aspects of the complexation reaction in this case: less than five minutes stirring was enough for complete interaction of the elements Hg(II), Ag(I), Au(III), and Pd(II), whereas Pt(IV) requires two hours for complete complexation. Introducing cationic moieties showed a significant change in the complexation behaviour of PEI. For instance, the methylated PEI sulfate (PMP) showed not only complexation abilities to bivalent ions such as Cu, Zn, Cd, Co, but also an interaction potential with anionic species [33, 43]. Chromium(VI) and selenium(IV) are quantitatively retained at pH 8.5 as chromate and selenite ions.

This ambivalent interaction behaviour can be explained by the multifunctional structure of this PEI derivative. On the one hand, the divalent metal ions can interact with the nitrogen atoms of the amino groups of the basis polymer, and one the other hand, the methylated moieties lead to an anion exchange character of the polymer. Thus, simultaneous complexation and retention of various elements are feasable with this polymer.

5 Heterogeneous Phase Complexation

One of the most abundant uses of chemically modified polymers is in the form of synthetic ion-exchange resins. There are different natural and synthetic products which show these properties. The organic resins are by far the most important ion exchangers. Their main advantages are high chemical and mechanical stability, high ion-exchange capacity, high exchange rate. Another advantage is the possibility of selecting the fixed ligand groups and the degree of cross-linking. In this context, PEI has been extensively used as matrix to obtain water-insoluble resins.

5.1 PEI Resins

Water-insoluble PEI resins are usually obtained by polymer-analogous cross-linking, alkylation of BPEI and LPEI or by introducing a polymerizable group in the side chain of PEI, which is then homo- or copolymerized.

5.1.1 Alkylated Resins

An alkylating procedure using 2,5-divinyl pyridine (DVP) was used by Druzin et al. [54] for the preparation of a PEI derivative with an unsaturated moiety in the side-chain:

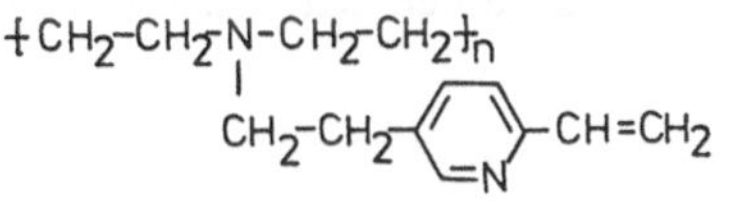

$$\begin{array}{c} +CH_2-CH_2-N-CH_2-CH_2\frac{1}{n} \\ | \\ CH_2-CH_2-\underset{=N}{\overbrace{}}-CH=CH_2 \end{array}$$

The product was homo- and copolymerized and investigated in respect to the transition metal sorption properties. The dynamic exchange capacity (DEC) was determined for Cu(II), Ni(II) and Co(II), from nitrate solutions in the presence of NaCl (Table 3).

Table 3. Sorption capacity of homo- and copolymers from poly[(ethyleneimine)-co-(2,5-divinylpyridine)] (PEI-DVP)

Resin	DEC [meq g^{-1}]		
	Cu(II)	Ni(II)	Co(II)
PEI-DVP	7.5	7.8	6.3
PEI-DVP-co-MMA[1]	4.45	4.40	2.98
PEI-DVP-co-DBI[2]	3.90	3.85	2.54

[1] MMA = Methyl methacrylate;
[2] DBI = Dibutyl itaconate

In addition, the PEI-DVP polymer was modified with chloroacetic acid in order to study the complexing properties. This derivative showed a lower sorption capacity and slower kinetics of sorption for Cu^{2+}, Ni^{2+}, and Co^{2+} compared to the unmodified copolymers [54].

5.1.2 Resins by Cross-Linking and N-Methylation of BPEI and LPEI

BPEI and LPEI were cross-linked with different dihalogen derivatives, formaldehyde, epichlorohydrin [55–73]. The cross-linking reaction allows to change the chemical and physical properties of the resins, e.g. the swelling capacity which is important for complex and adduct formation. Moreover, it is possible to vary the pore size which is also relevant in the ligand-metal complex formation.

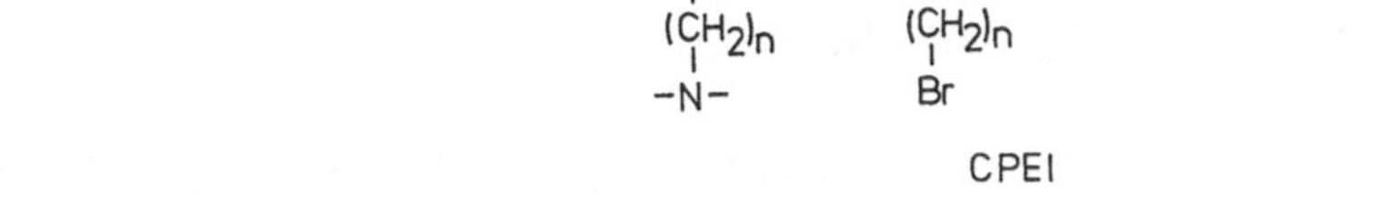

$$\begin{array}{c} +NH-CH_2CH_2\frac{1}{m} + Br(CH_2)_nBr \longrightarrow +N-CH_2CH_2\frac{1}{x}+N-CH_2CH_2\frac{1}{y}+NH-CH_2CH_2\frac{1}{z} \\ \qquad\qquad\qquad\qquad\qquad\qquad (CH_2)_n \qquad\quad (CH_2)_n \\ \qquad\qquad\qquad\qquad\qquad\qquad\quad -N- \qquad\qquad\quad Br \end{array}$$

CPEI

Poly(ethyleneimine) cross-linked (CPEI) with ethylenedichloride forms stable complexes with copper(II) as well as with cobalt(II). The RCl type of cross-linked poly(ethyleneimine) having an anion-exchange capacity of 6.8 meq g^{-1} retains copper from $CuSO_4$ and cobalt from 1 M aq. $CoCl_2$ solutions [55]. PEI is by itself a weak basic anion-exchange resin and forms stable complexes with anions and cations. The process is probably accompanied with chelate ring formation:

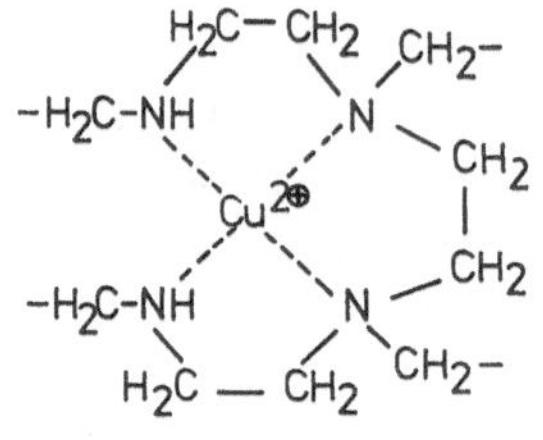

CPEI adsorbs also transition metals such as mercury(II) and nickel(II) [56].

Bartulin et al. [57–63] have used LPEI and BPEI which were cross-linked with different cross-linking agents. These resins were tested for to copper(II) and uranium(VI) using the batch method (Table 4).

Table 4. Adsorption of copper(II) and uranium(VI) by LPEI resins at pH 2

Cross-linker	Adsorption [%]	
	Cu(II)	U(VI)
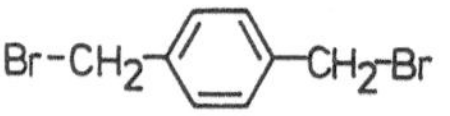 Br–CH$_2$–⟨ ⟩–CH$_2$–Br	1.2	71.8
$Cl-CH_2-C\equiv C-CH_2-Cl$	35.2	86.5
$Br-(CH_2-CH_2O)_2CH_2-CH_2-Br$	85.8	84.7

Cross-linkers with a more rigid structure like the 4,4′-dibromoxylene and the 1,4-dichloro-2-butyne lead to a lower retention capacity for copper compared to 1,2-bis(2-bromoethoxy)-ethane as cross-linking agent; the uranium, however, does not show dependence on the chemical structure of cross-linkers, probably because uranium forms adducts with the different amino groups and does not need the stereochemical arrangements of four nitrogen atoms as copper does. The adducts are preferentially formed with tertiary amino groups and ammonium groups [64].

Reaction with a weak basic resin:

$$2\,R_3N + H_2SO_4 \rightleftharpoons (R_3NH^+)_2SO_4^{2-}$$

$$(R_3NH^+)_2SO_4^{2-} + [UO_2(SO_4)_2]^{2-} \rightleftharpoons (R_3NH^+)_2[UO_2(SO_4)_2]^{2-} + SO_4^{2-}$$

Reaction with a strong basic resin:

$$(R_4N^+)_2SO_4^{2-} + [UO_2(SO_4)_2]^{2-} \rightleftharpoons (R_4N^+)_2[UO_2(SO_4)_2]^{2-} + SO_4^{2-}$$

BPEI and LPEI have been cross-linked with several different dihalogen compounds for comparison (Table 5) and then N-methylated with dimethyl sulfate to yield the corresponding resins which are insoluble both in water and in common organic solvents [65–73].

Table 5. Cross-linkers and abbreviations of the corresponding resins

Cross-linker	Resin from	
	BPEI	LPEI
1,4-Dibromo-2-butene	B-1	L-1
1,9-Dibromononane	B-2	L-2
1,10-Dibromodecane	B-3	L-3
1,3-Dibromo-2-propanol	B-4	L-4
1,7-Dibromoheptane	B-5	L-5
1,12-Dibromododecane	B-6	L-6
1,4-Dibromo-2,3-butanedione	B-7	L-7

The N-methylation reaction with the aim of quantitative conversion occurred according to:

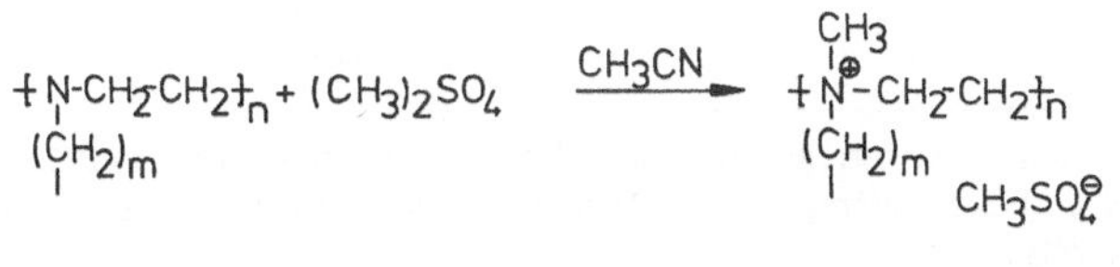

$$+N\text{-}CH_2CH_2\!\!+_n + (CH_3)_2SO_4 \xrightarrow{\;CH_3CN\;} +\overset{\overset{\displaystyle CH_3}{|}}{N}^{\oplus}\text{-}CH_2CH_2\!\!+_n$$
$$\underset{(CH_2)_m}{|} \qquad\qquad \underset{(CH_2)_m}{|}\; CH_3SO_4^{\ominus}$$

The sorption capacity for Cu^{2+} and UO_2^{2+} of all the resins was tested following the batch method at pH 2. Adsorption of a metal ion on a resin is expressed using the distribution coefficient (K_d) which is defined as [74]:

$$K_d = \frac{\text{amount of metal ion in the resin}}{\text{amount of metal ion in the solution}} \times \frac{\text{volumen of solution [ml]}}{\text{weight of resin [g]}}$$

The results are summarized in Table 6.

The resins from LPEI have a greater degree of alkylation than those prepared from BPEI with respect to the adsorption capacities, almost all the resins have a low K_d for copper(II) compared to UO_2^{2+}. For example, K_d (copper) for resin B-4 is 100 meq/g and K_d (uranium) is 19900 meq/g. The resins obtained from BPEI show a better adsorption capacity than those from LPEI, particularly for UO_2^{2+}.

Table 6. Relationship of the distribution coefficients K_d at pH 2 for Cu^{2+} and UO_2^{2+} with cross-linking and alkylation for resins from BPEI and LPEI[a]

Resin	Cross-linking [%]	Alkylation [%]	$K_d \cdot 10^{-2}$ [ml g^{-1}][b]	$K_d \cdot 10^{-2}$ [ml g^{-1}][c]
B-1	21.8	58.1	15.660	6.26
B-2	17.3	51.3	1.200	6.18
B-3	12.0	42.0	3.320	1.50
B-4	23.6	53.5	1.100	199.00
B-5	12.8	31.4	0.004	31.26
B-6	18.3	70.9	11.500	0.50
B-7	37.3	76.3	0.100	46.60
B-1M	–	–	0.100	15.66
B-2M	–	–	0.004	6.38
B-3M	–	–	0.002	15.66
B-4M	–	–	0.004	8.62
B-5M	–	–	0.400	9.12
B-6M	–	–	0.026	9.02
B-7M	–	–	0.082	2.66
L-1	32.1	83.7	11.840	11.80
L-2	28.2	88.9	0.150	0.37
L-3	31.0	92.6	0.240	0.29
L-4	28.8	66.7	3.540	8.10
L-5	23.7	82.0	2.320	0.92
L-6	16.0	63.5	11.500	0.50
L-7	46.5	98.0	0.410	9.19
L-1M	–	–	0.026	4.50
L-2M	–	–	0.004	1.10
L-3M	–	–	0.034	3.76
L-4M	–	–	0.266	5.50
L-5M	–	–	0.030	4.46
L-6M	–	–	0.020	1.76
L-7M	–	–	0.080	2.66

[a]   The names of the resins were assigned as follows: those from BPEI (B-) and from LPEI (L-). The L-M and B-M correspond to *N*-methylated resins from cross-linked LPEI and BPEI, respectively;

[b, c] correspond to distribution coefficients for copper(II) and uranium(VI)

Thus, K_d (uranium) for B-4 is 19 900 meq/g and for L-4 it is only 810 meq/g. In general, the *N*-methylated resins adsorb more UO_2^{2+} and less Cu^{2+} ions than the unmethylated resins. This may be due to the presence of tertiary amino and ammonium groups which are assumed to favor the adduct formation with UO_2^{2+}.

In order to recover the resin, sulfuric acid and sodium carbonate solutions were added. 85% of copper was eluted from loaded B-3 resin by 2 M aq. H_2SO_4 which is a good eluent for copper. Nevertheless, Cu^{2+} and UO_2^{2+} may be eluted in acid and basic medium. UO_2^{2+} is usually eluted with Na_2CO_3, due to a more stable complex $[UO_2(CO_3)_3]^{4-}$ which is formed. 98% of the uranyl ion has been eluted by 1 M aq. Na_2CO_3 from resin L-5.

The kinetics of the uranyl adsorption from acidic sulfate solutions with some of the anterior resins has been also reported [75, 76]. For example, BPEI cross-linked with epichlorohydrin, 1,3-dibromopropane, 1,2-dibromoethane and 1,4-bis(bromomethyl)benzene was investigated in terms of determining the rate controlling mechanism by the finite solution volume method [77] and shrinking core model of Gopala and Gupta [78] for testing film or pore diffusion control. The results showed that the ion exchange reactions involving uranyl sulfate and weak base resins are predominantly particle diffusion controlled.

5.1.3 Resins by Zwitterion Copolymerization of Ethyleneimine

Ethyleneimine has been also copolymerized in the absence of an initiator by zwitterion copolymerization with maleic anhydride [79] and methacrylic acid [80]. Both copolymers were checked following the batch method for copper(II), uranium(VI) and iron(III) and did not adsorb iron(III), but copper(II) and uranium(VI) at pH 2. Poly(ethyleneimine-co-methacrylic acid) adsorbed more than 95% UO_2^{2+}. The ions were almost quantitatively eluted by contact of the loaded resin with 1 M aq. H_2SO_4 [79, 80].

5.2 Graft PEI Resins and Other Derivatives

Saegusa et al. [81] reported on the grafting of poly(ethyleneimine) on poly(p-chloromethylstyrene) (PS-g-PEI).

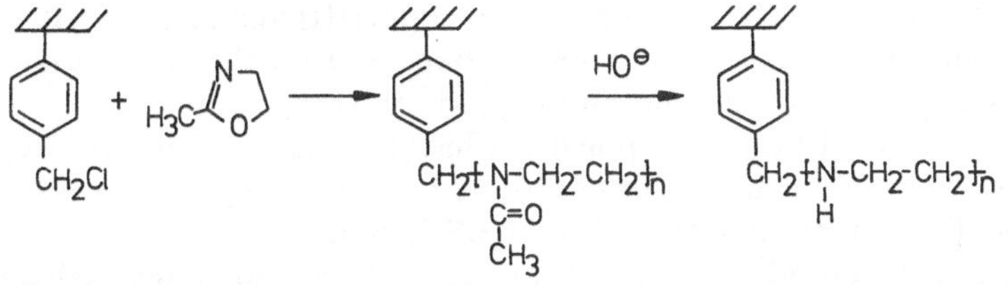

The interaction of this resin with Cu(II), Hg(II) and Cd(II) was investigated. In this study, the adsorption selectivity of this resin was: Hg(II) > Cd(II) > Cu(II).

It is also possible to obtain a resin with a high adsorption capacity for Ca(II) by carboxyalkylation of PS-g-PEI [82].

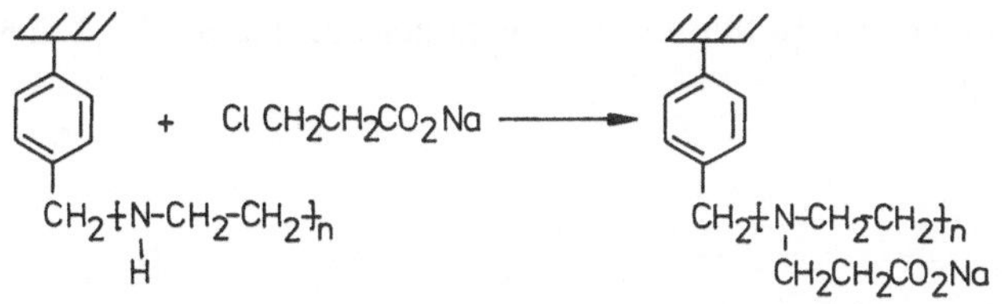

The calcium adsorption capacity of this resin was correlated with the carboxylate content.

A sulfur atom can be introduced into the side chain of PS-g-PEI by polymer-analogous reaction with ethylene sulfide [83].

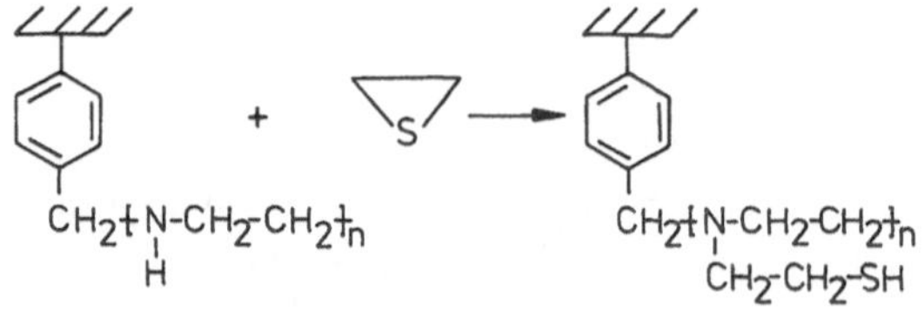

The resin adsorbs Cu(II), Cd(II) and Hg(II) and shows a selectivity depending on the pH. The thiol groups are stable towards oxidation in the atmosphere for two months.

It is well known that the secondary amines form dithiocarbamates with carbon disulfide, which are important complexing agents in analytical chemistry [84]. Therefore, PS-g-PEI was allowed to react with carbon disulfide to obtain a resin with a higher affinity to Hg(II) than the resin with thiol groups [83].

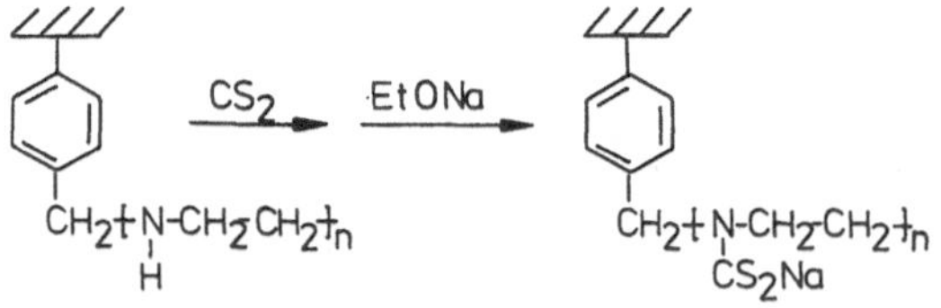

BPEI was cross-linked by N-alkylation with poly(vinylchloride), producing a polymer whose chelating properties to copper(II) and mercury(II) depend on the concentration of secondary amino groups and on the ion concentration of the solution but do not depend on the pH in the range of 0–3 [85].

Generally, PEI is a basis polymer which is easily accessible by several synthetic routes at low cost and exhibits outstanding complexing properties for many metal ions. The complexing capability of LPEI is in many cases lower than that of BPEI but is dependent on the type of metal. The reason for that is that the branched structure offers more favorable complexation sites than the linear polymer. In LPEI the coordination of metal ions is usually restricted by reason of a strong neighbouring interaction of the amino groups. In case of the cross-linked PEI resins the major influence on metal complex formation can be attributed to the degree of cross-linking and to the nature of the cross-linking agent.

Acknowledgement: B.L.R. thanks the Alexander von Humboldt Foundation for a grant.

6 References

1. Dick RC, Ham GE (1970) J Macromol Sci, Chem A4: 1301
2. Kindler WA, Swanson JW (1972) J Polym Sci, Part A-2 9: 853
3. Bloys van Treslong CJ (1978) Rec Trav Chim Pays Bas 97: 13
4. Schmidt K, Geckeler K (1974) Anal Chim Acta 71: 79

5. Geckeler KE, Eckstein H (1987) Analytische und präparative Labormethoden, Vieweg, Wiesbaden
6. Saegusa T, Ikeda H, Fujii H (1973) Macromolecules 6: 315
7. Saegusa T, Ikeda H, Fujii H (1972) Polym J 3: 35
8. Saegusa T, Ikeda H, Fujii H (1972) Macromolecules 5: 108
9. Tomalia DA, Sheetz DP (1966) J Polym Sci, Part A-1 4: 2253
10. Saegusa T, Ikeda H, Fujii H (1972) Macromolecules 5: 359
11. Saegusa T, Yamada A, Taoda H, Kobayashi S (1978) Macromolecules 11: 435
12. Litz M, Levy A, Herz J (1975) J Macromol Sci, Chem A9: 703
13. Kagiya T, Narisawa S, Maeda T, Fujii K (1966) J Polym Sci, Polym Lett Ed 4: 441
14. Kagiya T, Matsuda T (1971) J Macromol Sci, Chem A5: 1268
15. Levy A, Litt M (1967) J Polym Sci, Polym Lett Ed 5: 881
16. Bassiri TG, Levy A, Litt M (1967) J Polym Sci, Polym Lett Ed 5: 871
17. Levy A, Litt M (1968) J Polym Sci, Part A-1 1: 57
18. Bartulin J, Rivas BL, Rodriguez M, Angne U (1982) Makromol Chem 183: 2935
19. Rivas BL, Ananias SI (1987) Polym Bull 18: 189
20. Rivas BL, Zapata M (1990) Polym Bull 23: 571
21. Saegusa T, Ikeda H (1983) Macromolecules 6: 805
22. Tanaka R, Kataoka K, Ueoka T, Saito S (1985) Makromol Chem 186: 1787
23. Saegusa T, Yamada A, Kobayashi S Yamashida S (1979) J Appl Polym Sci 23: 2343
24. Berb WG (1955) J Chem Soc 2564
25. Jones GD (1944) J Org Chem 9: 125
26. Wolf F (1967) Plaste und Kautschuk 14: 85
27. Berb WG (1955) J Chem Soc 2577
28. Leibnitz E, Koenecke H-G, Gawelek G (1958) J Prakt Chem 6: 289
29. Salomon G (1949) Rev Trav Chim 68: 903
30. Bastian H (1950) An Chem 566: 210
31. Badische Anilin und Soda Fabrik AG (1954) Ger Pat 872-269
32. Saegusa T, Kobayashi S, Yamada A, Kashimura S (1979) Polymer J 11: 1
33. Geckeler KE, Bayer E, Shkinev VM, Spivakov BYa (1988) Naturwissenschaften 75: 198
34. Geckeler K, Weingärtner K, Bayer E (1979) Prepr Int Symp Polym Amin 131
35. Geckeler K, Lange G, Eberhardt H, Bayer E (1980) Pure Appl Chem (IUPAC) 52: 1883
36. Geckeler K, Weingärtner K, Bayer E (1980) In Goethe SE (Ed): Polymeric amines and ammonium salts, Pergamon, Oxford, p. 277
37. Bayer E, Geckeler K, Weingärtner K (1980) Makromol Chem 181: 585
38. Bayer E, Eberhardt H, Geckeler K (1981) Angew Makromol Chem 97: 217
39. Geckeler K, Pillai VNR, Mutter M (1981) Adv Polym Sci 39: 65
40. Bayer E, Spivakov BYa, Geckeler K (1985) Polym Bull 13: 307
41. Bayer E, Eberhardt H, Grathwohl P, Geckeler K (1985) Israel J Chem 26: 40
42. Spivakov BYa, Geckeler KE, Bayer E (1985) Nature 315: 313
43. Geckeler KE, Bayer E, Spivakov BYa, Shkinev VM, Vorobeva GA (1986) Anal Chim Acta 189: 285
44. Shkinev VM, Vorobeva GA, Spivakov BYa, Geckeler KE, Bayer E (1987) Sep Sci Technol 22: 2165
45. Geckeler KE, Shkinev VM, Spivakov BYa (1987) Angew Makromol Chem 155: 151
46. Geckeler K, Bayer E, Spivakov BYa, Shkinev VM, Zolotov YA (1987) Abstr Pittsburgh Conf Exp 1076
47. Geckeler KE, Bayer E, Spivakov BYa, Shkinev VM (1989) Abstr 6th Internat Symp, Synthetic membranes in science and industry, Tübingen, Sept, p 131
48. Novikov AP, Shkinev VM, Spivakov BYa, Myasoedov BF, Geckeler KE, Bayer E (1989) Radiochim Acta 46: 35
49. Shkinev VM, Spivakov BYa, Geckeler KE, Bayer E (1989) Talanta 36: 861
50. Shkinev VM, Spivakov BYa, Geckeler KE, Bayer E (1989) Solv Extr Ion Exch 7: 499
51. Geckeler KE, Bayer E, Shkinev VM, Spivakov BYa (1989) Fresenius Z Anal Chem 333: 763

52. Shkinev VM, Gomolitzki VN, Spivakov BYa, Geckeler KE, Bayer E (1989) Talanta 36: 861
53. Geckeler KE, Bayer E, Shkinev VM, Spivakov BYa (1990) Anal Chim Acta 230: 171
54. Druzin MI, Văkova IN, Schenniloa NI, Koroleva LI, Karapatyan LP, Valkova AK, Zaitseva IV, Pavlova ID (1974) J Polym Sci, Polym Symp 47: 369
55. Nonogaki S, Makishima S, Yonada Y (1958) J Phys Chem 62: 601
56. Shepherd EJ, Kitchener JA (1957) J Chem Soc 86
57. Bartulin J, Maturana HA, Rivas BL, Rodriguez MT (1982) An Quim 78: 221
58. Bartulin J, Maturana HA, Rivas BL, Rodriguez MT (1982) An Quim 78: 223
59. Bartulin J, Maturana HA, Rivas BL, Perich IM (1984) Polym Bull 12: 189
60. Bartulin J, Maturana HA, Rivas BL, Perich IM (1984) Bol Soc Chil Quim 29: 373
61. Bartulin J, Ramos ML, Rivas BL (1984) Polym Bull 12: 393
62. Bartulin J, Maturana HA, Rivas BL, Perich IM, Angne U (1985) Bol Soc Chil Quim 30: 3
63. Bartulin J, Rivas BL, Cardenas G, Perich IM, Angne U (1983) Polym Commun 24: 30
64. Barnes CD, da Silva Neves RD, Streat M (1974) J Appl Chem Biotechnol 24: 787
65. Rivas BL, Maturana HA, Perich IM, Angne U (1985) Polym Bull 14: 239
66. Rivas BL, Maturana HA, Perich IM, Angne U (1986) Polym Bull 15: 121
67. Rivas BL, Maturana HA, Bartulin J, Catalan RE, Perich IM (1986) Polym Bull 16: 299
68. Rivas BL, Maturana HA, Catalan RE, Angne U, Perich IM (1986) Polym Bull 16: 305
69. Rivas BL, Maturana HA, Angne U, Catalan RE, Perich IM (1988) Eur Polym J 24: 967
70. Rivas BL, Maturana HA, Catalan RE, Perich IM, Angne U (1988) Polym Bull 19: 609
71. Rivas BL, Maturana HA, Catalan RE, Perich IM (1989) J Appl Polym Sci 38: 801
72. Rivas BL, Maturana HA, Catalan RE, Perich IM (1988) Bol Soc Chil Quim 33: 151
73. Rivas BL, Maturana HA, Catalan RE, Perich IM, Casas S (1990) Bol Soc Chil Quim 35: 187
74. Kawabuchi K, Kanke M, Muraoka T, Yamauchi M (1976) Bunseki Kagaku 25: 213
75. Rivas BL, Klattenhoff D, Perich IM (1990) Polym Bull 23: 219
76. Rivas BL, Klattenhoff D, Perich IM (1990) Polym Bull 23: 315
77. Sano Y, Yamaguchi N, Adachi J (1974) J Chem Eng Japan 7: 255
78. Gopala M, Gupta AK (1982) AICH Symposium Series 78 N° 219, 103
79. Rivas BL, Canessa GS, Pooley SA, Maturana HA, Angne U (1985) Eur Polym J 21: 939
80. Rivas BL, Canessa GS, Pooley SA, Maturana HA (1989) Polym Bull 22: 189
81. Saegusa T, Kobayashi S, Yamada A (1975) Macromolecules 8: 390
82. Saegusa T, Kobayashi S, Yamada A (1977) J Appl Polym Sci 21: 2489
83. Saegusa T, Kobayashi S, Kobayashi K, Yamada A (1978) Polym J 10: 403
84. Hulanicki A (1967) Talanta 14: 137
85. Avny Y, Porath D (1976) J Macromol Sci, Chem A10: 1193

Editor H. Höcker
Received January 21, 1991

Oxidation of Hydrocarbon Polymers

Milan Lazár and Jozef Rychlý
Polymer Institute, Slovak Academy of Sciences,
842 36 Bratislava, Czechoslovakia

Oxidation of hydrocarbon polymers which is the main process causing deterioration in the useful properties of polymeric products is discussed with emphasis on the elementary reactions of the process. The principal difference between the oxidation of low and high molecular weight hydrocarbons which consists of the effect of catalyst residues, structural defects as well as in the existence of heterogeneous oxidized microdomains and morphological heterogeneities in the polymer is outlined. Various pathways of initiation of free radical oxidation such as those caused by atmospheric pollutants, ozone, singlet oxygen, ultraviolet and cosmic radiation and different reactive groups inherently present in the polymer together with a large variety of propagation free radicals which differ each from other in reactivity and in mobility lead to a complex process the general features of which have been elucidated but particular steps are still the subject to correction.

The present paper is a review of about one hundred references, of which about 60% are from last 5 years.

1 Introduction 190

2 Elementary Reactions 191

3 Initiation of the Chain Oxidation 193

4 Atmosphere as a Source of Free Radicals 193

5 Photoinitiation 196

6 Radiation Generation of Free Radicals 197

7 Thermally Induced Initiation 198

8 Propagation in Oxidation 202

9 Addition of Peroxyl Radicals to a C=C Bond 207

10 Transfer Reactions of Peroxyl Radicals 208

11 Degenerated Branching in Oxidation Reactions 209

12 Fragmentation of Radicals 211

13 Termination Reactions 215

14 Concluding Remarks 217

15 References 219

Advances in Polymer Science, Vol. 102
© Springer-Verlag Berlin Heidelberg 1992

1 Introduction

A common drawback of large-scale polymers is their tendency to oxidize. Most of them become less thermally stable when in contact with atmospheric oxygen which leads to such changes in the chemical structure causing deterioration of the properties of polymeric products.

The main problem with oxidation of many polymers is that embrittlement occurs at a rather low extent of reaction. Absorption of less than one oxygen molecule per one hundred carbon atoms can lead to such a reduction in mechanical properties that a failure of the material can take place while the molecular mass decrease does not exceed the factor 3.

Oxidation of polymers occurs as a radical chain reaction. It begins by the generation of free radicals effected by the light or almost by all kinds of radiation, by mechanical stresses due to deformation of macromolecules, and by chemical reactants coming from the surrounding medium. The secondary source of free radicals further accelerating the chain oxidation are reactions of intermediates of oxidation. In the propagation stage of a chain reaction, the initial structure of free radicals is changed while the number of radicals remains unchanged. It concerns mainly the reaction of radicals with oxygen, with surrounding molecules of hydrocarbon, or the fragmentation of free radicals. In conjunction with these reactions, the mutual interactions of free radicals occurs, leading to their decay.

Although the general features of polymer oxidation are known for a relatively long time and as such are accepted, the importance of particular elementary reactions is still the subject of discussions and corrections in opinions. Homogeneity and/or heterogeneity of the oxidation process throughout the polymer sample is another problem. Even though chemical reactions as such take place at a molecular level, their rates and a corresponding reaction channel may be determined by arrangement in supermolecular structure of a polymeric material. Nonuniformity in spatial organization of respective structural domains is displayed at three levels [1]. At the molecular level, microanisotropy of polymer system exists along and across the macromolecules. Topological irregularities on a polymer chain such as macromolecular junctions and entanglements, folds of the lamellar structure, end groups or segments linking macromolecules between particular crystallites represent more important structural heterogeneities. In the region of these topological defects, the density of a polymer is lower which brings about the higher solubility of oxygen and other low molecular mass compounds. Morphological heterogeneities of macromolecular crystallites belong to the third level of structural organization.

Within topological and morphological heterogeneities of a polymer, oxygen and residual impurities accumulate such as metals residues, peroxides, etc. If initiation of oxidation occurs in these microdomains [2], the oxidation products appear mainly there since the macroradicals which propagate the reaction are not capable of diffusing very far away [3]. The existence of heterogeneous oxidation zones is thus due to a large difference in solubility and diffusion coefficients of low and high molecular mass compounds in arranged amorphous regions. The heterogeneous course of reaction is accentuated, moreover, by an increased reactivity towards oxidation of intermediates accumulated in respective microregions of the polymer sample which resembles then a system of microreactors. Within these local isolated microregions where the oxidation of a polymer takes place, another kind of inhomogeneity may be observed which is quite usual in solid-state reactions. It is conditioned by the fact that at least one of the reacting particles of the polymer system has a reduced mobility. The different energetical states of reacting particles thus do not succeed to equilibrate, which may lead to the observation of the several rate constants for the reaction of identical reactants, the phenomenon which is called "polychromatic" kinetics [4].

When indicating the effect of different types of heterogeneities in the oxidation of polymers, the role of the interface of the solid polymer/air should be accounted for through which oxygen enters the polymer bulk where it is consumed. This gives rise to the appearance of a gradient in oxygen concentration and to the gradient of oxidation products. The effect of the surface becomes particularly evident when a relatively large concentration of free radicals is produced in the polymer as, e.g., during very intense irradiation or at high reactivity of oxidation reactant (O_3, $^1O_2$). In such a case, the surface of the polymer product acquires properties completely different from those of the polymer bulk.

2 Elementary Reactions

The sequence order and relative importance of elementary reactions in the process of chain oxidation depends markedly on temperature and microviscosity of a polymer medium. As some elementary reactions have different activation energies, the changes in temperature induce not only the changes in reaction rate but also in the composition of reaction products.

The efficiency of free radical generation from their corresponding precursors in polymer is lower when compared with that in liquid or gaseous systems. This is due to a larger propensity of a cage effect and more difficult rehybridization of macroradicals in a polymer medium, which affects the ratio of fragmentation reaction and, also the ratio of reaction products when compared with reactions of low-molecular mass analogs.

Specific feature of polymers which differentiates them from low molecular mass compounds is the existence of anomalous or defect structural units which are always a part of the chain of normal mers. The presence of the defect mers is a consequence of the polymer synthesis, where the products of side-reactions are integrated into the polymer chains as anomalous units usually without the possibility of their further separation.

The number of anomalous mers in a polymer chain depends on the difference between the Gibbs energy of formation of the regular and anomalous mers, on the type of reactive intermediates in the polymerization, and on the reaction conditions. Mers with a similar structure are incorporated into the chain preferably. The higher energy declination of an irregular mer from the regular structure, the lower will be its concentration in the polymer. A peculiarity of the chain reaction is that its rate may be affected by a small structural change which reduces the strength of the bond of some site in the molecule. Let the activation energy of a given reaction be a reactivity scale depending on the change of bond strength. The reduction of several percent in the bond strength and thus in activation energy is reflected in a multiple value of the reaction rate at the same temperature (Table 1) or in the large difference in temperatures at which both reactions proceed with the same rates (Fig. 1). This seems to be illustrative enough to understand why oxidizability of a polymer is being usually higher than that of low-molecular models. The same reasoning may be put forward when explaining the different oxidizability of apparently identical polymers synthesized from the same parent monomer but at different experimental conditions.

Table 1. Multiples of the rate constant of a given reaction for the decrease of activation energy by a value E calculated for 4 different temperatures

E, kJ/mol	1	2	4	8	16	32	64
20 °C	1.51	2.28	5.2	27.1	734	5.4×10^5	2.9×10^{11}
100 °C	1.38	1.91	3.65	13.4	178	3.2×10^4	1.0×10^9
200 °C	1.29	1.67	2.78	7.7	60	3.6×10^3	1.3×10^7
400 °C	1.2	1.43	2.05	4.2	18	3.1×10^2	9.8×10^4

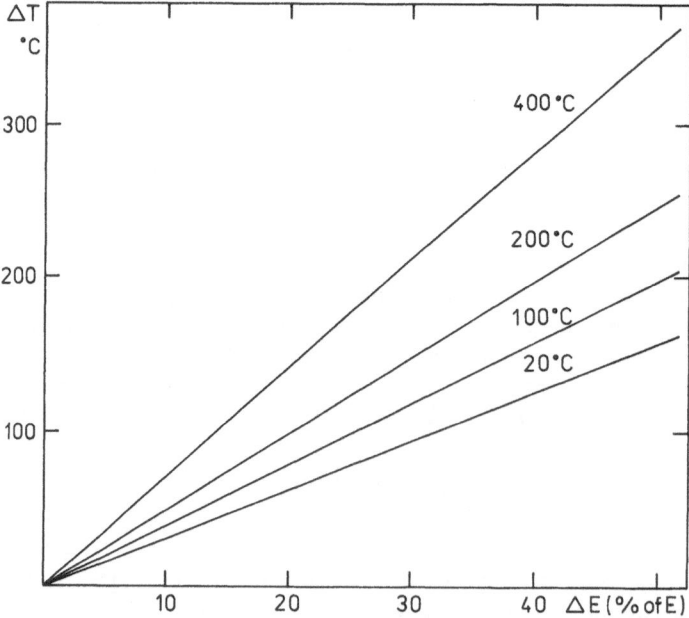

Fig. 1. The effect of the change of activation energy on the difference in temperatures at which given reactions proceed with equal rates

3 Initiation of the Chain Oxidation

The generation of free radicals in a polymer system occurs via various pathways, the resulting level of radicals being determined by the conditions of a polymer storage and use. Generally, the precursors of free radicals may be all compounds and substrates containing weaker bonds, the decomposition of which may be induced by radiation or reactive pollutants. Molecular oxygen or ozone which is always present in the air belong to particles which give rise to the easy appearance of free radicals in the polymer.

4 Atmosphere as a Source of Free Radicals

Photolysis of atmospheric pollutants by solar radiation results in an increase of ozone concentration in certain urban areas and is the cause of a sequence of oxidation reactions with polymers. Ozone reacts with practically all organic materials especially with alkenes. The rate of its reaction with alkene is several orders of magnitude higher than that with alkane. The ratio of the rate constants of ozone with ethene/ethane is 1.5×10^5, with propene/propane 1.6×10^6, and with butene-1/butane 1.1×10^6, at room temperature [5].

Reactivity of ozone with alkanes is still rather high and corresponds to radicals of medium reactivity. For instance, the rate constant for reaction of ozone with 2,3-dimethyl pentane has the value $0.3\,dm^3\,mol^{-1}\,s^{-1}$ at 20 °C [6]; the rate constant for linear alkanes is by one order lower, approximately.

In the ozonolysis of the aliphatic hydrocarbons the rate-controlling step is the abstraction of hydrogen atom, and the energy of the $C-H$ bond thus determines the overall rate constant [7].

For evaluation of the role of ozone in oxidation of hydrocarbons, the comparison of the reaction rates of ozone and oxygen with a certain hydrocarbon may be of use. The rate constant of this reaction is by 10^3 times higher [8], at minimum, for ozone and methane as reactants than that for oxygen and methane at the temperature as high as 200 °C. On the other hand, the concentration of ozone in the atmosphere is rather low under normal conditions.

The difference in reactivity may become more distinct at lower temperatures and may compensate the large difference in concentrations of ozone and oxygen. This is due to lower activation energy for the reaction of ozone with hydrocarbon, which is 61 kJ/mol, while that for oxygen is 230 kJ/mol. Thus, a conclusion can be put forward that ozone may play a significant role as an initiator of oxidation of polymers particularly when its concentration in the medium is somewhat higher than the average value. From, the possible initiation reactions of ozone, the likely seems that at which alkyl peroxyl and hydroxyl radicals are formed. It should be recalled that for the reaction of hydrocarbon and oxygen it is the reaction yielding $HO_2^{\cdot}$ radicals and alkyl radicals.

In the case of alkenes, the reaction with ozone starts on double bond $C=C$ as a cycloaddition:

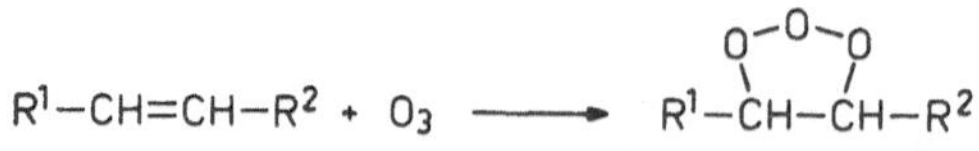

$$R^1-CH=CH-R^2 + O_3 \longrightarrow R^1-\underset{\underset{O}{\big|}}{CH}-\underset{\underset{O}{\big|}}{CH}-R^2$$

Trioxolane units (molozonide) are unstable and react to cyclic ozonides, branched polymeric peroxides, and tetroxanes.

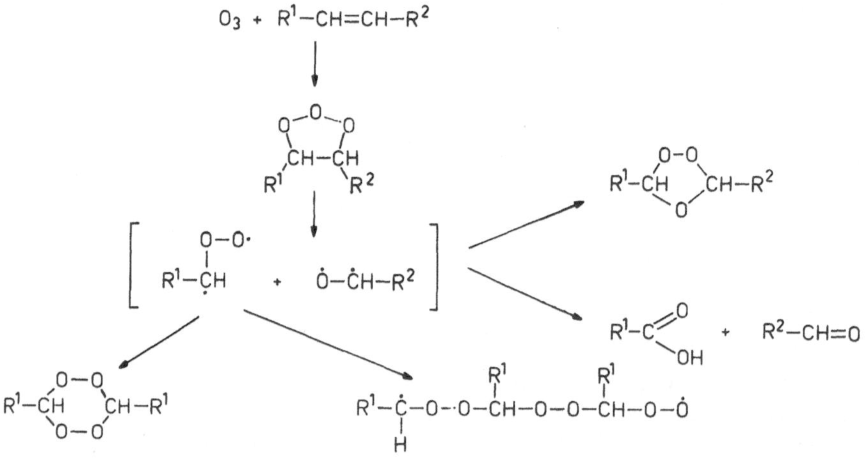

$$O_3 + R^1-CH=CH-R^2$$

Decomposition and transformations of molozonide

The ratio of respective reaction products depends on the conditions at which ozonolysis has been carried out [9].

Reactive intermediates may be captured when working with very high dilution of a reactant, in nonaqueous solvents and at a low temperature. It should be emphasized that at the reaction of ozone with polydienes, the main chain of a macromolecule is broken down ultimately at the site of the original double bond which leads to the deterioration of mechanical properties of rubbers.

Polydienes also can undergo oxidation with singlet oxygen. The present consensus is that the oxidation of polydienes with singlet oxygen occurs via a so-called ene-mechanism:

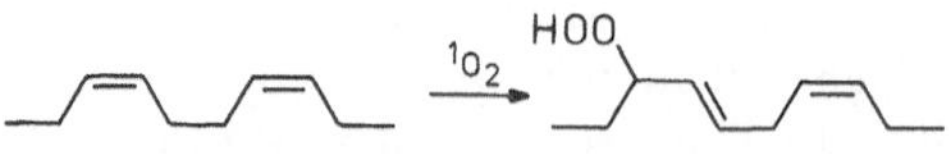

involving a shift of the double bond, its *cis-trans* isomerization, and formation of allyl hydroperoxide [10].

With natural rubber, several reaction pathways of the reaction with singlet oxygen exist.

H_3C H_3C
HOO
major product

H_2C H_3C
HOO
minor products

H_3C H_3C
HOO

H_3C H_3C
$+ \ ^1O_2$

In the polymer matrix, singlet oxygen may be generated in situ in several different ways: by chemical reactions, through the photosensitization of different chromophores as, e.g., carbonyl groups whose $n - \pi$ triplets are quenched by the ground state of oxygen with formation of singlet oxygen molecules.

One should bear in mind, however, that photodegradation of cis-1,4-polybutadiene in air carried out in the absence of a photosensitizer proceeds as a free radical process [11].

In the free radical oxidation of polybutadienes, both the peroxides and carbonyl groups are formed typically. Consequently, the presence or absence of carbonyl groups in the reaction products of the photooxidation of polydiene may be an indication of the oxidation mechanism.

The efficiency of the initiation effect of triplet oxygen depends on the hydrocarbon structure, i.e., on the strength of the attacked C−H bond. For instance, the ratio of the rate constants of the reaction of oxygen with formaldehyde and methane is 1.3×10^9 at 100 °C [8]. This indicates that intermediates of oxidation may be more sensitive towards oxidation than the original substrate which may contribute to the appearance of heterogeneous regions where the oxidation takes place preferably.

Peroxy nitric acids and organic peroxy nitrates are another precursors of free radicals, which may be introduced into polymers from polluted atmosphere. They produce both peroxy radicals and reactive nitrogen oxides (NO and NO_2) on decomposition. With alkyl peroxynitrates, decomposition proceeds via OO−N bond fission having activation energy 87 kJ/mol, their half-life being several seconds at 0 °C [12].

In initiation of oxidation, the important role may also be played by reactive bonds in a polymer. With polyethylene which is the most simple structurally, such reactive sites may be small concentrations of double bonds or more numerous sites of branching of a main chain to side alkyls. Investigation of the oxidation reaction of different types of polyethylene has, however, revealed that the degree of polyethylene branching from 0 to 20 branches for each 1000 carbon atoms of the main chain does not affect the induction period of oxidation [13].

On the other hand, a considerably lower level of $C=C$ bonds brings about a significant shortening of the induction period of oxidation; at a concentration of $C=C$ bonds below 0.1/1000 C the induction period is about 100 min while at 0.6 $C=C$/1000 C twenty min only at 160 °C. This leads to the suggestion that allyl hydrogen or double bond $C=C$ should be taken seriously as initiators of the first stages of polymer oxidation.

With high-density PE, unsaturation content is even more important than the metal impurities in determining the rate of oxidation. Samples with initial unsaturation level 0.385 mol% containing over twice the amount of Ti and 18 times of Al than those with a higher level of unsaturations (0.481 mol%) are still more stable both thermally and photochemically [14].

The higher unsaturation content has also a significant effect on the degree of crosslinking which is higher for samples with higher initial level of unsaturations.

On the other hand, with polypropylene, metal impurities play by far a more significant role [15] which is probably due to a considerably higher concentration of metal ions in this polymer (particularly in samples prepared from the gas phase), lower content of unsaturation, and also to the fact that polypropylene does not crosslink when oxidized.

As it was demonstrated by staining the oxidized polypropylene and its observation by UV microscope, a high degree of inhomogeneity at the micron level was observed even for the most thoroughly annealed samples. In most cases this is clearly associated with the catalyst residues. An observation of microdomains in the oxidized polymer in which the degree of oxidation is by far pronounced than in the rest of polymer may be explained by the effect of higher oxidation state of transition metal ions M which interact directly with polymer:

$$M^{(n+1)+} + PH \rightarrow P^{\cdot} + H^{+} + M^{n+}$$

$$P^{\cdot} + O_2 \rightarrow PO_2^{\cdot}$$

$$PO_2^{\cdot} + PH \rightarrow POOH + P^{\cdot}$$

$$POOH + M^{n+} \rightarrow PO^{\cdot} + {}^{-}OH + M^{(n+1)+} .$$

Obviously, the above sequence of reactions, which may be repeated many times concentrates the oxidative attack to the close vicinity of catalyst residue particle [16]. The question is whether the same effect can be obtained when traces of peroxides are decomposed from which only one or two free radicals can be formed at one site.

5 Photoinitiation

Pure hydrocarbon polymers not having proper chromophores should not absorb incident ultraviolet light from the surrounding atmosphere. If, in spite of this, photodegradation occurs in saturated polymers, it has to be attributed to the presence of chromophore groups absorbing UV light of wavelengths above 295 nm, such as aromatic polyconjugated structures, carbonyl groups, $C-T$ complexes polymer/molecular oxygen, and catalytic residues of metals which are evenly present in any polymer either bound or as admixtures. In the photoinitiation step, these inherent or admixed chromophores absorb the light and produce primary free radicals. The visible light has still sufficient energy to cleave the labile $O-O$ bond but this bond usually does not absorb in this range of wavelengths. The photolysis of hydroperoxidic groups by visible light occurs indirectly by an energy transfer mechanism. The necessary condition for the energy transfer to occur is the short distance between the excited chromophore group and peroxidic bond.

Experiments demonstrated that photolysis of secondary hydroperoxides in polyethylene does not lead to the initiation of chain oxidation of the polymer [17]. Hydroperoxides are decomposed here either by a molecular mechanism or by free radicals. If formed, they disappear during transformation within a cage reaction from which only a small part of free radicals succeeds to escape.

Using nanosecond laser flash spectroscopy, it has been shown that excited ketones can abstract hydrogen of hydroperoxide groups and generate free radicals as follows [18]:

$$[\text{>C=O}]^* \ + \ \text{POOH} \ \longrightarrow \ \text{>\.C-OH} \ + \ \text{P\.O}_2$$

Although ketone groups can produce free radicals in polymers also by other photochemical pathways, their role in initiation seems to be considerably less during initial stages of oxidation than that of hydroperoxides.

Hydroperoxides are much more efficient than ketones for initiating photooxidation of ethylene-propylene copolymers [19]. This fact was confirmed by the results from photolysis of low-molecular model compounds and isotactic polypropylene [20].

Despite the numerous papers devoted to photooxidation of hydrocarbon polymers [21], the initiation step has not been clearly established yet even for polyethylene or polystyrene which were the most studied [22, 23]. Difficulties which follow from solution of this problem consist in the necessity of analysis of small amounts of decomposing unstable structures and products which are thereby formed. Moreover, photoinitiation does not include one reaction only but the overall complex of many chemical and physical processes, which importance depends on experimental conditions.

6 Radiation Generation of Free Radicals

By the effect of ionization irradiation side substituents are abstracted from excited macromolecules in parallel to the scission of main chains. Elimination of side substitutions such as hydrogen and halogen atoms may be the main primary process. For instance, in polyethylene, although the $C-H$ bond is stronger than the $C-C$ bond, excited macromolecules eliminate hydrogen atoms preferentially through a $C-C$ bond cleavage of a macromolecular backbone. Part of the excitation energy may be transformed to the kinetic energy of dissociation products. This is a route by which hot hydrogen atoms of high reactivity are formed. The hydrogen atoms formed react fast with the CH_2 groups of a surrounding polymer chain or with the parent macroradicals and form secondary alkyl radical or unsaturation and molecular hydrogen. Radiation yield (number of species formed by absorption of energy of 100 eV) of formation of macroradicals [24] is 2–4.

Hydrogen atoms may react also with oxygen dissolved in the polymer and hydrogen peroxyl radicals are thus formed. Part of the hot hydrogen atoms may, however, react to give hydroxyl radicals and oxygen atoms. This should explain the relatively high yield of radicals [25] formed during initial phases of polyolefin irradiation.

The formation of oxygen anion radicals and molecules of ozone also should be counted with at the ionization initiation of oxidation [26]. Initiation reaction caused by oxygen anion radicals may play an important role within the polymer bulk while the effect of ozone forming in the surrounding air atmosphere will include only the formation of radicals on the polymer surface. The latent effect of ionization initiation on polymer oxidation which is very distinct may be documented on a relatively fast increase of concentration of carbonyl groups, observed over 1 year after irradiation crosslinking of polyethylene [27].

The ionizing radiation combined with light causes synergetic effects and contributes to intense chemical changes in polymers. Photoradiation effect can result in a sharp increase of free radical yield which may accelerate oxidation. The synergetic effect of both the gamma irradiation and light on free radical yield [28] depends on the polymer structure [29]. An increase of free radical concentration may be explained by a photochemically induced increase in the escape of free radicals from the cage of radical pairs formed during ionization irradiation. Without the effect of light, these radicals should recombine.

Taking into account the effect of radiation on initiation of oxidation, one thing should be born in mind, namely that the cosmic radiation present everywhere may be one of the reasons of oxidation which occurs more or less slowly in any organic material.

7 Thermally Induced Initiation

Voluminous literature data exists on the thermal degradation of polymers. It was, e.g., established that characteristic parameters of the polymer degradation are determined by the way of preparation of a particular sample. It has been postulated that the less stable samples contain weak links in the macromolecule. However, the nature of these links is not well defined; it is known that the weakest links in polymer chains are peroxidic bonds. In hydrocarbon polymer stored in air, a certain concentration of peroxides will always be present as well as of free radicals derived from the decomposition of peroxides.

In spite of low concentration of these reactive particles, they may start the initiation of thermo-oxidation. The dependence of the course of the polymer oxidation on the residual amount of these initiating particles may be exemplified by experiments where a standard sample of polypropylene was annealed at 150 °C for a certain time in nitrogen atmosphere and subsequently oxidized in oxygen. The induction time τ of oxidation at 150 °C increased with the time t_a of sample annealing, accordingly, indicating that the primary initiator was partially consumed (Fig. 2). Assuming that the initiation of oxidation is due to hydroperoxides decomposing by the first-order mechanism, the following equation may be derived for induction time of oxidation in dependence on annealing time t_a:

$$\tau = 1/k_1 \left[2 * \ln 2 \left(1 - \sqrt{POOH_0} \cdot \sqrt{k_1 k_6}/k_3 \cdot PH \cdot e^{-\frac{k_1}{2} \cdot t_a} \right) - 1 \right],$$

where k_1 is the rate constant of hydroperoxide decomposition, k_6 is the rate constant of peroxy radical recombination, k_3 is the rate constant of the transfer of peroxy radical towards the polymer molecules PH, and $POOH_0$ is the initial concentration of hydroperoxides in the polymer.

Optimization of parameters of these equations for a set of data from Ref. [30] yields values of $k_1 = 9.5 \times 10^{-4}$ min^{-1} and $\sqrt{k_6 ROOH_0}/k_3 = 100$ mol s$^{1/2}$ kg^{-1}. It may be of interest that the rate constant k_1 is more than 4 times higher than that for decomposition of *tert*-butyl hydroperoxide at 150 °C (2×10^{-4} min^{-1}) (inert solvent).

Concentration of peroxides, the rate of their decomposition, and their structure depends on conditions of storage of a polymer and on its structure. It is, e.g., well known that the rate constant of decomposition of polypropylene peroxides

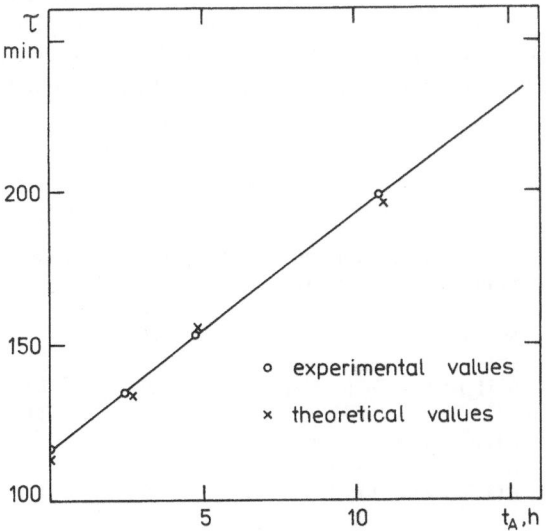

Fig. 2. The plot of induction time of oxidation of polypropylene (oxygen, 150 °C) on time t_A of the polymer annealing in nitrogen at 150 °C

prepared in advance by an initiated oxidation at higher temperatures (5.9×10^{-3} min^{-1} for 150 °C) [31, 32, 33] is considerably higher than that obtained by simple storage of the polymer.

This is obviously due to the aggregation of hydroperoxides formed in local microdomains of the polymer. Accumulation of hydroperoxides in a certain microregion brings about their faster decomposition by an assisted bimolecular reaction and the domain increases its sizes.

Peroxidic oxygen may be incorporated into the polymer already during synthesis. The problems arising from more or less perfect removal of oxygen during the polymer synthesis may be documented on the study of methyl methacrylate polymerization carried out without any initiator in an inert medium [34], from which one can deduce that the "thermal" polymerization of methyl methacrylate is initiated by adventitious peroxides.

Overview articles on the formation and decomposition of polymer peroxides [35, 36, 37] confirm the formation of weak bonds during the polymer synthesis following from various experimental data.

The values of dissociation energy of O−O bonds in various molecules (Table 2) corresponding temperatures of decomposition are consistent with the fact that organic peroxides are generally unstable. The second important fact which ensues is a large temperature interval within which free radical decomposition of peroxides may occur dependent on their structure.

Ozonization of the polypropylene powder creates the peroxidic species in the polymer, as well. The activation energy [41] of the thermal decomposition of these peroxides is 100 kJ/mol. In the decomposition of peroxides more than one type of radicals was trapped. Moreover, the three exotherms (peak at 40, 90, and 130 °C) were observed on DSC thermograms of ozonized sample which also indicates the presence of several types of peroxides. Besides the peroxidic bonds in polymer, selective thermal decomposition may occur also with such bonds in the polymer as, e.g., with end groups containing the initiator moieties [42]. This, however, takes place at higher temperatures than it corresponds to usual temperatures at which the thermo-oxidation starts.

Table 2. Dissociation energy of O−O bonds and approximative temperatures for some peroxides at which half-life of their decomposition in inert solvent is 10 h [38–40]

Molecule	kJ/mol	°C	Molecule	kJ/mol	°C
HO−OH	213	330	$(CH_3)_3CO−OC(CH_3)_3$	159	125
.O−OH	276	−	$CH_3CH_2O−OCH_2CH_3$	156	119
$HO−OC(CH_3)_3$	194	170	$CH_3C(O)−O−O−C(CH_3)_3$	134	68
$CH_3O−OCH_3$	159	126	$(CH_3)_3CO−O−O−OC(CH_3)_3$	73	−93
$.O−OCH_3$	245	− *	RO−O−O−OR**	68	−112

* During fragmentation of free radicals, primary process is not the dissociation of O−O bond but rather H−O$_2$. bond (205 kJ/mol), or .O−CH$_3$ (103 kJ/mol).
** R = $(CH_3)_3C(CH_3)_2C−$

Organometallic compounds with dissociation energy of the metal-carbon bond between 50 and 300 kJ/mol belong among thermally unstable compounds, too. For structurally similar compounds, the values of this dissociation energy depends on the position of the central metal atom in the periodic system increasing in a given transition period from left to right and within a group on going from top to bottom.

The low values of dissociation energy of a bond metal-carbon are characteristic particularly for alkyl transition-metal compounds. Some of them are the key intermediates of Ziegler-Natta low-pressure polymerization of alkenes. Outside the laboratory or chemical plant [43, 44]. Nature gives us also important representatives of such compounds namely vitamin B_{12} or co-enzyme B_{12}.

The role of metals in the oxidation of polymers is manifold; it does not include the only the initiation step and the generation of free radicals following the homolysis of relatively unstable bond metal-carbon but they may also enter propagation and termination reactions.

On conditions of low-temperature oxidation the current bond of a polymer chain may be cleaved into radicals when it is under simultaneous effect of mechanical stress. The scission of the chemical bond by the effect of mechanical forces may be understood as the process of mechanically activated thermal decomposition. Primarily formed radicals are usually transformed to more stable ones. For example, the scission of polyoxy methylene macromolecules gives at first two kinds of radicals, namely $RCH_2O.$ and $ROCH_2^{.}$. These primary particles abstract hydrogen atom from unbroken parts of macromolecular chains and change thus to one type of secondary radicals, $RO\dot{C}HOR$. The radicals are formed here mainly by ruptures of the tie-molecules linking crystalline regions [45]. The formation of free radicals depends on the extent of deformation. The high concentration of radicals may be received even before the macroscopic rupture of the sample is observed. This fact explains the chemical reasons for the fatigue and ageing of organic polymers during their exposure to mechanical forces. The alkyl radicals thus formed are prone to react with atmospheric oxygen and the original structure and properties of the polymer system are changed accordingly.

Free radicals formed in polymers due to thermomechanical stress appear not only during the polymer use but also during the polymer processing and shaping to final products [46]. The kind of initiation which prevails in a certain polymer depends not only on initial conditions of oxidation but also on the extent of a previous oxidation as well as on the occurrence of additional interactions among oxidation products. Increasing extent of oxidation is usually characterized by higher concentration of hydroperoxides which are secondary sources of initiation. The products of oxidation formed may alter the kinetics and mechanism of hydroperoxide decomposition so that the rate of initiation is the result of several mutually coupled processes.

The initiation effect of ultrasound generating hydroxyl radicals in reaction medium as, e.g., during sonochemical oxidation of aldehydes has been widely studied but rarely utilized.

8 Propagation in Oxidation

The reaction of alkyl radicals with molecular oxygen should formally be considered as an addition reaction:

$$R^{\cdot} + O_2 \rightleftharpoons RO_2^{\cdot}.$$

As the ground state of oxygen is triplet, the above reaction may also be interpreted as the recombination. This is in correspondence with the high value of the frequency factor as well as with almost zero value of the activation energy observed for this reaction. The activation energy of the backward process is considerably higher.

The stability of $RO_2^{\cdot}$ radicals may be characterized by a ceiling temperature at which the rate of the forward and backward reaction are equal. For isopropyl and isopropyl peroxy radicals it is about 350 °C.

The β-scission process with peroxy radicals that are highly resonance stabilized, such as with triphenyl methyl peroxyls may be observed even at a markedly lower temperature. The regeneration of alkyl radicals in a reaction system may however proceed through numerous fragmentation and transfer reactions which makes the correct determination of ceiling temperature of peroxy radicals backward decomposition more complicated.

The rate constants for the reactions of carbon-centered radicals with oxygen are not very sensitive to the extent of spin delocalization. For example, the rate constant (in $dm^3 \, mol^{-1} \, s^{-1}$) for *tert*-butyl is 4.9×10^9, benzyl 2.4×10^9, cyclohexadienyl 1.6×10^9, 4-nitrobenzyl 9×10^8, and diphenyl methyl 7.5×10^8, respectively.

In the gas phase, the high-pressure limiting rate constants for methyl [52] and for ethyl radical [53] have the value 1.3×10^9, resp. 2.6×10^9 which are comparable with the rate constants in the liquid medium.

Oxygen atom reacts with alkyl radicals by two orders of magnitude faster than molecular oxygen, the rate constant being the highest for alkyl radicals with the lowest ionization potentials [54] (Table 3).

Table 3. Rate constants for the reaction R. + O

Radical	Ionization potential eV	k $10^{-10} \, cm^3 \, molecule^{-1} \, s^{-1}$
$(CH_3)_3C^{\cdot}$	6.9	8.7
$C-\dot{C}_6H_{11}$	7.4	2.7
$(CH_3)_2CH^{\cdot}$	3.5	4.9
$CH_3\dot{C}O$	8.1	3.2
$CH_3CH_2^{\cdot}$	8.5	2.2
$H\dot{C}O$	8.6	2.1
$\dot{C}H_3$	9.8	1.4

The high-pressure limiting rate constants also correlate with the ionization potential of the alkyl radical; the lower the ionization potential of the alkyl radicals, the higher is their rate constant with molecular oxygen [55, 56].

It is of interest that the carbonyl radicals react more slowly with molecular oxygen than do the alkyl radicals, both having similar ionization potentials. Even higher declination from the correlation between ionization potential and the rate constant of the reaction $R^{\cdot} + O_2$ may be seen on the case of methyl radicals and hydrogen atoms. Despite the higher ionization potential, hydrogen atoms react with a higher rate constant $(4.5 \times 10^{10} \, mol^{-1} \, dm \, s^{-1})$ than methyl radicals. The trend in the correlation is not due to the sterical effect. The radical center in the *tert*-butyl radical is shielded by three methyl groups, and yet the rate of its reaction with oxygen is 10 times faster than for the reaction of methyl radicals with oxygen.

The low effect of the structure of oligomer alkyl radicals on the rate constant of their reaction with oxygen also follows from the review article on the copolymerization of vinyl monomers with oxygen in the liquid phase.

radical	$-CH_2-\overset{\cdot}{C}-(CH_3)-COOCH_3$	$-CH_2-\overset{\cdot}{C}H-CH=CH_2$	$-CH_2-\overset{\cdot}{C}H-C_6H_5$
k (dm^3 mol^{-1} s^{-1})	1×10^7	1×10^7	3×10^7

Comparing these data with the rate constants in the gas or even in a liquid we may say that the above constants for oligomer radicals are considerably lower. This may be related to the conformation effects causing more difficult accessibility of oxygen to a radical site or to a lower solubility of oxygen in the environment of these radicals than it is the average value in the system as a whole.

The rate constant for the reaction of ethyl radicals and oxygen increases with the increasing concentration of inert helium [57]. This indicates the possible participation of the molecules in the process of stabilization of peroxy radicals formed. The excited alkyl peroxy radicals do not fragmentate only to original reactants but rearrange to products through separate pathways [58].

Besides the splitting molecular oxygen out with the rate constant k_a, ethyl peroxyl radicals may fragment (k_b) to hydrogenperoxyl and ethylene. The ratio of rate constants k_a/k_b for these two alternatives is about 6 for excited ethyl peroxyl radicals at the value of $k_a \, 10^6 \, s^{-1}$. The rate constant for the deactivation of excited radicals with the third body M:

$$C_2H_5OO^* + M \xrightarrow{k_7} C_2H_5OO^{\cdot} + M$$

is however rather high (for $M = He$, $k_7 = 6 \times 10^{10} \, dm^3 \, mol^{-1} \, s^{-1}$) which indicates that reactions of excited radicals are marginal.

The reduction in the mobility of radicals and in the accessibility of oxygen to a radical site has a decelerating effect on addition reaction. Such a case may be observed [59] during the oxidation of carbon-centered radicals derived from triglycerides of some fatty acids such as arachidic $(CH_3-(CH_2)_{18}-COOH$, m.p. 348 K), oleic $(CH_3-(CH_2)_7-CH=CH(CH_2)_7COOH$, m.p. 290 K), linoleic $(CH_3-(CH_2)_4-CH=CH-CH_2-CH=CH-(CH_2)_7-COOH$, m.p. 268 K), linolenic $(CH_3-CH_2-CH=CH-CH_2-CH=CH-CH_2-CH=CH-(CH_2)_7$ $-COOH$, m.p. 262 K). While triarachidin is polycrystalline, triolein and other unsaturated triglycerides tend to form a glass at low temperatures. In triarachidin samples (m.p. 350 K) irradiated at 77 K, peroxyl radical formation begins at 200 K and at 210 K only $RO_2^{\cdot}$ species are detectable. On the other hand, in unsaturated triglycerides the conversion of carbon-centered radicals to peroxyl radicals occurs already near 125 K for each sample. The difference in the temperature at which the addition reaction with oxygen occurs is given by the difference in the physical state of the triglyceride matrix. The production of peroxyl radicals is explained by the release in the diffusion of oxygen and by its migration to carbon-centered radicals which remain immobile. This may be due to a relatively high viscosity at a given temperature. As a consequence, a very high conversion of $R^{\cdot}$ to $RO_2^{\cdot}$ (near 90%) without termination of free radicals is observed.

Reactivity of carbon-centered radicals may remain high even in crystalline state of the matrix when the access of oxygen towards unpaired electrons is not hindered by the sterical arrangement of surrounding molecules. This is, e.g., the case of peroxyl radicals formed during irradiation of cholesterol [60]. Oxygen reacts with alkyl radicals derived from cholesterol already at 125 K which is far below the melting point of the matrix (423 K). Two peroxyl radicals of cholesterol were observed (Scheme).

The former, having oxygen attached to the tertiary carbon atom of alkyl side chain, prevails at low temperature of oxidation while the latter with oxygen on allyl carbon of cycloalkane appears at room temperature. This result is somewhat contradictory to the reactivity of C−H bonds attacked. With regard to a lower dissociation energy of C−H bond in allyl $(D(C-H)_{allyl} = 362$ kJ/mol) than that in tertiary alkyl $(D(C-H)_{alkyl} = 385$ kJ/mol), one should expect that it will be the allyl hydrogen which is more reactive at lower temperature. The fact that it is not so may be related to a different mobility of functional groups at different temperatures. The more rigid cyclic structure with allyl hydrogens has obviously higher activation energy of conformation changes than the side alkyl.

The kinetics of an addition of oxygen to macroradicals in polymer systems performs the stepwise character which is caused by the kinetic non-equivalency of reactants in the solid state [4]. Achieving a certain conversion, the process ceases and may be reinitiated by a temperature increase. The carbon macroradicals begin to react already at 90 K. The resulting extent of transformation towards peroxyl radicals does not depend on the way how the temperature increase has been performed (i.e. whether all at once or in steps), which indicates that the stepwise character of a reaction may be attributed to the existence of discrete subsystems of free radical sites in a polymer which have relaxation times from several seconds to hundred of hours. The kinetic curves obtained for a certain reaction under conditions of subsequent de-icing of the polymer matrix thus submit an information on distribution of identical reactants of different reactivity. The stepwise trans- formation of radicals is characterized by different rate constants for identical reaction occurring in structurally non-equivalent parts of the system. This fact is displayed namely in a value of activation energy.

From a variety of measurements, the large differences in the rate constant for addition of oxygen to macroradicals may be seen for the same polymers and a certain temperature [1] (Table 4). The different rate constants cannot be assigned to a different reactivity of a radical site. This follows not only from the estimations of electron density on the carbon atoms of a radical centre and from the rate constant of the corresponding reaction in the liquid phase but also from the fact that identical radicals which were generated through different methods have different rate constants because of different conditions of mobility, other conditions being comparable. Reactivity of chemically identical radicals in a given polymer depends markedly on the physical structure of the solid phase (degree of crystallinity, orientation of macromolecules, sizes of lamellas in crystallites, mobility of chain segments in the surroundings of a radical center, etc.). The kinetic parameters of addition reaction of oxygen on macroradicals occurring in solid polymer thus reflect more the parameters of ordering in the assembly of macromolecules than the chemical properties of a radical site. The importance of the factor of accessibility and mobility of the radical center arises not only in the reactions taking place in the same polymer system but also in explanation of variations in kinetics of peroxyl radicals formation in different polymers.

The effect of freezing of various molecular motions of reactants on reactivity of alkyl radicals may be exemplified by comparable values of the rate con-

Table 4. The values of rate constants k of transformation of alkyl radicals to peroxyl radicals in polymer matrix at 200 K and data of minimum (min) and maximum (max) activation energy E

Matrix	Generation of R.[a]	Temperature K	k $dm^3 \, mol^{-1} \, s^{-1}$		E, kJ/mol	
					min	max
polytetrafluorethylene	r	273–305	4×10^{-5}		41.9	
poly(methyl methacry-late	m	133–203	9×10^{-3}		33.1	
poly(vinylalcohol)	m	193–253	4×10^{-1}		24.7	
polystyrene	m	123–178	4×10^{-2}		22.2	
	r	90–130	4×10^3	1×10^5	29.3	35.2
	p	95–180	3×10^4	8×10^5	22.6	28.1
i-polypropylene	r	100–140	4×10^2	9×10^3	19.3	24.7
isopropylbenzene	r	90–130	6×10^4		23	

[a] The route of free radicals generation: r — radiation, m — mechanical degradation, p — photochemically

stant and activation energy for addition of oxygen on cumyl radical in the solid cumene.

Worth of noticing is the difference about 10 orders of magnitude for the rate constant of reaction of oxygen with polytetrafluoroethylene and polystyrene radicals generated by ionization irradiation (Table 4). The deceleration of the reaction of oxygen with perfluoroalkyl radicals is mostly due to a sterical shielding of unpaired electrons by surrounding substituents similarly as in persistent radicals which are formed following the irradiation of perfluoroisopropyl-4-methyl cyclohexane [61] or addition reactions of radicals [62] on perfluoro-4-methyl pentene-2.

Following the irradiation of polyethylene, allyl-type radicals accumulate in the system at room temperature. In contact with oxygen they are converted to peroxyl radicals. The rate of oxidation of radicals in crystalline regions of low-density polyethylene is by two orders higher than in high-density polyethylene [63]. This may partially be due to the higher rate of oxygen diffusion in the lower density crystalline regions. (The density of crystalline lattice in low-density polyethylene is by 1.5% lower than that in HDPE.) The difference in density of crystalline phases of HDPE and LDPE is however too small to explain a marked difference in kinetics and mechanism of oxidation reaction. It was, e.g., ascertained that in LDPE, peroxyl radicals ($RO_2^•$) while in HDPE oxyl ($RO^•$) radicals are formed.

The ratio of $RO_2^•/RO^•$ in HDPE upgrades with increasing temperature and with increasing proportion of side branches on the polymer chain [64]. For analogous reactions performed in the gas phase, the ratio $RO_2^•/RO^•$ decreases with increasing temperature particularly as a consequence of higher activation energy for the formation of oxyl radicals when compared with peroxyls. In the solid polymer, the effect of this difference in activation energies is overlapped by the counteracting effect of temperature on the structure of the solid matrix. Increasing temperature decreases the density of packing of polymer chains. The enlargement of free volume in polymer crystallites is more favorable for the formation of peroxyl than oxyl radicals.

The overall activation energy for both processes is about 60 kJ/mol which is one half of the value for the activation energy of oxygen diffusion into the crystalline region of HDPE [65]. The lower values of activation energy for oxidation of allyl radicals show on the participation of reactions of such macroradicals which are in the close proximity of amorphous regions.

The formation of oxyl radicals in reaction of alkyl radicals with oxygen is important for the possible occurrence of branching of oxidation reaction due to oxygen atoms. This aspect is not, however, obvious from the experiments performed. It seems that oxyl radicals in HDPE are not formed in propagation reaction of alkyl radicals and oxygen but as the product of termination of two peroxyl radicals.

$$2 RO_2^• \rightarrow 2 RO^• + O_2 .$$

The reaction of alkoxy radicals, as the intermediates of hydrocarbon oxidation, with molecular oxygen takes place in the case of primary and secondary radicals by disproportionation route:

$$RCH_2O^{\cdot} + O_2 \rightarrow RCHO + HO_2^{\cdot}$$

which has a relatively low activation energy (Table 5).

Table 5. Arrhenius parameters and rate constants for disproportionation of alkoxyl radicals and oxygen

Radical	A $dm^3\,mol^{-1}\,s^{-1}$	E kJ/mol	k $dm^3\,mol^{-1}\,s^{-1}$	Ref.
$CH_3O^{\cdot}$	7.6×10^7	11.2	8.1×10^5	[66]
$CH_3O^{\cdot}$	—	—	3×10^7	[67]
$C_2H_5^{\cdot}$	1.8×10^8	7.7	8.1×10^6	[68]
$C_3H_7O^{\cdot}$	1.7×10^8	7.3	9.1×10^6	[68]
$(CH_3)_2CHCH_2O^{\cdot}$	1.9×10^8	6.9	1.2×10^7	[68]

9 Addition of Peroxyl Radicals to a C=C Bond

Unsaturated hydrocarbons have two important pathways of transformation in oxidation reaction. One alternative consists of the transfer of reactive hydrogen:

$$RO_2^{\cdot} + R'CH_2CH=CH_2 \rightarrow ROOH + R'\overset{\cdot}{C}HCH=CH_2$$

with formation of hydroperoxides and allyl radical, the second elementary reaction is the addition of a peroxyl radical to double bond C=C.

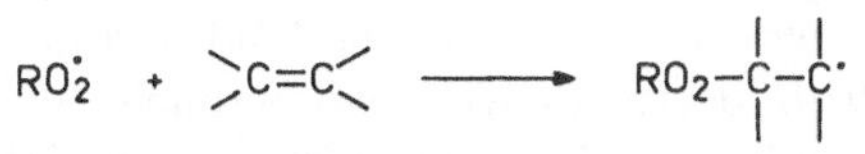

The addition of the peroxyl radical to the double bond is governed by the electron density in the alkene bond and by electrophility of the radical. The rate constants of addition reactions increase with an increase of electron density on the double bond and with the increase of the electrophilic character of a radical (Table 6). The considerably larger electrophility of acyl peroxy radical ($CH_3CO_3^{\cdot}$, $C_6H_5CO_3^{\cdot}$) may explain by 5 orders faster addition of acyl peroxyl radicals [69] to α-methyl styrene at 20 °C. Electrophility of radicals leads to the marked reduction of activation energy of addition to alkenes: methyl peroxyl radical has 47 kJ/mol, while acetyl peroxyl radical has 19 kJ/mol [70].

Alkyl radical containing peroxy group in the vicinity of a radical site which was formed in an addition reaction is unstable and decomposes rapidly to oxirane and alkoxyl radical.

Table 6. Rate constants $(dm^3 \, mol^{-1} \, s^{-1})$ for the reaction of oligomer peroxyl radicals with some monomers at 50 °C

Radical Alkene	$-CH_2(C_6H_5)(H)CO_2^{\cdot}$	$-CH_2(CH_3)(CH_3OCO)CO_2^{\cdot}$	$-CH_2(CH_3)(C_6H_5)CO_2^{\cdot}$
$(C_6H_5)(CH_3)C=CH_2$	204	43	28
$(C_6H_5)(H)C=CH_2$	98	23	13
$(CH_3)(CH_3OCO)C=CH_2$	12	1.8	1.8

10 Transfer Reactions of Peroxyl Radicals

It seems obvious to assume a priori that the facility of the transfer reaction depends on the strength of the broken and newly formed bond between radical and hydrogen atom transferred. The dependence between the activation energy E and dissociation energy D of the broken bond $R'-H$:

$$E = \alpha D + \beta \qquad \text{(Evans-Polanyi rule)}$$

may be approximated by the straight line only within the series of compounds of the same type. The constant α, which depends on the kind of attacking radical, is from the interval 0.2 to 0.9. It decreases with exothermicity of the transfer reaction. The constant α may simultaneously be taken as a rating of the selectivity of a radical at the abstraction of hydrogen from hydrocarbons; the radicals with higher α are more selective. Peroxyl radicals have the value of α about 0.4. With respect to a small difference in the values of the frequency factor for the transfer reaction of a given radical with the same type of chemical bond, the rate constants may be correlated with the values of the dissociation energy [71, 72].

The rate constant of a transfer reaction will therefore be the higher, the weaker $C-H$ bond is attacked by a peroxyl radical. From this it is obvious that the maximum rate of oxidation of polyethylene will increase with increasing number of tertiary hydrogens in the polymer [13]. Since the process includes the interaction of a macroradical with a macromolecule which both are of restricted translational mobility, the maximum rate of oxidation does not depend on the low content of reactive allyl hydrogens in polyethylene.

The reactivity of peroxyl radicals increases with electron-withdrawing power of their substituents [73].

In the solid state of a polymer there exists a connection between the degree of motional freedom of a peroxyl radical and its reactivity [74].

The transfer of hydrogen to peroxyl radicals may proceed intra- or inter-molecularly. Intramolecular transfer reaction (isomerization) of peroxyl macrora-dicals of polypropylene occurs during the oxidation of the polymer in a solution of inactive solvent [75] while the intermolecular transfer is preferred during the oxidation in reactive solvent or in the crystalline state [76].

11 Degenerated Branching in Oxidation Reactions

Oxidation reactions of hydrocarbons have a typical course. From the low rates, the reaction accelerates successively due to the consecutive formation of another source of free radicals which increases the rate of the primary initiation reaction. The amplification of the number of reactive free radicals is caused mainly by the decomposition of alkyl hydroperoxides, dialkyl and diacyl peroxides and peracids which are formed as intermediates in the oxidation reaction.

Selectivity in the transformation of hydrocarbon towards a specific hydroper-oxide is usually low which is due to the parallel decomposition of the latter and to side reactions of peroxyl radicals with oxidation products. In the case of polymers this is moreover enhanced by the presence of reactive defect structures in a polymer, which are the sites of the oxygen attack in the first stages of the reaction.

At low concentrations of oxygen in hydrocarbon, not all carbon-centered radicals react with oxygen but part of them attack hydroperoxides formed. At higher concentration of oxygen, a more quantitative transformation of alkyl radicals to peroxyl radicals occurs, which leads to a higher yield of hydroperoxides [77].

In the presence of molecular oxygen, the oxidation of initially pure paraffins and cycloparaffins takes place at temperatures from 120 to 160 °C. In parallel with other products, there are formed acids, aldehydes, and ketones which induce the decomposition of hydroperoxides. This is documented by kinetic curves of hydroperoxide accumulation which show a maximum [78].

The appearance of the maximum on the kinetic curves of hydroperoxide formation may be affected also by a successive consumption of reactive defect structures in the case of polymers.

For most polymers, the yield of hydroperoxides is relatively low even in the presence of oxygen excess. The relatively high values were, e.g., obtained during oxidation of atactic polypropylene [79]. In the initial phases of oxidation, the yield of hydroperoxide related to 1 mol of oxygen absorbed is 0.6 at 130 °C; when passing the maximum concentration it decreases considerably. In isotactic polypropylene, the maximum yield of hydroperoxides attains the value 0.2, only [80]. This may be probably related with a local accumulation of hydroperoxides in domains of defects in the crystalline structure which leads to an increased ratio of participation of hydroperoxide groups in the chain reaction of an oxidation process (induced decomposition of hydroperoxides) and finally to a lower yield of hydroperoxides per oxygen consumed. This is in accordance with the fact that the rate of decomposition of many polymeric hydroperoxides is more affected by the method of preparation than by the type of a polymer itself [33]. At the same time, a significant change in the rate of decrease of polymer hydroperoxide groups when different previously oxidized polymers are heated in an inert atmosphere may be assigned mainly to a different propensity of a chain radical-induced decomposition of hydroperoxide while the changes in dissociation energy of $O-O$ bond play a minority role here. The different rate in disappearance of hydroperoxide groups observed for polyvinyl acetate and polyacrylic acid dissolved in different solvents may be interpreted obviously in the same way [81].

If the decomposition of hydroperoxides takes place in hydrocarbon solvents, the oxidation products are formed either from radicals of hydroperoxide (see scheme below) or from radicals which arise in transfer reaction of the former with the solvent.

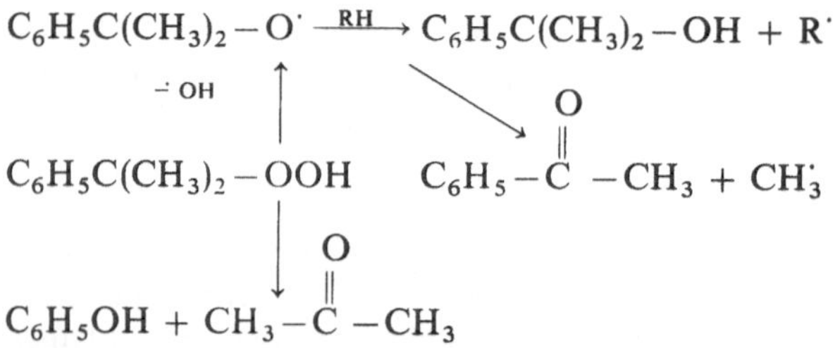

$$C_6H_5C(CH_3)_2-O^{\cdot} \xrightarrow{\text{RH}} C_6H_5C(CH_3)_2-OH + R^{\cdot}$$

$$C_6H_5C(CH_3)_2-OOH \qquad C_6H_5-\overset{\overset{\text{O}}{\|}}{C}-CH_3 + CH_3^{\cdot}$$

$$C_6H_5OH + CH_3-\overset{\overset{\text{O}}{\|}}{C}-CH_3$$

Example of the formation of decomposition products in a solvent from cumene hydroperoxide

Oxidation products from the solvent are formed even in the absence of molecular oxygen in a reaction system. Alcohols derived from the solvent molecules arise as products of an induced decomposition of hydroperoxides:

$$R_2\dot{C}H + R'OOH \rightarrow R_2CH-OH + R'O^{\cdot}$$

while the ketones result from interaction of solvent radicals and alkyl peroxyl radicals.

$$R_2\dot{C}H + R'OO^{\cdot} \rightarrow R_2C=O + R'OH$$

The amount of alcohols and ketones formed depends on the nucleophile character of alkyl radicals (Table 7). As documented by analytical data, increasing concentration of cumene hydroperoxide (CHP) reduces the extent of attack of hydrocarbon by primary radicals. In decomposition of CHP in heptane at 125 °C, the relative amount of heptanols (in mmol/100 mmol CHP) formed decreases (21.7, 18.0, 13.6) with an increasing concentration of CHP (0.3, 0.6 and 1 mol·dm^{-3}) while the amount (in mol%) of heptanones does not change (10.9, 11.7, 11.0). Obviously, the oxygen-containing radicals formed during the

Table 7. Reaction products obtained after 80 h of decomposition of cumene hydroperoxide (CHP) ($0.6\ mol \cdot dm^{-3}$) in some solvents at $125\ °C$[a]

Solvent	2-Phenylpropan-2-ol	Acetophenone	Phenol (acetone)	Alcohols	Ketones
				from solvent	
	mmol/100 mmol CHP				
Cyclohexane	66.4	15.3	4.0	15.4	10.9[b]
Heptane	78.9	19.5	–	18.0	11.7
cis-Pinane	88.5	11.5	–	32.0[c]	–

[a] Table constructed from the results of Lauterbach et al. [82].
[b] Uncertain result because of the possible confusion with α-methyl styrene.
[c] Pinan-2-ol 3 mol%, α-terpineol 29 mol%; induced decomposition of CHP is preceded by isomerization of pinanyl radicals

decomposition of the CHP react with the initial hydroperoxide than with the solvent. With increasing concentration of CHP in the system, a relative ratio of $R_2CH^{\cdot}$ radicals decreases at the expense of $RO^{\cdot}$ radicals and therefore, the ratio of heptanol and heptanone decreases, as well.

12 Fragmentation of Radicals

Before the abstraction of hydrogen atom from surrounding molecules of hydrocarbon or any other reaction with other reactants, peroxy radical may split into a saturated molecule and a new radical. The most discussed was the fragmentation leading to alkyl radical and oxygen. This reaction affects not only the oxidation kinetics but also directly influences the composition of reaction products. Variations in product composition will be predominantly due to a different ratio of peroxyl and alkyl radical concentrations and thus to different rates of competition reactions. The changes in the equilibrium constant of propagation of alkyl radicals with oxygen and a reverse fragmentation induced by reaction conditions (concentration of oxygen, temperature, the pressure of surrounding gaseous phase) have, therefore, an important impact on the course of polymer oxidation.

Of importance may be also another fragmentation giving rise to hydrogenperoxyl radicals:

$$R-CH_2-CH(R')-OO^{\cdot} \rightarrow R-CH=CH-R' + {}^{\cdot}OOH$$

and double bonds on a polymer chain. The better the conjugation of the formed double bond with the system of bonds in a molecule moiety as, e.g., during the oxidation of 1,4-polydienes, the more probable the above fragmentation occurs [83, 84].

The fragmentation reaction of peroxy alkyl radicals is the key step in the formation of oxiranes during the oxidation of alkenes. This reaction may be understood as an intraradical decomposition of peroxides. The more nucleophile the alkyl radical, the more quickly fragmentation occurs. Thus the rate constant of fragmentation of β-peroxyalkyl radical:

$$
\begin{array}{c}
O-O-C(CH_3)_3 \\
| \\
R-CH-\overset{\cdot}{C}H-R
\end{array}
\longrightarrow
R-CH\overset{\displaystyle O}{\overset{\displaystyle \diagup\diagdown}{-}}CH-R \;+\; {}^{\cdot}O-C(CH_3)_3
$$

increases, if the alkyl radical has several more electron donor methyl groups $(R = CH_3)$ instead of hydrogens $(R = H)$ [85].

For $R = H$, the rate constant is e.g. $4 \times 10^3 \text{ s}^{-1}$, while that for $R = CH_3$ reaches up to $2 \times 10^6 \text{ s}^{-1}$ at 298 K. An increase in the nucleophile character of the corresponding radical of the same type enhances the attack on the peroxidic bond rather significantly. The rate constant of the interaction of perfluorotriphenyl verdazyl radical with $O-O$ bond of dibenzoyl peroxide in benzene is $8 \times 10^{-3} \text{ dm}^3 \text{ mol}^{-1} \text{ s}^{-1}$ at 20 °C, whereas the triphenyl verdazyl radicals which are more nucleophilic react with peroxide by 500 times faster [86].

β-Peroxyalkyl radicals are formed in addition reactions of peroxyl radicals to $C = C$ double bonds. Radicals which have a larger distance between peroxidic bond and a radical site are formed in intraradical transfer of hydrogen to oxygen of peroxyl radical.

$$
\begin{array}{c}
OO^{\cdot} \\
| \\
R-CH-CH_2-CH_2-CH_2-R'
\end{array}
\left\{
\begin{array}{c}
\quad\quad OOH \\
\quad\quad | \\
\longrightarrow R-CH-CH_2\overset{\cdot}{C}H-CH_2R' \\
\\
\quad\quad OOH \\
\quad\quad | \\
\longrightarrow R-CH-CH_2-CH_2-\overset{\cdot}{C}HR'
\end{array}
\right.
$$

Such reactions of hydroperoxide alkyl radicals are of importance during the oxidation of hydrocarbons in the gaseous phase where they give a relatively high yield of cyclic ethers.

A fragmentation reaction on the oxygen atom of the peroxyacid group:

$$
R^{\cdot} + HOO-\overset{\displaystyle O}{\overset{\displaystyle \|}{C}}-R \rightarrow ROH + RCO_2^{\cdot}(R^{\cdot} + CO_2)
$$

takes place at the decomposition of peroxy acids [87].

The yield of alcohol ROH depends on the nucleophility of the radical as well as on the degree of the delocalization of an unpaired electron. If the unpaired electron is not delocalized, the reactivity of the π-alkyl radical increases with its nucleophility which may be correlated with the ionization potential. On the other hand, if the unpaired spin is delocalized, the orbital overlap of the radical and the peroxidic oxygen becomes an important reactivity factor. A radical seemingly more nucleophilic on the basis of the ionization potential then proves to be less reactive than a nondelocalized π-alkyl radical. This is the case for the benzyl radicals.

σ-Radicals like phenyl, cyclopropyl, or 1-bicyclo-[2,2,1]-heptyl are unreactive with regard to peracids and give only a small amount of alcohol. The electrophilic alkoxy radical does not react on the O−O bond but abstracts a H atom of the peroxyacid group.

A common feature of fragmentation reactions of alkyl peroxyl radicals is the fact that the peroxide group decomposes into an inactive compound and only one free radical is formed per one free radical consumed. Thus the overall number of free radicals potentially available for initiation reaction is reduced.

A particular type of fragmentation is assumed for secondary alkyl peroxyl radicals [88].

$$\overset{\displaystyle OO^{\cdot}}{\underset{\displaystyle \sim CH_2-CH-CH_2\sim}{|}} \rightarrow \sim CH_2COOH + {}^{\cdot}CH_2\sim \, .$$

This reaction has been put forward to explain the observed fact that the number of chain scissions corresponds to the number of carboxyl groups formed in the oxidation of polyethylene. Activation energy of both processes is 140 kJ/mol. The mechanism of such an elementary fragmentation reaction remains however uncertain. The reactions of a chain scission are likely to precede the isomerization of original secondary alkyl peroxy radicals.

A great tendency towards fragmentation is displayed by alkoxyl radicals which are formed as primary species of hydroperoxides or peroxides decomposition or during self-interaction of peroxyl radicals.

$$2\,RO_2^{\cdot} \rightarrow 2\,RO^{\cdot} + O_2 \, .$$

Fragmentation of alkoxyl radicals plays an important role in the decrease of molecular mass of a polymer. The chain-centered alkoxy radicals are cleaved in the β-bond (related to a radical site).

$$\underset{\displaystyle \underset{\displaystyle O^{\cdot}}{|}}{\sim CH_2-CH-CH_2\sim} \rightarrow \sim CH_2CHO + \sim CH_2^{\cdot} \, .$$

This reaction was extensively studied and most studies on polymer oxidation take it into account.

Decarbonylation of acyl radicals is another kind of fragmentation reaction:

$$R\overset{\cdot}{C}O \rightarrow R^{\cdot} + CO \, .$$

Carbonyl radicals in a polymer arise through a radical oxidation of aldehyde groups. The comparison of the rate of transfer reaction of methyl radical to tertiary alkanes and to aldehydes is more favorable for aldehydes which are by one to two orders more reactive [89]. The concentration of aldehyde groups in an oxidized polymer is usually considerably lower than that of alkane bonds $C-H$. The course of decarbonylation becomes therefore significant only in such a case when further oxidation occurs in highly oxidized microregions. As a random reaction taking place in the polymer bulk it is less probable.

At the oxidation of halogen-containing polymers, the release of halogen atoms which are in β-position to a radical site of chloroalkyl radicals should be considered as well. Worth of noting here is the fact that abstraction of a halogen atom may occur also in α-position [90].

Calculated dissociation constants for the fragmentation reaction predict a lifetime that is less than $10^{-10}\,s^{-1}$ for the $^{.}OCCl_{3-x}F_x$ radicals.

When increasing the temperature, the fragmentation reaction:

$$RO_2^{.} \rightarrow R_1^{.} + \text{products}$$

decreases the relative importance of the transfer reaction of peroxy radicals to polymer:

$$RO_2^{.} + RH \rightarrow ROOH + R^{.},$$

where R_1 are either hydroxy radicals or other low-molecular radicals. The low-molecular-mass free radicals play a significant role in oxidation of a polymer at increased temperatures and possibly also at relatively low temperatures. This is conditioned by their high mobility in the polymer medium. It is documented by a complex dependence of the oxidation rate on the sample thickness [92, 93]. On decreasing the sample thickness, the oxidation rate increases slowly, then, in a certain interval of thickness a sharp increase appears followed by a rapid decrease, tending to a zero oxidation rate at zero thickness. An attempt has been made to explain this dependence by a combined effect of oxygen diffusion into the sample and the escape of small free radicals out of it. If there are at least two types of small radicals differing in their mobility and reactivity like $^{.}OH$ and $HO_2^{.}$, they may act in different zones around the site of their appearance. The more active the radical, the shorter is the distance which it survives in the polymer. Provided that this distance is comparable with the sample thickness, the loss of reactive free radicals by a simple evaporation will cause the rate of oxidation for very thin samples to be low. A sharp increase in the rate when increasing the thickness of the polymer sample is apparently due to the effect of reactive low-molecular-mass radicals such as hydroxy radicals which remain in the polymer mass and enter the chain propagation reaction. This is supported also by the escape of less active radicals from the sample participating in the chain termination mainly, which may lead to a marked increase of the reaction rate, as well.

13 Termination Reactions

The decay of free radicals taking part in oxidation of a polymer may occur as a recombination or disproportionation of alkyl radicals, alkyl and peroxyl radicals, or peroxyl radicals:

$$R^{\cdot} \; + \; R^{\cdot} \longrightarrow R{-}R \qquad \left(RH \; + \; {\scriptstyle >}C{=}C{\scriptstyle <} \right)$$

$$R^{\cdot} \; + \; RO_2^{\cdot} \longrightarrow R{-}O{-}O{-}R$$

$$RO_2^{\cdot} \; + \; RO_2^{\cdot} \longrightarrow R{-}O{-}O{-}R \; + \; O_2 \qquad\qquad \text{(tert. alkyl peroxyl radicals)}$$

$$\longrightarrow R^1R^2C{=}O \; + \; R^1R^2CH{-}OH \; + \; O_2 \quad \text{(sec. alkyl peroxyl radicals)}$$

It was shown that the relative contribution of the above three termination steps at oxygen pressures from O to 760 torr is markedly dependent upon the structure of a radical. For instance, for oxidation of model hydrocarbon, 2,6-dimethylhepta-2,5-diene the ratio of $RO_2^{\cdot} + RO_2^{\cdot}$ and $RO_2^{\cdot} + R^{\cdot}$ reactions is only 1 : 1 even at the oxygen pressure 760 torr. For aliphatic alkyl radicals (primary, secondary, or tertiary), the rate of reaction with oxygen is very fast and for the most of industrially produced polymers the quadratic termination step will include almost exclusively the reaction of two peroxy radicals. In a bulk of a polymer, however, the restricted diffusion of oxygen may bring about that the reaction $R^{\cdot} + RO_2^{\cdot}$ may become decisive.

In polymers, two distinct stages in the decay of free radicals may be differentiated, namely that of a mutual diffusional and migrational approach and the chemical reaction itself. Free radical site may be transferred from place to place either chemically as a sequence of reactions:

$$\dot{R}^1 + R^2H \rightarrow R^1H + \dot{R}^2$$

or (in the presence of oxygen):

$$R^1O_2^{\cdot} + R^2H \rightarrow R^1OOH + \dot{R}^2 .$$

The track of the motion of a free radical center is labelled by the sequence of OOH groups formed.

The second route of getting free radical center near is mediated by low-molecular free radicals or compounds which are either present in the polymer or are gradually formed there by fragmentation reactions. While tertiary peroxy radicals propagate only by abstraction of hydrogen atom from surrounding C$-$H bonds, secondary peroxy radicals may easily cleave to hydroxy radicals and ketones as follows:

$$
\begin{array}{c}
\qquad\qquad\quad \overset{\displaystyle |}{C}H_2 \qquad\qquad\qquad\qquad\quad \overset{\displaystyle |}{C}H_2 \\[2pt]
\qquad\qquad\quad | \qquad\qquad\qquad\qquad\qquad\quad | \\[2pt]
H_3C{-}\overset{\displaystyle |}{\underset{\displaystyle |}{C}}{-}O{-}O^{\cdot} + RH \rightarrow H_3C{-}\overset{\displaystyle |}{\underset{\displaystyle |}{C}}{-}O{-}OH + R^{\cdot} \\[2pt]
\qquad\qquad\quad CH_2 \qquad\qquad\qquad\qquad\quad CH_2 \\[2pt]
\qquad\qquad\quad | \qquad\qquad\qquad\qquad\qquad\quad |
\end{array}
$$

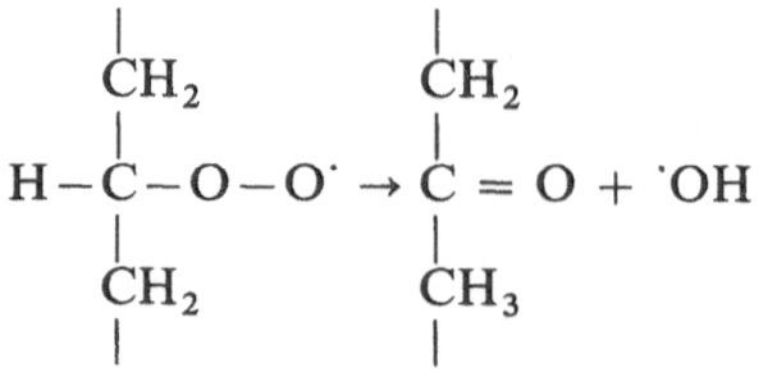

$$
\begin{array}{ccc}
\overset{\displaystyle |}{CH_2} & & \overset{\displaystyle |}{CH_2} \\
| & & | \\
H-C-O-O^{\cdot} & \rightarrow & C=O + {}^{\cdot}OH \\
| & & | \\
\overset{\displaystyle CH_2}{|} & & \overset{\displaystyle CH_3}{|}
\end{array}
$$

The latter reaction which has the activation energy about 85 kJ/mol seems to be more probable at higher temperatures; at lower temperatures hydroxy radicals may appear in other reaction pathways as in the monomolecular decomposition of hydroperoxides, etc. Hydroxyl which are very mobile and very reactive propagate the reaction site only to a short distance (the rate constant of transfer reaction of $HO^{\cdot}$ radicals with CH_3 group of ethane is equal 10^8 dm^3 mol^{-1} s^{-1}, approximately at 20 °C).

In initiation step of polymer oxidation:

$$PH + O_2 \rightarrow P^{\cdot} + HO_2^{\cdot}$$

there appear also hydrogen peroxy radicals which could play the role of chain terminators. Their reaction with secondary alkyl peroxy radicals according to the concerted Russell's cyclic mechanism:

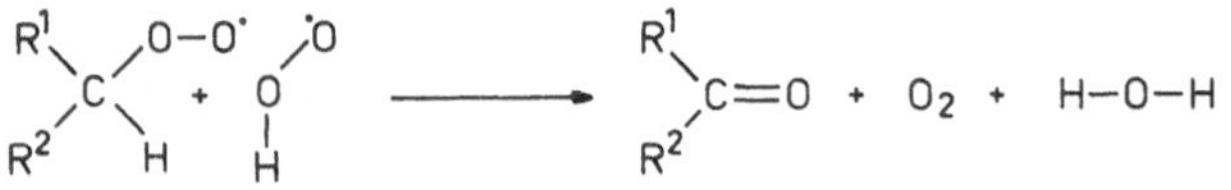

represents one of the most exothermic elementary process in hydrocarbon oxidation which provides an excited ketone, water, and singlet oxygen.

Although this mechanism appears to be a very attractive explanation for the observed low activation energy and high value of the rate constant of self-reaction of secondary peroxy radicals, it is not out of the question that ionic or at least very polar reaction steps specific to a condensed system are also involved.

The dissociation energy of $O-O$ bond in the intermediate tetroxide structure is about 35 kJ/mole. Comparing the stability of tetroxides with respect to peroxides [95] by the value of temperature at which in 1 mol of a corresponding compound at least 1 molecule is dissociated, we receive the value -200 °C against 30 °C. From this it is obvious that the conception of tetroxides in oxidation mechanisms at ambient and higher temperatures is unconceivable.

There exists another pathway of self-reaction of two secondary or primary alkyl peroxy radicals which is even more favorable from the viewpoint of exothermicity than it is Russell's mechanism. It is the reaction in which two molecules of ketone and hydrogen peroxide (or hydrogen and oxygen) are formed as follows:

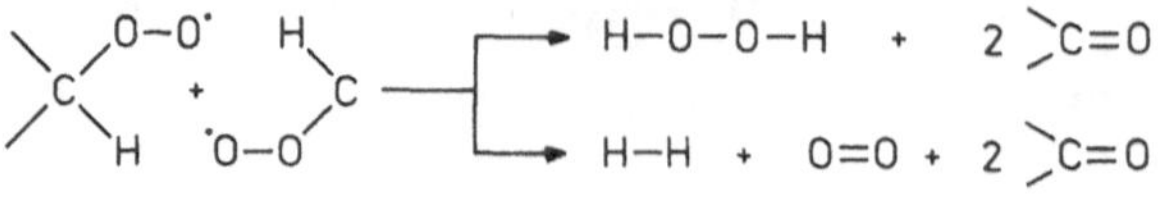

(Hydrogen was observed among reaction products of primary hydroperoxide decomposition).

One thing should be born in mind, namely that the oxidation of a polymer system in which radicals of different reactivity and mobility are formed, occurs more as co-oxidation than oxidation. Sometimes a following scheme may start to operate in such systems:

$$\dot{R}^1 + O_2 \rightarrow R^1 O_2^{\cdot},$$
$$\dot{R}^1 + R^2 H \rightarrow \dot{R}^2 + R^1 H,$$

where R^2 radicals react slowly with oxygen but rapidly with peroxy radicals. This leads to a practically complete stop of the consumption of $R^1 H$ component in the system. An example may be put forward concerning the rate of oxidation of cumene which decreases remarkably by the addition of triphenyl methane [96] or other polynuclear aromatic hydrocarbon.

14 Concluding Remarks

The aim of all studies of polymer degradation is to understand the mechanism of the reaction in detail to be able to develop a suitable stabilizing system and to predict the lifetime of the product under conditions of use.

Degradation of a hydrocarbon polymer starts at or before the processing stage and in many aspects this history of the sample is the least studied parameter of the degradation. The combination of high temperature, low oxygen concentration, and shear stress as it occurs, e.g., in extruders is very difficult to simulate in the laboratory. For this reason, the development of melt stabilizers is still very empirical, the problem being considerably less studied than that in the case of other stabilizers.

The most sensitive indicator of melt degradation is the molecular mass of the polymer which may either increase or decrease depending on the balance between crosslinking and chain scission. For instance, when subjected to oxygen at 100 °C in an oxygen uptake apparatus, samples of unstabilized HDPE films prepared by compression moulding had lost all elongation [97] at a point when only one chain scission on average had taken place. At that point, the M_n had halved but M_w had dropped to less than one tenth of the original value (Table 8). It is of interest that the functional dependence $1/M_n^2$ or $1/M_w^2$ against the amount of absorbed oxygen fits the above results the best (Fig. 3); since $1/M_n$ or $1/M_w$ may be taken as a relative scale of the amount of macromolecules in the polymer sample, it seems therefore that one molecule of oxygen as a carrier of two potential radical sites is also capable of cleaving two macromolecules. A theoretical dependence $M_{w,n} = M_{w,n,o}/\sqrt{a[O_2] + 1}$, where $[O_2]$ is the amount (in moles) of oxygen absorbed in oxidation per one mole of polyethylene, yields values of $a = 21.5$ for M_w and 1.13 for M_n.

On continuing oxidation, the molecular mass distribution is getting narrower with the shift of the maximum to lower values of molecular mass.

Table 8. Changes of molecular mass of high-density polyethylene with the amount of oxygen absorbed at 100 °C

mol O_2/mol PE	M_w	$M_{w\,theor.}$	M_n	$M_{n\,theor.}$
0	151000	151000	8270	7430
0.04	137000	90170	6450	7310
0.76	29100	25380	5610	5480
1.24	15900	19980	4620	4320
3.06	13800	12790	3700	3540
3.64	10400	11730	3200	3300
6.28	8460	8940	2670	2620
8.51	7080	7690	2390	2290

Theoretical values were obtained by optimization of parameters M_{wo} and a in M_w vs. $M_{wo}/\sqrt{a * [O_2] + 1}$ dependence

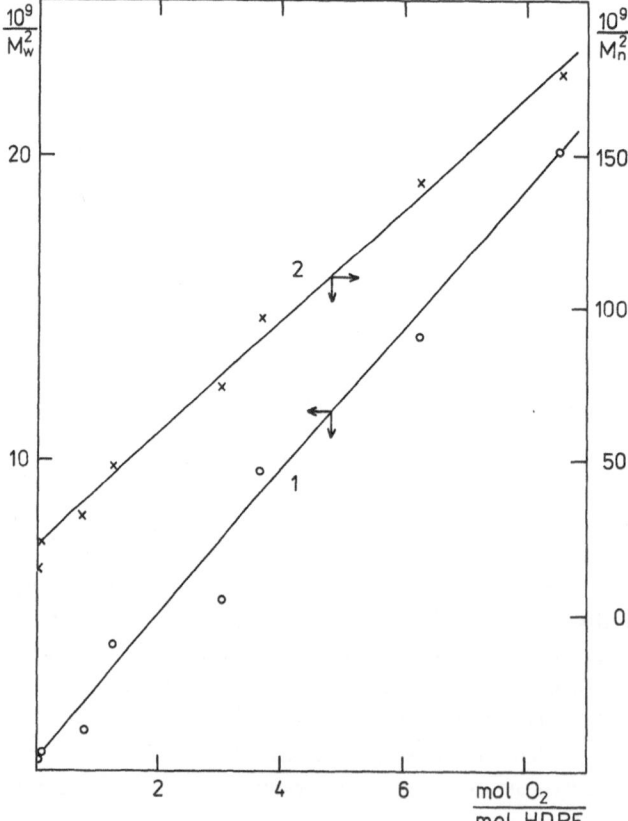

Fig. 3. Plot of $1/M_w^2$ (1) and $1/M_n^2$ (2) on the amount of oxygen absorbed at 100 °C in polyethylene

The difference in behavior between M_w and M_n shows that the large molecules are more susceptible to chain scissions, undergoing several scissions per molecule while the smaller molecules do not break at all. This may have a practical impact on the polymer stabilization where the stabilizer should be either mobile enough or its contact with higher molecular mass fractions should be in some way ensured to obtain an appropriate effect.

Studies of thermal oxidation of polymers are usually carried out as accelerated by raising temperature of the sample so that the oxidation can take place in a considerably short time scale. This makes any extrapolation to lower temperatures unreliable since many processes are involved in degradation and several of these may become rate-limiting as the temperature is changed. This is particularly valid when such extrapolation is performed through the melting range.

As for the mechanism of the process, there exist indications that ion-radical chain reaction may occur under certain circumstances together with purely free radical mechanism involving transfer of oxygen atom [98]:

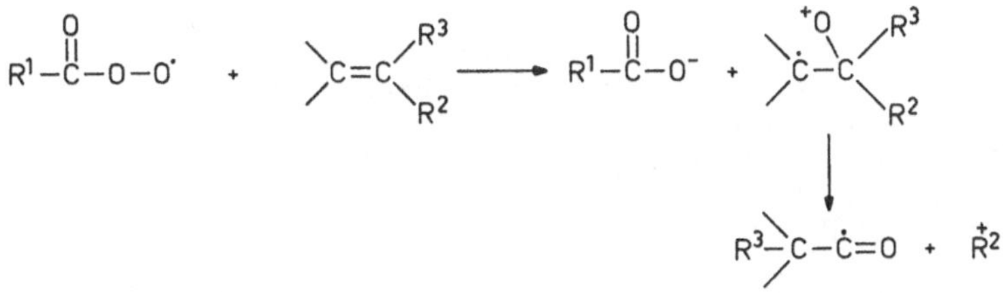

These reactions could take place in oxidized microregions of a polymer. However, experimental data are needed to find out whether and when the proposed scheme is valid.

15 References

1. Shlyapnikov YA, Kiryushkin SG, Marin AL (1986) Antioxidative stabilisation of polymers (in Russian). Publ House Khimiya, Moscow
2. Billingham NC, Calvert PD (1983) In: Degradation and stabilisation of polyolefins, Allen NS (ed) Applied Sci Publ, London, p 1
3. Weir NA (1987) In: Developments in polymer degradation 7, Grassie N (ed) Elsevier, Amsterdam, p 193
4. Emanuel NM, Buchachenko AL (1982) Chemical physics of aging and stabilisation of polymers. Nauka, Moscow, p 55
5. Atkinson R, Carter WPL (1984) Chem Rev 84: 437
6. Razumovskii SD, Rakovski SK, Shopov DM, Zaikov GE (1983) Ozone and its reactions with organic compounds (in Russian). Publ House of Bulg. Acad of Sci, Sofia
7. Rakovski S, Cherneva D (1990) Int J Chem Kinet 22: 321
8. Gesser HD, Hunter NR, Prakasch B (1985) Chem Rev 85: 235
9. Griesbaum K, Volpp W, Huh TS, Jung IC (1989) Chem Berichte 122: 141
10. Shopov I, Kassalova N, Kossmehl G (1989) Poly Degrad Stab 25: 31
11. Rabek JF, Ranby B (1979) Photochem Photobiol 30: 133
12. Zabel F, Reimer A, Becker KH, Fink EH (1989) J Phys Chem 93: 5500
13. Iring M, Tudos F (1990) Prog Polym Sci 15: 217

14. Chirinos-Padron AJ, Hernandez PH, Chavez E, Allen NS, Vasilious C, De Poortere M (1987) Europ Polym J 23: 935
15. Allen NS, Fatinikun KO, Henman TJ (1983) Europ Polym J 19: 551
16. Billingham NC, Makromol Chem (1989) Makromol Symp 28: 145
17. Ginhac JM, Gardette JL, Arnaud R, Lemaire J (1981) Makromol Chem 182: 1017
18. Stewart LC, Carlson DJ, Wiles DM, Scaiano JC (1983) J Am Chem Soc 105: 3605
19. Geuskens G, Debre F, Kabamba MS, Nedelkos G (1984) Polym Photochem 5: 313
20. Faucitano A, Buttafava A, Martinotti F (1986) Proc Int Conf Advances in the Stab and Controlled Degrad of Polymers, Luzern, Switzerland, May 1986, p 19
21. Rabek JF (1987) Mechanism of photochemical processes and photochemical reactions in polymers. Wiley, Chicester
22. Gugumus F (1988) Proceedings of the Internat Conf on Advances in the Stabilisation and Controlled Degradation of Polymers, p 43–80, Luzern
23. Weir NA, Whiting K (1989) Europ Polym J 25, 291
24. Perez E, Vanderhart DL, J Polym Sci (1988) B 26, 1979
25. Babic D (1989) Proc. 7th Bratislava IUPAC Conf on Modified Polymers 1988: Makromol Chem, Makromol Symp 28: 231
26. Clough RL, Gillen KT (1984) J Phys Sci, Polym Chem A 27: 2312
27. Birkinshaw C, Bugy M, Daly S (1988) Poly Degrad Stab 22: 285
28. Shelukhov IP, Zhdanov GS, Klinshpont ER, Milinchuk VK (1986) Radiat Phys Chem 28: 617
29. Bobyleva AV, Berlyant SM, Zhdanov GS, Milinchuk VK (1987) Vysok Soed 29: 839
30. Zlatkevich L (1985) J Polym Sci, Polym Phys Ed, 23: 2633
31. Zolotova NV, Denisov ET (1971) J Polym Sci A 19: 3311
32. Van Sickle DE (1972) J Polym Sci A1, 10: 355
33. Lazár M (1983) Solid state reactions of peroxides. In: The chemistry of functional groups. Peroxides. Patai S (ed) Wiley, New York p 777
34. Lehrle RS, Shortland A (1988) Europ Polym J 24: 425
35. Makundan T, Kishore, K (1987) Mascromolecules, 20: 2382
36. Terwiesch B (1982) J Macromol Sci Chem A17: 1081
37. Mogilevich (1979) Usp Khimii 48: 362
38. Cremer D (1983) General and theoretical aspects of the peroxide group, In: Chemistry of Peroxides. Patai S (ed), Wiley London, p 1
39. Lazár M, Rychlý J, Klimo V, Pelikán P, Valko L (1989) Free radicals in chemistry and biology. CRC Press, Boca Raton, p 62
40. Plesnicar B (1983) Organic polyoxides. In: Chemistry of Peroxides. Patai S (ed) Wiley, London, p 483
41. Catiore B, Verney V, Michel A (1989) Makromol Chem Makromol Symp 25: 199
42. Krstina J, Moad G, Solomon DH (1989) Europ Polym J 25: 767
43. Halpern J (1983) Pure Appl Chem 55: 1059
44. Dowd P, Shapiro M (1984) Tetrahedron 40: 3063
45. Takeuchi Y, Yamamoto F, Konaka T, Nakagawa K (1986) J Polym Sci B, Polym Phys Edit 24: 1067
46. La Mantia FP (1988) Proc Tenth Ann Int Conf on Advances in the Stabilisation and Controlled Degradation of Polymers, Luzern, Switzerland, 25–27 May, p 123
47. Starchevski VL, Mohryi EN, Margolis MA (1984) Zh Phys Khim 58: 845
48. Einhorn C., Einhorn J, Luche JL (1989) Synthesis 787
49. Stagle IR, Ratajczak E, Heaven MC, Gutman D, Wagner AF (1985) J Amer Chem Soc 107: 1838
50. Neta P, Huie RE, Mosseri S, Shastri LV, Mittal JP, Maruthamuthu P, Steenken S (1989) J Phys, Chem 93: 4099
51. Maillard B, Ingold KV, Scaiano JC (1983) J Amer Chem Soc 105: 5095
52. Cobos CJ, Hippler H, Luther K, Ravishankara AR, Troe J (1985) J Phys Chem 89: 4332
53. Kaiser EW, Rimai L, Wallington TJ (1989) J Phys Chem 93: 4094
54. Miyoshi A, Matsui H, Washida N (1989) J Phys Chem 93: 5813
55. Ruiz RP, Bayes KD (1984) J Phys Chem 88: 2592

56. Wu D, Bayes KD, (1986) Int J Chem Kinetics 18: 547
57. Plumb IC, Ryan KR (1981) Int J Chem Kinetics 13: 1011
58. Saito K, Ito R, Kakumoto T, Imamura A (1986) J Phys Chem 90: 1422
59. Yanez J, Sevilla CL, Becker D, Sevilla MD (1987) J Phys Chem 91: 487
60. Sevilla CL, Becker D, Sevilla MD (1986) J Phys Chem 90: 2963
61. Allayarov SR, Barkalov IM (1988) Khimyia vyssh energyi 22: 207
62. Allayarov SR, Gordon DA, Gumina IV, Barkalov IM, Mikhailov AI (1988) Izv Akad Nauk ZSSR, Ser Khim 184
63. Maksimov VL, Agnivtseva TG , Khaikin SY, Pukshanskii MD (1986) Vysok Soed A28: 106
64. Maksimov VL, Agnivtseva TG (1987) Vysok Soed B29: 920
65. Hori Y, Fukunaga Z, Shimada S, Kashiwabara H (1979) Polymer 20: 181
66. Cox RA, Derwent RG, Kearsey SW, Batt L, Patrick KG (1980) J Photochem 13: 149
67. Gesser HD, Hunter NR, Prakasch C (1985) Chem Rev 85: 235
68. Zabarnick S, Heicklen J (1985) J Chem Kinet 17: 455
69. Sawaki Y, Ogata Y (1984) J Org Chem 49: 3344
70. Sway MJ, Waddington DJ (1983) J Chem Soc, Perkin Trans II, 139
71. Jolly GS, Paraskevopoulos G, Singleton DL (1985) Int J Chem Kinetics 17: 1
72. Baulch DL, Campbell IM, Saunders SM (1985) J Chem Soc, Farad Trans I, 81: 259
73. Alfassi ZB, Mosseri S, Nata P (1989) J Phys Chem 93: 1380
74. Becker D, Yanez J, Sevilla MD, Alonso-Amigo MG, Schlick SJ (1987) J Phys Chem 91: 492
75. Boss RC, Jabloner H, Vendenberg EJ (1972) Polymer Lett 10: 915
76. Shimada S, Hori Y, Kashiwabara H (1985) Macromolecules 18: 170
77. Ladygin BY (1981) DAN USSR 257: 1401
78. Guseva LN, Mikheev YuA, Sukhareva SV, Toptygin DY (1988) Vysok Soed A30: 9888
79. Monakhova TV, Bogaevskaya TA, Shlyapnikov YuA (1989) Vysok Soed A31: 636
80. Shlyapnikov YuA, Bogaevskaya TA, Kiryushkin SG, Monakhova TV (1979) Europ Polym J 15: 737
81. Tsvetkov NS, Dutka VS, Markovskaya RF (1981) Ukrainskii khim Zh 47: 411
82. Lauterbach G, Pritzkow W, Tien TD, Voerckel V (1988) J Prakt Chem 330: 933
83. Lala D, Rabek JF (1981) Europ Polym J 17: 7
84. Abdel-Razik EA (1989) J Polymer Sci, Polym Chem A27: 343
85. Bloodworth AJ, Courtneidge JL, Davies AG (1984) J Chem Soc Perkin Trans II, 523
86. Ryabokov IG, Polumbrik OM, Markovski LN (1984) DAN USSR 49
87. Lefort D, Fossy J, Gruselle M, Nedelec JY (1985) Tetrahedron 41: 4237
88. Iring M, Tudos F, Fodor Z, Kelen T (1980) Polym Degrad Stab 2: 143
89. Choudhury TK, Sanders WA, Lin MC (1989) J Chem Soc, Farad Trans 2, 85: 801
90. Kleindienst TE, Shepson PB, Nero CM (1989) Int J Chem Kinetics 21: 863
91. Li Z, Francisco JS (1989) J Am Chem Soc 111: 5660
92. Serenkova IA, Gorelov EP, Shlyapnikov YA (1982) Europ Polym J 18: 5
93. Shlyapnikov YA (1990) Proc Microsymposium Degradation and Stabilisation of Polymers, Stará Lesná, High Tatras, June 1990
94. Russell GA (1957) J Am Chem Soc 79: 3871
95. Lazár M, Bleha T, Rychlý J (1989) Chemical reactions of natural and synthetic polymers. Ellis Horwood and Publ House Alfa, Chicester, Bratislava
96. Mahoney LR (1965) J Am Chem Soc 87: 1089
97. Klemchuk PP, Li Horng P (1984) Polym Degrad Stab 7: 131
98. Shilov AE (1990) React Kinet Catal Lett 41: 223

Editor: H.-J. Cantow
Received February 2, 1991

Author Index Volume 101–102

Author Index Vols. 1–100 see Vol. 100

Améduri, B. and *Boutevin, B.*: Synthesis and Properties of Fluorinated Telechelic Monodispersed Compounds. Vol. 102, pp. 133–170.

Barshtein, G. R. and *Sabsai, O. Y.*: Compositions with Mineralorganic Fillers, Vol. 101, pp. 1–28.
Boutevin, B. and *Robin, J. J.*: Synthesis and Properties of Fluorinated Diols. Vol. 102, pp. 105–132.
Boutevin, B. see Améduri, B.: Vol. 102, pp. 133–170.

Friedman, M. L. see Terent'eva, J. P.: Vol. 101, pp. 29–64.

Geckeler, K. E. see Rivas, B.: Vol. 102, pp. 171–188.
Grubbs, R., Risse, W. and *Novac, B.*: The Development of Well-defined Catalysts for Ring-Opening Olefin Metathesis. Vol. 102, pp. 47–72.

Hall, H. K. see Penelle, J.: Vol. 102, pp. 73–104.

Kulichikhin, S. G. see Malkin, A. Y.: Vol. 101, pp. 217–258.

Lazár, M. and *Rychlý, J.*: Oxidation of Hydrocarbon Polymer. Vol. 102, pp. 189–222.

Malkin, A. Y. and *Kulichikhin, S. G.*: Rheokinetics of Curing. Vol. 101, pp. 217–258.

Novac, B. see Grubbs, R.: Vol. 102, pp. 47–72.

Okada, M.: Ring-Opening Polymerization of Bicyclic and Spiro Compounds. Vol. 102, pp. 1–46.

Padias, A. B. see Penelle, J.: Vol. 102, pp. 73–104.
Penelle, J., Hall, H. K., Padias, A. B. and *Tanaka, H.*: Captodative Olefins in Polymer Chemistry. Vol. 102, pp. 73–104.
Pospisil, J.: Functionalized Oligomers and Polymers as Stabilizers for Conventional Polymers. Vol. 101, pp. 65–168.

Risse, W. see Grubbs, R.: Vol. 102, pp. 47–72.
Rivas, B. and *Geckeler, K. E.*: Synthesis and Metal Complexation of Poly(ethyleneimine) and Derivatives. Vol. 102, pp. 171–188.
Robin, J. J. see Boutevin, B.: Vol. 102, pp. 105–132.
Rychlý, J. see Lazár, M.: Vol. 102, pp. 189–222.

Sabsai, O. Y. see Barshtein, G. R.: Vol. 101, pp. 1–28.
Singh, R. P. see Sivaram, S.: Vol. 101, pp. 169–216.
Sivaram, S. and *Singh, R. P.*: Degradation and Stabilization of Ethylene-Propylene Copolymers and Their Blends: A Critical Review. Vol. 101, pp. 169–216.

Tanaka, H. see Penelle, J.: Vol. 102, pp. 73–104.
Terent'eva, J. P. and *Fridman, M. L.*: Compositions Based on Aminoresins. Vol. 101, pp. 29–64.

Subject Index

ABA-triblock copolymers 56
Abrasion resistance 157
Acid anhydride 139
− fluoride 135, 154, 161
Acrylate 149, 150
−, telomerization 140
Acrylic 140, 149
Activated monomer mechanism 25, 29
Activation energy, decomposition of peroxides 200
Acyclic olefins, disproportionation 48
Acyl-oxygen scission 23
α-Acylamidoacrylates 84, 87
Acyllactam, aminolysis 25
Aldehyde 141
Aldol group transfer 54
Alkene 137, 158, 160
Alkoxyl radicals, disproportionation 207
Alkyl radicals, rate constant with oxygen 203
Alkyl-oxygen scission 21, 24
Alkylation 180, 181, 183
Alkylidene 47, 63
−, transfer from phosphoranes 61
Allyl glycidyl ether 159
− acetate, telomerization 140
Amines 112
1,6-Anhydro cellobiose 16
1,4-Anhydro-α-D-galactopyranose 15
1,4-Anhydro-α-D-glucopyranose 15
1,4-Anhydro-α-D-lyxopyranose 15
1,4-Anhydro-α-D-ribopyranose 15
1,4-Anhydro-α-D-xylopyranose 15
1,4-Anhydro-α-L-arabinopyranose 15
1,6-Anhydro-β-D-galactopyranose derivative 9
1,4-Anhydro-2,3-di-O-benzyl-α-D-xylopyranose 14
1,6-Anhydro-2,3,4-tri-O-benzyl-β-D-glucopyranose 11, 12
1,6-Anhydro-2,3,4-tri-O-benzyl-β-D-allopyranose 5
1,4-Anhydro-2,3,6-tri-O-benzyl-α-D-glucopyranose 12
1,4-Anhydro-2,3,6-tri-O-benzyl-β-D-galactopyranose 14

Anhydrosugar derivatives, cationic polymerization 11−16
− −, oxacarbenium ion mechanism 14
− −, trialkyloxonium ion mechanism 13
Anion exchange separation 177
Anomeric effect 21
Antiperiplanar rule 4, 14
Asymmetric copolymerization 10
2-Azabicyclo[2.2.1]hept-5-en-3-one 31

Back-biting 19
− secondary isomerizations 53
Benzonorbornadiene 54
7,8-Benzotricyclo[4.2.2.0]deca-3,7,9-triene 58
Benzoxazoles 118
Benzvalene 66
Bicyclic acetals 3−11, 32
− lactams 24, 25, 29−32
− lactones 17−47
− orthoesters 32−35
Bisbenzoxazoles 151
Bisphenol, hexafluorinated 145
Bistelomerization 140
Bis(titanacycle) 56
Bistrichloromethyl compounds 107
Block copolymers 54, 55
− −, synthesis 162
Bond-forming initiation 92−95, 100
Branching, oxidation reactions 209
Brominated compounds 140
4-Bromo-6,8-dioxabicyclo[3.2.1]octan-7-one 20
4(a)-Bromo-6,8-dioxabicyclo[3.2.1]octane 8, 9
α-tert-Butylthioacrylonitrile 77, 78, 81

Carbenes 49
Carbenium ion, nonclassical-type 41
Carbon-centered radicals, reactivity 204
Catalysts, copper salts 140
−, ruthenium complex 140
−, tertiary amines 151
Cationic oligomerization 20, 27, 35
Captodative, acrylamides 83

Captodative, acrylates 82−89, 99
−, acrylonitriles 81, 85, 86, 88, 98
−, olefins 75
−, radicals 75, 80, 91
Ceiling temperature 25, 81, 83
Chain extenders 144
− transfer 85
− − agents 63
Charge-transfer polymerization 93
Chemical resistance 135, 164
Chlorhydrine 112, 113
Chlorination 119
Chlorocarbonylation 143
Chlorofluorinated chain 137, 154
Chlorotrifluoroethylene (CTFE) 137−141, 154
Comb-shaped branched polysaccharides 15
Composites 145
Coupling 138, 148
Crosslinking 180, 182
Cumene hydroperoxide, decomposition 210 to 211
Cumyl acetate, telomerization 140
Cyclic oligomer 49
Cyclobutane adducts 95, 96, 100
−, polymerization 62
Cycloparaffins, oxidation 209
Cyclopropanes 96, 100

Decarbonylation of radicals 213−214
Decarboxylation 138, 139
Defect units 190
Degradation, molecular mass dependence 217−218
Dehalogenation 136
Dehydrohalogenation 136, 155
Depolymerization 8
Deslongchamps' theory 33
Diacids 106, 108, 115, 135, 136, 139−141
Diadduct 140
Dialcoholate 122
Dialdehyde 135, 141
Diamines 124, 135, 139, 142, 146, 151, 155, 156
2,3-Dicarbomethoxy-norbornadiene 63
Dichlorides 107, 110, 121
exo-Dicyclopentadiene 54
Diels Alder, retro- 66
Dienes, acyclic, polymerization 61
−, butadiene 152
−, nonconjugated 140, 152
Diepoxy 135, 145, 157
Diesters 109, 115
Difluorides 114, 115
Difluoromethylene 135, 141
Dihalogenide 122, 123

Dihydrate 141
Diiodides 138−140
Diisocyanates 135, 141−143
Dimerization, reductive 152
3,3-Dimethylcyclopropene 53
Diols 106, 108−111, 115, 116, 123
2,6-Dioxabicyclo[2.2.1]heptane 32
2,7-Dioxabicyclo[2.2.1]heptane 4, 32
2,5-Dioxabicyclo[2.2.2]octan-3-one 22
2,6-Dioxabicyclo[2.2.2]octan-3-one 21
6,8-Dioxabicyclo[3.2.1]octane 4, 10
6,8-Dioxabicyclo[3.2.1]octan-7-one 17, 18
Dioxacarbenium ion 9
− −, cyclic 33
1,3-Dioxolan-2-ylium ion 33
Dipotassium ruthenium penta chloride 69
Diradicals 75, 96, 98
Dissociation energy, metal-carbon bond 201
− −, O-O bond 200
Distribution coefficient 183
Dynamic exchange capacity 181

Elastomer 144
Electrophilicity of radicals, effect on addition to alkenes 207
Enantiomer selection 10, 11, 19
Epimerization 22, 28
Epoxides 109, 111−113
Epoxy-alcohol 145
Equilibrium polymerization 25, 36
ESR 84, 91, 98, 99
Esters 164
Ethers 163
Ethylene 140, 161
−, telomerization 140
α-Ethylthioacrylate 85, 98
α-Ethylthioacrylonitrile 84−86
Evans-Polanyi rule 208

Filtration factor 178, 179
Fluorinated chain 137, 141, 150, 153, 164
Fluorination 115, 118, 119, 135, 162
−, electrochemical 135, 153
Fragmentation, free radicals 202, 211
Free radicals, excited 203
− −, thermomechanical stress 201

(1→6)-β-D-Galactopyranan 9
Graft copolymers 56
Grafts 109, 126, 128

Halogenation 139
Hexafluoroacetone 152−154, 161
Hexafluoroisopropanol 154
Hexafluoropropene 139, 147
Hydrogen atoms, reactions with oxygen 198

Hydrogenperoxyl radicals, reactivity 214
Hydrolysis, acidic 159, 164
—, stability to 151
—, with oleum 139
Hydroperoxides, decomposition 210
—, initiation of photooxidation 197
—, yield 209
Hydrophobic properties 159
2,2-bis(2-Hydroxyhexafluoro-2-propyl)-
    benzene 147
Hydroxyl radicals, reactivity 214

Initiation 80, 84, 92—96, 99, 138, 154, 155,
    158
Initiators 120—122
Interpolymers 175
Iodide 157, 158
Iodinated compounds 107
Iodination 147
Ion transport 19
Isocyanurates 117

Jacobson-Stockmayer theory 17

Ketones 136, 161

Lewis acids 63, 90—92
Ligands, selective 179
Living polymerization 52
— — catalysts 60

Macrodiols 109
Malonate, alkyl 160
Membrane filtration 177
Metal carbenes, mechanism 48
— complexation 173, 177
Metallacyclobutanes 50
Metals, seperation 179
Metathesis catalysis 32
4-Methoxycarbonyl-2,6-dioxabicyclo[2.2.2]-
    octan-3-one 22
6-Methylbenzonorbornadiene 54
4-Methylene-1,3-dioxolanes 43
α-Methylthioacrylonitrile 77, 81, 84
1-Methyl-2,6,7-trioxabicyclo[2.2.2]-heptane
    33
2-Methyl-1,4,6-trioxaspiro[4.5]decane 35
2-Methyl-1,4,6-trioxaspiro[4.4]nonane 35
Methylundecylenate 140
Molozonide 194
Molybdenum 62
Monoacetate 140
Monoaddition 158
Monoepoxy 145

Natural rubber, oxidation 195
Nitrile 116, 117
Norbornene spiro orthocarbonate 40

Olefin, allyl 140
—, cyclic 136
—, difunctional 158
—, perfluorinated 154
Oleophobic properties 159
Oligoesters, cyclic 17
Oligomer alkyl radicals, rate constant with
    oxygen 203
—, CTFE 137—141, 154
Oligomerization, cationic 20, 27, 35
—, ionic 154
Osmium 67
2-Oxa-5-azabicyclo[2.2.2]octan-6-one 29,
    30
Oxacarbenium ion mechanism 14
Oxadiazole 118
Oxane 161
7-Oxanorbornene 67
7-Oxanorbornene-2,3-dicarboxylic acid 68
Oxazine 174
Oxazoline 174
Oxethanes 119, 152, 153, 160—162
Oxidation 106, 107
—, catalytic 136
—, permanganic 136, 140, 159
—, pure paraffins/cycloparaffins 209
Oxolane 161
Oxonium exchange reaction 6, 8
Oxygen addition, rate constant 205
Ozone, initiation of oxidation 193

Paraffins, oxidation 209
Peracetic acid 136
Perfluorinated acid 136
— chain 153, 157, 159
Perfluoroacetone see Hexafluoroacetone
Perfluoroalkyl iodides 153
Perfluoroepoxides 113
Perfluoroolefins 113
Perfluoropinacol 152
Periodic acid 136
Peroxides, activation energy in decomposi-
    tion 200
Peroxy alkyl radicals, fragmentation 212,
    213
— nitrates 195
— nitric acids 195
— radicals, disproportionation 207
— —, termination 206
— —, transfer reaction 208
Phosphoranes, alkylidene transfer 61
Photooxidation 113
Polyacetals, stereoregular 10
Polyacetylene 58, 64—66
Polyamide 145, 156
—, amphiphilic 24

Polyaziridine 173, 176
Polybenzoxazoles 151
Polycarbonate 39, 145, 160
Polychelatogens 177
Polycondensate 135
Polycyclooctatetraene 65
Polydienes, oxidation of 194
Polydispersities 53
Polyelectrolyte 173, 176
Polyesters 21, 110, 116, 144, 150, 160, 162, 164
−, isomerization 17
Polyether ester 35
− carbonate 39
Polyethers 113−115, 118, 123, 127, 144
Polyethylene 62
−, irradiation 206
−, unsaturations 196
Polyethyleneimine 173−176, 183, 186
Polyimide 145, 146, 150
Polyisocyanurate 151
Polyketone 43
Polymer-analogous reaction 180
Polypropylene, metal impurities 196
−, rate constant of hydroperoxide decomposition 199
Polysaccharides 5, 9, 11−17
−, synthetic 11−17
Polyurethanes 106, 113, 116, 143−145, 148, 156, 160, 163, 164
Polyvinyl alcohol 55
Prepolymer 135
Propagation 6, 21

Radicals, chemically identical, reactivity 205
−, decarbonylation 213−214
−, electrophilicity 207
Recyclability 68
Redox catalysis 107
− −, telomerization 140
Reduction, catalytic 141
Resins 180, 181
Resistance, abrasion 157
Ring-opening metathesis 52
ROMP 49, 67
− catalysts, mildly active 63
Rubber, oxidation of natural 195
Russell's mechanism 216
Ruthenium 67

Silicone 150
Solar cells 66
Solubility 145
Spacer 141, 164
Spiro orthocarbonates 39−43
− orthoesters 35−38

− tetrathioorthocarbonates 41
Stereoelectronic effect 31
Stereospecific polymerization 10
Steric hindrance 81, 84, 99
Strain energy 24, 36
Sulfur tetrafluoride 162
Sulfuryl halogenides 108
Surface properties 135, 144, 157, 164
Synperiplanar interactions 23

Tantalacyclobutane 59
Tantalum alkylidene 60
Taxogen 138
−, fluorinated 138
Tebbe 50
Telechelic polymers 135−164
Telogens 137, 155, 158
−, functional 124
−, halogenated 138
Telomerization 124, 125, 137−140, 153 to 155, 160
Telomers 125
−, chlorofluorinated 140
Termination, quadratic 215
Tetrafluoroethylene 136, 153, 154
Tetrakis(titanacyclobutane) 57
Tetramethylene 93, 98
1,4,6,9-Tetraoxaspiro[4.4]nonane 39
1,4,6,9-Tetrathiaspiro[4.4]nonane 41
Thermal resistance 135, 144, 145, 148, 151, 161, 162, 164
Thermostability 161
Thiol 157, 158
Titanacyclobutane 51, 53
Titanium metallacyclobutane 60
Transacetalization 6, 21
Transamidation 27
Trialkoxycarbenium ion 40
Trialkyloxonium ion 5
− − mechanism 13
Triazine 155
Triazoles 118
Triepoxy 149
Trimerization 146
Trimethylsilylcyclooctatetraene 65
2,6,7-Trioxabicyclo[2.2.2]octane 34
1,4,6-Trioxaspiro[4.4]nonane 35
Tris(alkylthio)carbenium ion 42

Ultrasound, initiation 201
UV radiation 114, 115, 125, 126, 128

Vinyl fluoride 161
Vinylidene fluoride 161

Wittig-type catalysts 54

Ziegler Natta catalysts 58